WAVE PROPAGATION AND SCATTERING IN RANDOM MEDIA

VOLUME 2
Multiple Scattering, Turbulence, Rough Surfaces, and Remote Sensing

Wave Propagation and Scattering in Random Media

AKIRA ISHIMARU

Department of Electrical Engineering
University of Washington
Seattle, Washington

VOLUME 2
Multiple Scattering, Turbulence,
Rough Surfaces, and Remote Sensing

ACADEMIC PRESS New York San Francisco London 1978
A Subsidiary of Harcourt Brace Jovanovich, Publishers

ACADEMIC PRESS, INC.
111 Fifth Avenue, New York, New York 10003

United Kingdom Edition published by
ACADEMIC PRESS, INC. (LONDON) LTD.
24/28 Oval Road, London NW1 7DX

Library of Congress Cataloging in Publication Data

Ishimaru, Akira, Date
Wave propagation and scattering in random media.

Bibliography: v. 1, p. v. 2, p.
CONTENTS: v. 1. Single scattering and
transport theory.--v. 2. Multiple scattering,
turbulence, rough surfaces and remote sensing.
1. Waves. 2. Scattering (Physics) I. Title.
QC157.I83 531'.1133 77-74051
ISBN 0-12-374702-3 (v. 2)

PRINTED IN THE UNITED STATES OF AMERICA

To Joyce, Jim, Jane, John
Yuko
Yumi and Shigezo Ishimaru

CONTENTS

PART IV □ WAVES IN RANDOM CONTINUUM AND TURBULENCE

CHAPTER 16 □ SCATTERING OF WAVES FROM RANDOM CONTINUUM AND TURBULENT MEDIA

CHAPTER 17 □ LINE-OF-SIGHT PROPAGATION OF A PLANE WAVE THROUGH A RANDOM MEDIUM—WEAK FLUCTUATION CASE

PART V □ ROUGH SURFACE SCATTERING AND REMOTE SENSING

CHAPTER 21 □ ROUGH SURFACE SCATTERING

CHAPTER 22 □ REMOTE SENSING AND INVERSION TECHNIQUES

APPENDIX A □ SPECTRAL REPRESENTATIONS OF A RANDOM FUNCTION

APPENDIX B □ STRUCTURE FUNCTIONS

APPENDIX C □ TURBULENCE AND REFRACTIVE INDEX FLUCTUATIONS

APPENDIX D □ SOME USEFUL MATHEMATICAL FORMULAS

PREFACE

The problem of wave propagation and scattering in the atmosphere, the ocean, and in biological media has become increasingly important in recent years, particularly in the areas of communication, remote-sensing, and detection. These media are, in general, randomly varying in time and space so that the amplitude and phase of the waves may also fluctuate randomly in time and space. These random fluctuations and scattering of the waves are important in a variety of practical problems. Communication engineers are concerned with the phase and amplitude fluctuations of waves as the waves propagate through atmospheric and ocean turbulence and with the coherence time and coherence bandwidth of waves in such a medium. Waves scattered by turbulence may be used for beyond-the-horizon communication links. The detection of clear air turbulence by a scattering technique contributes significantly to safe navigation. Geophysicists are interested in the use of wave fluctuations that occur due to propagation through planetary atmospheres in order to remotely determine their turbulence and dynamic characteristics. Bioengineers may use the fluctuation and scattering characteristics of a sound wave as a diagnostic tool. Radar engineers may need to concern themselves with clutter echoes produced by storms, rain, snow, or hail. Electromagnetic and acoustic probing of geological media requires the knowledge of scattering characteristics of inhomogeneities that are statistically distributed. Finally, the emerging field of radio oceanography is the study of ocean characteristics by scattered radio waves. Central to this technique is the knowledge of wave characteristics that have been scattered by rough surfaces.

All of these problems are characterized by the statistical description of waves and media. Because of this fundamental similarity it should be possible to develop basic formulations common to all these problems. In this book we will present the fundamental formulations of wave propagation and scattering in random media in a unified and systematic manner. This is not an easy task, but it is hoped that the readers can discern some common threads through various formulations and varieties of topics covered in this book. It should be emphasized that, because of the diverse nature of the problems, it is necessary to employ various approximations to obtain useful results. Therefore we will present a systematic exposition of useful approximation techniques applicable to a variety of different situations.

Remote sensing of geophysical and meteorological parameters by wave propagation and scattering techniques has become increasingly important as a useful tool with which to study the structure and dynamics of the atmosphere. This knowledge is used in improving weather forecasting, suggesting means for weather modification, studying pollution, and increasing air traffic safety. An up-to-date account of recent research in this area serves as an introduction to more general remote-sensing techniques.

This book, then, is intended for engineers and scientists interested in optical, acoustic, and microwave propagation and scattering in atmospheres, oceans, and biological media, and particularly for those involved in communication through such media and remote sensing of the characteristics of these media. This book is an introduction to the fundamental concepts and useful results of the statistical wave propagation theory. The theory includes a systematic exposition of radiative transfer and transport and multiple scattering theories that are also the concerns of chemists, geophysicists, and nuclear engineers. Prerequisites include some familiarity with solutions to wave equations, Maxwell's equations, vector calculus, Fourier series, and Fourier integrals.

Topics covered in this book may be grouped into three categories: "waves in random scatterers," "waves in random continua," and "rough surface scattering." Random scatterers are random distributions of many particles. Examples are rain, fog, smog, hail, ocean particles, red blood cells, polymers, and other particles in a state of Brownian motion. Random continua are the media whose characteristics vary randomly and continuously in time and space. Examples are clear air turbulence, jet engine exhaust, tropospheric and ionospheric turbulence, ocean turbulence, and biological media such as tissue and muscle. Rough surface examples are the ocean surface, planetary surfaces, interfaces between different biological media, and the surface roughness of an optical fiber.

Volume 1 (Parts I and II) deals with single scattering theory and transport theory. The single scattering theory is applicable to the waves in a tenuous distribution of scatterers. This covers many practical situations, including radar, lidar, and sonar applications in various media. Because of its relatively simple mathematical formulations it is possible to develop without undue complications a variety of fundamental concepts such as coherence bandwidth, coherence time, temporal frequency, moving scatterers, and pulse propagation. We also include some numerical values of the characteristics of particles in the atmosphere, the ocean, and in biological media. The transport theory, which is also called the radiative transfer theory, deals with transport of intensities through a random distribution of scatterers. This applies to many optical and microwave scattering problems in the atmosphere and in biological media. We present several approximate solutions, including diffusion theory, Kubelka-Munk theory, the plane-parallel problem, isotropic scattering, and forward scattering theory.

Volume 2 (Parts III, IV, and V) contains many recent developments in the theory of waves in random media. Part III is devoted to the theory of multiple scattering of waves by randomly distributed scatterers. Part IV covers theories of weak and strong fluctuations of waves in random continua and turbulence. Part V deals with fundamentals of rough surface scattering and remote sensing of the characteristics of random media, including important inversion techniques.

ACKNOWLEDGMENTS

This book is based on a set of lecture notes prepared for a first-year graduate course on waves in random media, given in the Department of Electrical Engineering at the University of Washington for the past several years. The material in this book has also been used as a text for a one-week UCLA short course offered in 1973 and 1974 on the theory and application of waves in random media.

The author wishes to express his appreciation to his colleagues and graduate students who have made many valuable comments. In particular, he wishes to thank R. A. Sigelmann and F. P. Carlson at the University of Washington, Richard Woo at the Jet Propulsion Laboratories, Cavour Yeh at UCLA, Curt Johnson at the University of Utah, Victor Twersky at the University of Illinois, Chicago Circle, N. Marcuvitz, Polytechnic Institute of New York, and J. B. Keller at New York University, Courant Institute, for discussions, helpful suggestions, and encouragement. A large part of the work included in this book was supported by the Deputy for Electronic Technology, formerly the Air Force Cambridge Research Laboratories, the National Science Foundation, and the National Institute of Health. The author is also grateful to Mrs. Eileen Flewelling for her expert typing and her help in organizing the manuscript.

CONTENTS OF VOLUME 1

SINGLE SCATTERING AND TRANSPORT THEORY

PART III □ MULTIPLE SCATTERING THEORY

CHAPTER 14 □ MULTIPLE SCATTERING THEORY OF WAVES IN STATIONARY AND MOVING SCATTERERS AND ITS RELATIONSHIP WITH TRANSPORT THEORY

As we discussed in the introduction to Chapter 7, there are two general approaches to the problem of wave propagation in randomly distributed particles: analytical theory and transport theory. Chapters 7–13 are devoted to the transport theory in which the propagation of intensities in a random medium is investigated using the transport equation.

In analytical theory, which is also called multiple scattering theory, we start with fundamental differential equations governing field quantities and then introduce statistical considerations (see Crosignani *et al.*, 1975; the excellent review paper by Barabanenkov *et al.*, 1971; Frisch, 1968). Early studies on multiple scattering include those by Ryde (1931), Ryde and Cooper (1931), Foldy (1945), Lax (1951), and Snyder and Scott (1949). These were extended by Twersky who obtained consistent sets of integral equations. Twersky's theory gives a clear physical picture of various processes of multiple scattering, and therefore the first portion of this chapter is devoted to a derivation of Twersky's integral equations. See Twersky (1964), Keller (1964), Twersky (1967, 1970a,b, 1973), Ishimaru (1975, 1977a), Beard (1962, 1967), and Beard *et al.* (1965).

The diagram method gives a systematic and concise formal representation of the complete multiple scattering processes based on an elementary use of Feynman diagrams (Frisch, 1968; Marcuvitz, 1974; Tatarski, 1971, Chapter 5). This leads to the diagram representation of the Dyson equation for the average field and the Bethe–Salpeter equation for the correlation function. It is noted, however, that it is impossible to obtain the explicit exact expressions of the operators in these integral equations, and it is necessary to resort to approximate representations. The simplest and the most useful is called the first order smoothing approximation. This approximation can be shown to be equivalent to the Twersky integral equations (Ishimaru, 1975).

The relationships between multiple scattering theory and transport theory have been investigated in recent years, and many studies dealing with this approach have been reported (Bremmer, 1964; Dolin, 1966; Barabanenkov, 1968, 1969; Barabanenkov *et al.*, 1968a,b; Fante, 1973, 1974; Tatarski, 1971;

Ishimaru, 1975; Furutsu, 1975; Bugnolo, 1960, 1961, 1972; Watson, 1969, 1970; Stott, 1968; Feinstein, 1969; Feinstein *et al.*, 1972; Gnedin and Dolginov, 1964; Granatstein *et al.*, 1972; Kalaschnikov, 1966). In this chapter we also discuss the relationships between Twersky's theory and the transport theory presented in Chapter 7.

If the scatterers are in motion, the fields become functions of time and we need to consider the correlation of the fields in time as well as in space. This problem is also analyzed in this chapter and the fundamental equations are derived. The solutions that exhibit the field fluctuation in time and space are described in the following chapter.

14-1 MULTIPLE SCATTERING PROCESS CONTAINED IN TWERSKY'S THEORY

Let us consider a random distribution of N particles located at $\mathbf{r}_1, \mathbf{r}_2, \ldots, \mathbf{r}_N$ in a volume V. The particles need not necessarily be identical in shape and size. We consider a scalar field ψ^a at $\mathbf{r}_a$, a point in space between the scatterers, which satisfies the wave equation†

$$(\nabla^2 + k^2)\psi = 0, \tag{14-1}$$

where $k = 2\pi/\lambda$ is the wave number of the medium surrounding the particles. Let us designate by ϕ_i^a the incident wave in the absence of any particles at $\mathbf{r}_a$.‡ The field ψ^a at $\mathbf{r}_a$ is then the sum of the incident wave ϕ_i^a and the contributions U_s^a from all N particles located at $\mathbf{r}_s$, $s = 1, 2, \ldots, N$ (see Fig. 14-1):

$$\psi^a = \psi_i^a + \sum_{s=1}^{N} U_s^a \tag{14-2}$$

U_s^a is the wave at $\mathbf{r}_a$ scattered from the scatterer located at $\mathbf{r}_s$, and can be expressed in terms of the wave ϕ^s incident upon the scatterer at $\mathbf{r}_s$, and the scattering characteristic (u_s^a) of the particle located at $\mathbf{r}_s$ as observed at $\mathbf{r}_a$. We write (Fig. 14-2)

$$U_s^a = u_s^a \Phi^s. \tag{14-3}$$

Note that, in general, $u_s^a \Phi^s$ does not mean the product of u_s^a and Φ^s. $u_s^a \Phi^s$ is only a symbolic notation to indicate the field at $\mathbf{r}_a$ due to the scatterer at $\mathbf{r}_s$

† ψ may be a pressure wave in the case of acoustics, or a rectangular component of the electric or magnetic field.

‡ The superscript for a field such as ϕ_i^a denotes the location at which the field is observed, and the subscript denotes the origin of that field.

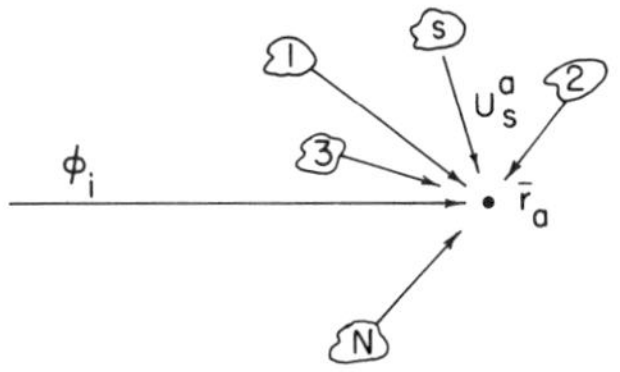

FIG. 14-1

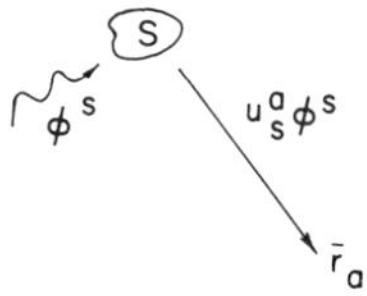

FIG. 14-2

FIG. 14-1 Field at $\mathbf{r}_a$ is the sum of the incident wave and the contributions from all N particles. (Vectors are expressed by the letters with an overbar in figures, while they are boldface in the text. Matrices have a double overbar in figures while they are bold face with an overbar in the text. Unit vectors have a caret.)

FIG. 14-2 The contribution from the particle s when the effective field Φ^s is incident upon it.

when the wave Φ^s is incident upon it. Only when Φ^s can be approximated by a plane wave propagating in the direction of a unit vector $\hat{\mathbf{i}}$,

$$\Phi^s = e^{i\mathbf{k}\cdot\mathbf{r}} \qquad \text{with} \quad \mathbf{k} = k\hat{\mathbf{i}}, \tag{14-4}$$

and when the distance between $\mathbf{r}_s$ and $\mathbf{r}_a$ is large, we can use the following far field approximation of $u_s{}^a$:

$$u_s{}^a \simeq f(\hat{\mathbf{0}}, \hat{\mathbf{i}}) \frac{e^{ikr}}{r} \qquad \text{with} \quad r = |\mathbf{r}_a - \mathbf{r}_s|, \tag{14-5}$$

where $\hat{\mathbf{0}}$ is a unit vector in the direction of $\mathbf{r}_a - \mathbf{r}_s$, and $f(\hat{\mathbf{0}}, \hat{\mathbf{i}})$ is the scattering amplitude of the scatterer.

We call the wave Φ^s incident upon the scatterer at $\mathbf{r}_s$ the "effective field." It consists of the incident wave $\phi_i{}^s$ and the wave scattered from all the particles except the one at $\mathbf{r}_s$. Thus we write (Fig. 14-3)

$$\Phi^s = \phi_i{}^s + \sum_{t=1,\, t\neq s}^{N} U_t{}^s. \tag{14-6}$$

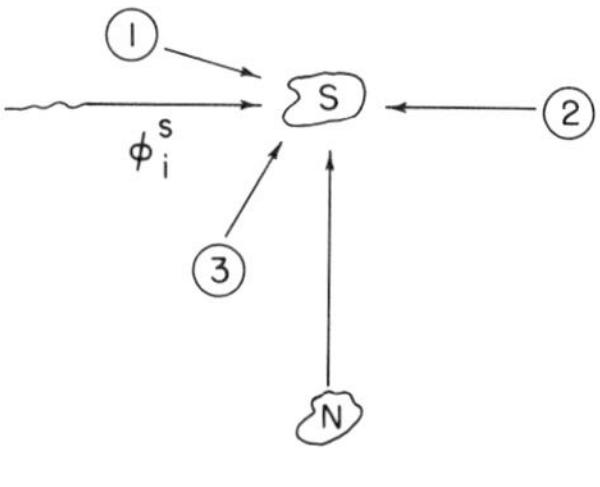

FIG. 14-3 The effective field on the particle s consists of the incident field and the contributions from all the particles except the particle s.

Equations (14-2) and (14-6) constitute the following fundamental pair of equations:

$$\psi^a = \phi_i{}^a + \sum_{s=1}^{N} u_s{}^a \Phi^s \tag{14-7a}$$

$$\Phi^s = \phi_i{}^s + \sum_{t=1,\,t\neq s}^{N} u_t{}^s \Phi^t. \tag{14-7b}$$

In principle we can eliminate Φ from these two equations and obtain the solution ψ^a for a given incident wave ϕ_i. This can be done by substituting (14-7b) into (14-7a) and iterating this process in the following manner:

$$\begin{aligned}\psi^a &= \phi_i{}^a + \sum_{s=1}^{N} u_s{}^a \left(\phi_i{}^s + \sum_{t=1,\,t\neq s}^{N} u_t{}^s \Phi^t\right)\\ &= \phi_i{}^a + \sum_{s=1}^{N} u_s{}^a \phi_i{}^s + \sum_{s=1}^{N} \sum_{t=1,\,t\neq s}^{N} u_s{}^a u_t{}^s \phi_i{}^t\\ &\quad + \sum_{s=1}^{N} \sum_{t=1,\,t\neq s}^{N} \sum_{m=1,\,m\neq t}^{N} u_s{}^a u_t{}^s u_m{}^t \phi_i{}^m + \cdots.\end{aligned} \tag{14-8}$$

Let us consider each term of (14-8). The first term is the incident wave $\phi_i{}^a$. The next term in this series

$$\sum_{s=1}^{N} u_s{}^a \phi_i{}^s \tag{14-8a}$$

represents all the single scattering (see Fig. 14-4a). The next summation,

$$\sum_{s=1}^{N} \sum_{t=1,\,t\neq s}^{N} u_s{}^a u_t{}^s \phi_i{}^t, \tag{14-8b}$$

represents all the double scattering (Fig. 14-4b).

The third summation is triple, but the terms $t = s$ and $m = t$ are excluded, while the term $s = m$ is *not* excluded. Therefore, we may write this summation so as to include the terms containing the separate s, t, and m, plus the terms corresponding to $s = m$:

$$\begin{aligned}&\sum_{s=1}^{N} \sum_{t=1,\,t\neq s}^{N} \sum_{m=1,\,m\neq t}^{N} u_s{}^a u_t{}^s u_m{}^t \phi_i{}^m\\ &= \sum_{s=1}^{N} \sum_{t=1\,t\neq s}^{N} \sum_{m=1,\,m\neq t,\,m\neq s}^{N} u_s{}^a u_t{}^s u_m{}^t \phi_i{}^m + \sum_{s=1}^{N} \sum_{t=1,\,t\neq s}^{N} u_s{}^a u_t{}^s u_s{}^t \phi_i{}^s.\end{aligned} \tag{14-8c}$$

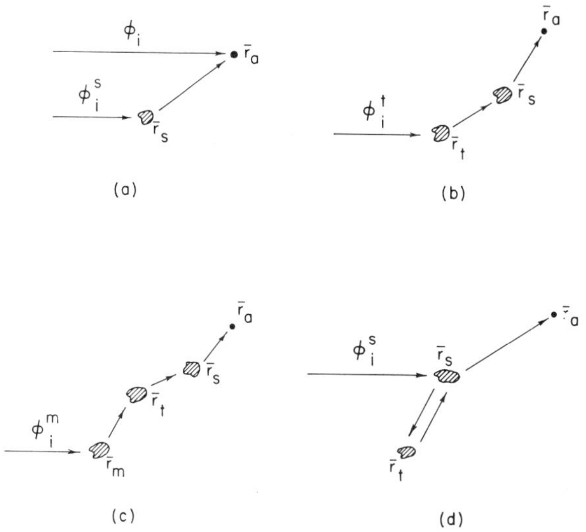

FIG. 14-4 (a) Single scattering, (b) double scattering, (c) triple scattering through different particles, and (d) triple scattering when the propagation path goes through the same particle more than once.

The first triple summation is pictured in Fig. 14-4c. The second summation of (14-8c) involves only two scatterers at $\mathbf{r}_s$ and $\mathbf{r}_t$ and is pictured in Fig. 14-4d.

In general, then, the complete field ψ^a at $\mathbf{r}_a$, which is composed of the incident wave and all the multiple scattered waves, can be divided into two groups:

1. One group, represented by the first summation in (14-8c), consists of all the multiply scattered waves which involve chains of successive scattering going through different scatterers. This is illustrated in Fig. 14-5a. Note that s is for all the scatterers and thus there are N terms for s; t should be for all the scatterers except s, and thus there are $N - 1$ terms for t. In a similar manner, there are $N - 2$ terms for m.

2. The other group of terms represented by the second summation in (14-8c) contains all the paths which go through a scatterer more than once, such as shown in Fig. 14-5b.

Twersky's theory includes all the terms belonging to this first group (Fig. 14-5a), but neglects the terms corresponding to the second group (Fig. 14-5b). Obviously, the first group takes care of almost all the multiple scattering, and so Twersky's theory should give an excellent result when the backscattering is insignificant compared with the scattering in other directions.

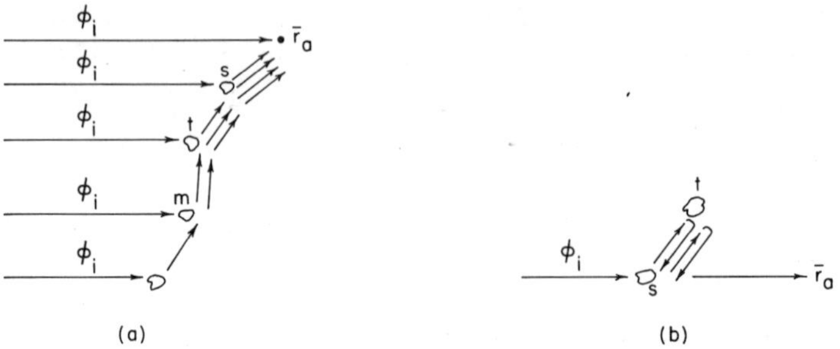

FIG. 14-5 (a) Chains of successive scattering going through different scatterers, and (b) chains of scattering paths which go through the same scatterer more than once.

Mathematically, Twersky's theory is based on the form of the field

$$\psi^a = \phi_i{}^a + \sum_{s=1}^{N} u_s{}^a \phi_i{}^s + \sum_{s=1}^{N} \sum_{t=1,\, t\neq s}^{N} u_s{}^a u_t{}^s \phi_i{}^t + \sum_{s=1}^{N} \sum_{t=1,\, t\neq s}^{N} \sum_{m=1,\, m\neq t,\, m\neq s}^{N} u_s{}^a u_t{}^s u_m{}^t \phi_i{}^m + \cdots. \tag{14-9}$$

Table 14-1 compares the number of terms involved in the exact [Eq. (14-8)] and the Twersky [Eq. (14-9)] multiple scattering processes. It is clear that as N becomes large, the difference between the exact process and the Twersky process becomes very small.

TABLE 14-1

	E = Exact (14-8)	T = Twersky (14-9)	Difference $\dfrac{E-T}{E}$
$\phi_i{}^a$ incident	1	1	0
Single scattering	N	N	0
Double scattering	$N(N-1)$	$N(N-1)$	0
Triple scattering	$N(N-1)^2$	$N(N-1)(N-2)$	$\dfrac{1}{N-1}$
Quadruple scattering	$N(N-1)^3$	$N(N-1)(N-2)(N-3)$	$\dfrac{3N-5}{(N-1)^2}$
	⋮	⋮	⋮

Equation (14-9), which is called the expanded representation by Twersky, is useful for understanding the scattering processes involved, but is impractical for the evaluation of important quantities. Instead, Foldy and Twersky developed the consistent integral equations shown in the following sections.

14-2 STATISTICAL AVERAGES FOR DISCRETE SCATTERERS

Let us consider a random function f which depends on all N scatterers. It may be a field quantity ψ or a product of field quantities. We consider the ensemble average of this function f. The average is given in terms of a probability density function $W(\underline{1}, \underline{2}, \underline{3}, \ldots, \underline{N})$:

$$\langle f \rangle = \iint \cdots \int f W(\underline{1}, \underline{2}, \ldots, \underline{s}, \ldots, \underline{N})\, d\underline{1}\, d\underline{2} \cdots d\underline{s} \cdots d\underline{N}. \tag{14-10}$$

The variable $\underline{s}$ designates all the characteristics of the scatterer s such as location $\mathbf{r}_s$, shape, orientation, and dielectric constant. Therefore, we may write

$$d\underline{s} = d\mathbf{r}_s\, d\xi_s \tag{14-11}$$

where $d\mathbf{r}_s$ designates the volume integral $dx_s\, dy_s\, dz_s$, and $d\xi_s$ represents all the other characteristics of the scatterer:

$$d\xi_s = d(\text{shape of scatterer } s)\, d(\text{orientation of } s)\, d(\text{size of } s) \cdots.$$

Now we consider an important case in which the particle density is low and the particle size is much smaller than the separation between particles. In this case, we can neglect the finite size of particles and we can assume that the location and characteristics of each scatterer are independent of the locations and characteristics of other scatterers. This also means that all particles are considered as point particles, and the effect of size appears only in scattering characteristics. Under this assumption, we have

$$W(\underline{1}, \underline{2}, \underline{3}, \ldots, \underline{s}, \ldots, \underline{N}) = w(\underline{1})w(\underline{2})w(\underline{3}) \cdots w(\underline{s}) \cdots w(\underline{N}). \tag{14-12}$$

Next, we assume that all scatterers have the same statistical characteristics. Then writing

$$w(\underline{s}) = w(\mathbf{r}_s, \xi_s), \tag{14-13}$$

we can perform integration with respect to all the characteristics ξ_s. Thus, we get

$$\langle f \rangle = \iint \cdots \int [f]_\xi w(\mathbf{r}_1) w(\mathbf{r}_2) \cdots w(\mathbf{r}_s) \cdots w(\mathbf{r}_N)\, d\mathbf{r}_1 \cdots d\mathbf{r}_N, \tag{14-14}$$

where $[f]_\xi$ represents the average of f corresponding to the average characteristics of the scatterer (shape, orientation, etc.).

The probability density function $w(\mathbf{r}_s)$ may be interpreted as follows:

$$\begin{aligned} w(\mathbf{r}_s)\, d\mathbf{r}_s &= \text{probability of finding the scatterer } s \text{ within an incremental volume } d\mathbf{r}_s \\ &= \frac{\text{number of scatterers within } d\mathbf{r}_s = dx_s\, dy_s\, dz_s}{\text{total number of scatterers in } V} \\ &= \frac{\rho(\mathbf{r}_s)\, d\mathbf{r}_s}{N}, \end{aligned} \tag{14-15}$$

where $\rho(\mathbf{r}_s)$ is the "number density" defined by the number of scatterers per unit volume. Thus, we get

$$w(\mathbf{r}_s) = \rho(\mathbf{r}_s)/N. \tag{14-16}$$

Note also that if the density $\rho(\mathbf{r}_s)$ is uniform throughout the volume V, then

$$\rho = N/V \qquad \text{and} \qquad w(\mathbf{r}_s) = 1/V.$$

The average is now given by

$$\langle f \rangle = \iint \cdots \int [f]_\xi \frac{\rho(\mathbf{r}_1)\rho(\mathbf{r}_2)\cdots\rho(\mathbf{r}_N)}{N^N}\, d\mathbf{r}_1\, d\mathbf{r}_2 \cdots d\mathbf{r}_N. \tag{14-17}$$

If $[f]_\xi$ depends only on the location of the scatterer(s) and not on the location of other scatterers, then writing $[f]_\xi = f(\mathbf{r}_s)$, we integrate (14-17) over all $\mathbf{r}_1, \ldots, \mathbf{r}_N$, except $\mathbf{r}_s$. Noting

$$\int w(\mathbf{r}_1)\, d\mathbf{r}_1 = \int \frac{\rho(\mathbf{r}_1)\, d\mathbf{r}_1}{N} = 1,$$

we get

$$\langle f(\mathbf{r}_s) \rangle = \int f(\mathbf{r}_s) \frac{\rho(\mathbf{r}_s)}{N}\, d\mathbf{r}_s. \tag{14-18}$$

If $[f]_\xi$ depends on the locations of two different scatterers s and t, then writing $[f]_\xi = f(\mathbf{r}_s, \mathbf{r}_t)$, we obtain

$$\langle f(\mathbf{r}_s, \mathbf{r}_t) \rangle = \iint f(\mathbf{r}_s, \mathbf{r}_t) \frac{\rho(\mathbf{r}_s)\rho(\mathbf{r}_t)}{N^2}\, d\mathbf{r}_s\, d\mathbf{r}_t. \tag{14-19}$$

Expressions (14-18) and (14-19) may be generalized to any number of scatterers. These expressions will be used to obtain the statistical averages in the following sections.†

† For dense distributions, the pair-distribution function needs to be introduced (Beard *et al.*, 1967; Twersky, 1975).

14-3 FOLDY–TWERSKY'S INTEGRAL EQUATION FOR THE COHERENT FIELD

Let us consider a field ψ^a at $\mathbf{r}_a$ in a random medium. In general, ψ^a is a random function of $\mathbf{r}_a$ and time, and can be divided into the average field $\langle\psi^a\rangle$ and the fluctuating field $\psi_f{}^a$.

The average field $\langle\psi^a\rangle$ is also called the coherent field, and the square of its magnitude $|\langle\psi^a\rangle|^2$ is called the coherent intensity. The fluctuating field $\psi_f{}^a$ is also called the incoherent field, and the average of the square of its magnitude $\langle|\psi_f{}^a|^2\rangle$ is called the incoherent intensity. The total intensity is the average of the square of the magnitude of the total field $\langle|\psi^a|^2\rangle$ and is equal to the sum of the coherent intensity and the incoherent intensity:

$$\langle|\psi^a|^2\rangle = \langle|\langle\psi^a\rangle + \psi_f{}^a|^2\rangle = |\langle\psi^a\rangle|^2 + \langle|\psi_f{}^a|^2\rangle. \tag{14-20}$$

As an example, consider a plane wave normally incident on a semi-infinite region containing random particles (see Fig. 14-6). We give the following

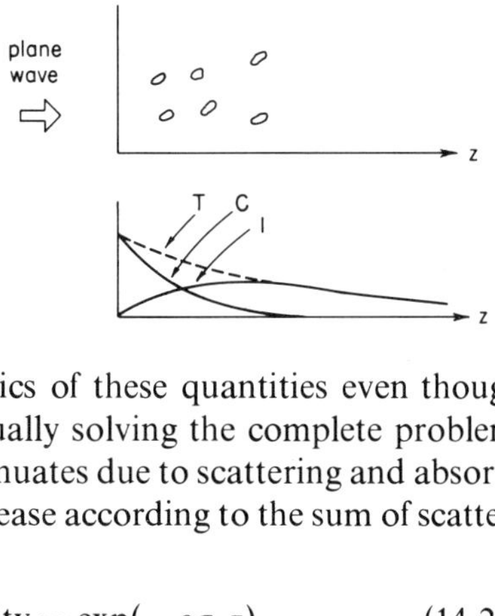

FIG. 14-6 Plane wave incident on a semi-infinite medium, and the coherent intensity C, the incoherent intensity I, and the total intensity T.

approximate picture of the characteristics of these quantities even though details cannot be obtained without actually solving the complete problem. For example, the coherent intensity attenuates due to scattering and absorption, and thus we expect that it will decrease according to the sum of scattering and absorption cross sections:

$$C = \text{coherent intensity} = \exp(-\rho\sigma_t z). \tag{14-21}$$

On the other hand, the power scattered is the incoherent power and, therefore, it contributes to the total intensity. As a result, the total intensity T is substantially dependent on the absorption only, and thus

$$T = \text{total intensity} \approx \exp(-\rho\sigma_a z). \tag{14-22}$$

The incoherent intensity I is, therefore, approximately equal to

$$I \approx \exp(-\rho\sigma_a z) - \exp(-\rho\sigma_t z). \tag{14-23}$$

These equations are only approximate, but they serve to give some quantitative overall characteristics of the field, as pictured in Fig. 14-6. We also note that the coherent intensity and incoherent intensity correspond to the reduced incident intensity and diffuse intensity, respectively, in the transport theory.

We now examine the coherent field $\langle\psi^a\rangle$ using Twersky's theory. We start with the expanded representation (14-9) of Section 14-1:

$$\langle\psi^a\rangle = \phi_i^{\,a} + \sum_{s=1}^{N} \langle u_s^{\,a}\phi_i^{\,s}\rangle + \sum_{s=1}^{N}\sum_{t=1}^{N} \langle u_s^{\,a}u_t^{\,s}\phi_i^{\,t}\rangle + \sum_{s=1}^{N}\sum_{t=1,\,t\neq s}^{N}\sum_{m=1,\,m\neq t,\,m\neq s}^{N} \langle u_s^{\,a}u_t^{\,s}u_m^{\,t}\phi_i^{\,m}\rangle + \cdots. \qquad (14\text{-}24)$$

Using (14-18) and (14-19) of Section 14-2, we get in the limit $N \to \infty$,

$$\langle\psi^a\rangle = \phi_i^{\,a} + \int u_s^{\,a}\phi_i^{\,s}\rho(\mathbf{r}_s)\,d\mathbf{r}_s + \iint u_s^{\,a}u_t^{\,s}\phi_i^{\,t}\rho(\mathbf{r}_s)\rho(\mathbf{r}_t)\,d\mathbf{r}_s\,d\mathbf{r}_t + \iiint u_s^{\,a}u_t^{\,s}u_m^{\,t}\phi_i^{\,m}\rho(\mathbf{r}_s)\rho(\mathbf{r}_t)\rho(\mathbf{r}_m)\,d\mathbf{r}_s\,d\mathbf{r}_t\,d\mathbf{r}_m + \cdots. \qquad (14\text{-}25)$$

To obtain (14-25), we used

$$\sum_{s=1}^{N} \langle u_s^{\,a}\phi_i^{\,s}\rangle = \sum_{s=1}^{N} \int (u_s^{\,a}\phi_i^{\,s})\frac{\rho(\mathbf{r}_s)}{N}\,d\mathbf{r}_s = \int u_s^{\,a}\phi_i^{\,s}\rho(\mathbf{r}_s)\,d\mathbf{r}_s,$$

and

$$\sum_{s=1}^{N}\sum_{t=1,\,t\neq s}^{N} \langle u_s^{\,a}u_t^{\,s}\phi_i^{\,t}\rangle = \sum_{s=1}^{N}\sum_{t=1,\,t\neq s}^{N} \iint u_s^{\,a}u_t^{\,s}\phi_i^{\,t}\frac{\rho(\mathbf{r}_s)\rho(\mathbf{r}_t)}{N^2}\,d\mathbf{r}_s\,d\mathbf{r}_t$$

$$= \frac{(N-1)}{N} \iint u_s^{\,a}u_t^{\,s}\phi_i^{\,t}\rho(\mathbf{r}_s)\rho(\mathbf{r}_t)\,d\mathbf{r}_s\,d\mathbf{r}_t, \qquad (14\text{-}26)$$

which in the limit $N \to \infty$ becomes

$$\iint u_s^{\,a}u_t^{\,s}\phi_i^{\,t}\rho(\mathbf{r}_s)\rho(\mathbf{r}_t)\,d\mathbf{r}_s\,d\mathbf{r}_t.$$

We note that this expanded form (14-25) is identical to the Foldy–Twersky integral equation

$$\langle\psi^a\rangle = \phi_i^{\,a} + \int u_s^a\langle\psi^s\rangle\rho(\mathbf{r}_s)\,d\mathbf{r}_s. \qquad (14\text{-}27)$$

The equivalence between (14-25) and (14-27) can be established by iterating (14-27).

The integral equation (14-27) is the *basic equation* for the coherent field in Twersky's theory. This equation was obtained by Foldy as an approximation, but Twersky has clearly established its physical significance as stated above. In essence, then, $\langle\psi^a\rangle$ as given by the integral equation (14-27) is the average of the field ψ^a pictured in Fig. 14-5a.

14-4 TWERSKY'S INTEGRAL EQUATION FOR THE CORRELATION FUNCTION

Twersky obtained an integral equation for the intensity consistent with the Foldy–Twersky integral equation (14-27) for the coherent field. In this section, we do not derive Twersky's integral equation. Instead, we start with the integral equation, show that it is consistent with the preceding results, and explain its physical significance.

Twersky's integral equation can be stated as

$$\langle\psi^a\psi^{b*}\rangle = \langle\psi^a\rangle\langle\psi^{b*}\rangle + \int v_s{}^a v_s^{b*}\langle|\psi^s|^2\rangle\rho(\mathbf{r}_s)\,d\mathbf{r}_s, \tag{14-28}$$

where $v_s{}^a$ satisfies the integral equation

$$v_s{}^a = u_s{}^a + \int u_t{}^a v_s{}^t \rho(\mathbf{r}_t)\,d\mathbf{r}_t. \tag{14-29}$$

These two integral equations (14-28) and (14-29) constitute the basic equations for the correlation function $\langle\psi^a\psi^{b*}\rangle$.

Let us consider the physical significance of these equations by expanding them in a manner similar to that of the preceding sections. First, we consider (14-29). We iterate (14-29) and obtain

$$v_s{}^a = u_s{}^a + \int u_t{}^a u_s{}^t \rho(\mathbf{r}_t)\,d\mathbf{r}_t + \int u_t{}^a u_m{}^t u_s{}^m \rho(\mathbf{r}_t)\rho(\mathbf{r}_m)\,d\mathbf{r}_t\,d\mathbf{r}_m + \cdots. \tag{14-30}$$

The first term $u_s{}^a$ represents the scattering characteristic of a scatterer s at $\mathbf{r}_a$ (Fig. 14-7). The second term, in the limit $N \to \infty$, becomes

$$\int u_t{}^a u_s{}^t \rho(\mathbf{r}_t)\,d\mathbf{r}_t = \sum_{t=1,\,t\neq s}^{N} \langle u_t{}^a u_s{}^t\rangle_s, \tag{14-31}$$

where $\langle\ \rangle_s$ represents the average with respect to the scatterer t, keeping the particle s fixed. Equation (14-31), therefore, represents the wave originating at s, scattered by t, and reaching $\mathbf{r}_a$. The third term represents the scattering from s to m to t to $\mathbf{r}_a$. Thus, $v_s{}^a$ represents all the scattering process from s to a going through various scatterers as shown in Fig. 14-7.

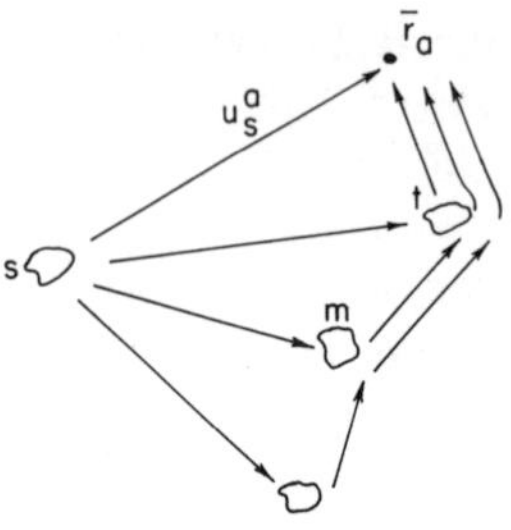

FIG. 14-7 Scattering processes for $v_s{}^a$.

Now we consider the integral equation (14-28) and iterate it in the same manner. We obtain

$$\langle \psi^a \psi^{b*} \rangle = \langle \psi^a \rangle \langle \psi^{b*} \rangle + \int v_s{}^a v_s^{b*} \,|\langle \psi^s \rangle|^2 \rho(\mathbf{r}_s)\, d\mathbf{r}_s$$

$$+ \int v_s{}^a v_s^{b*} v_t{}^s v_t^{s*} \,|\langle \psi^t \rangle|^2 \rho(\mathbf{r}_s)\rho(\mathbf{r}_t)\, d\mathbf{r}_s\, d\mathbf{r}_t$$

$$+ \int v_s{}^a v_s^{b*} v_t{}^s v_t^{s*} v_m{}^t v_m^{t*} \,|\langle \psi^m \rangle|^2 \rho(\mathbf{r}_s)\rho(\mathbf{r}_t)\rho(\mathbf{r}_m)\, d\mathbf{r}_s\, d\mathbf{r}_t\, d\mathbf{r}_m + \cdots. \tag{14-32}$$

Let us consider each term of (14-32). The first term, $\langle \psi^a \rangle \langle \psi^{b*} \rangle$, is the product of the coherent field at a and the complex conjugate of the coherent field at b. Since $\langle \psi^a \rangle$ represents the average of the field ψ^a, which is the sum of all the multiple scattering shown in Fig. 14-5a, we may illustrate this term by Fig. 14-8a.

The next term, $\int v_s{}^a v_s^{b*} \,|\langle \psi^s \rangle|^2 \rho(\mathbf{r}_s)\, d\mathbf{r}_s$, represents the wave at a generated through $v_s{}^a$ (Fig. 14-7) from the coherent field at s, and the wave at b generated through v_s^{b*} from the complex conjugate of the coherent field. This is illustrated in Fig. 14-8b.

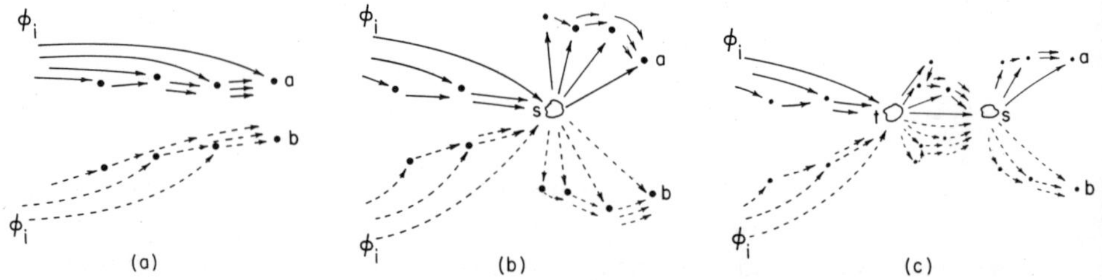

FIG. 14-8 Scattering processes for (a) the first term, (b) the second term, and (c) the third term of (14-32).

In a similar manner, the third term is pictured in Fig. 14-8c.

Continuing this process, it is clear that Twersky's integral equation can be generated by the average of the product of ψ^a and ψ^{b*} as given by the basic chains of scattering processes (14-9) illustrated in Fig. 14-5a. In Appendix 14A, we show an example of these processes for $N = 3$.

Thus, both Foldy–Twersky's integral equation for the coherent field and Twersky's integral equation for the intensity are originated by the same basic processes described in (14-9), and are, therefore, consistent with each other.

It should be mentioned that these integral equations are consistent with the first order smoothing approximation to the more rigorous Dyson and Bethe–Salpeter equations. Derivation of the latter equations can be made using the diagram method (Frisch, 1968). For more rigorous formulation, the readers are referred to Marcuvitz (1974) and Furutsu (1975).

14-5 COHERENT FIELD

Let us consider a slab of thickness d containing a large number of scatterers. A plane wave is incident normally upon the slab (see Fig. 14-9).

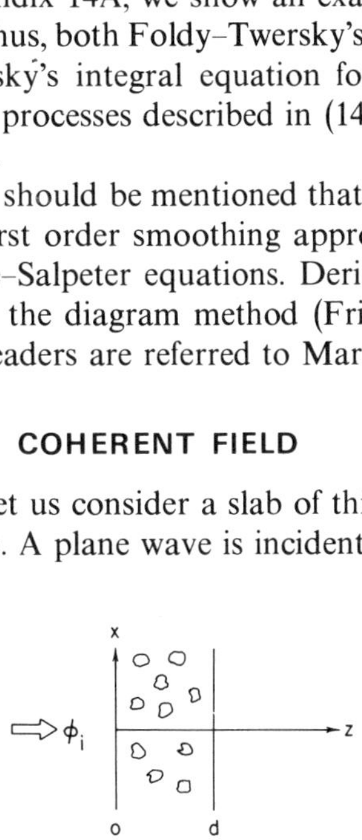

FIG. 14-9

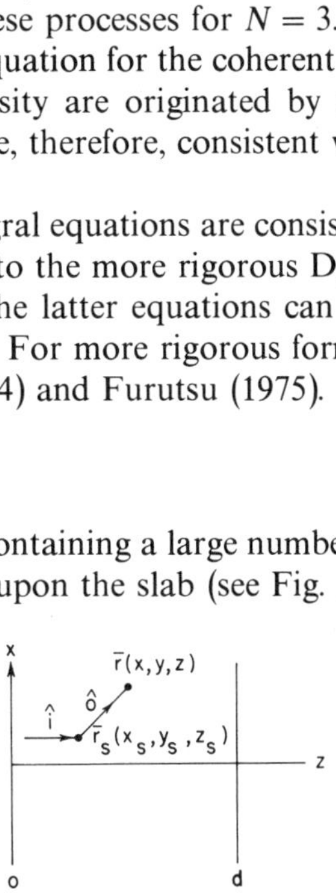

FIG. 14-10

FIG. 14-9 Plane wave incident on a slab of thickness d.
FIG. 14-10 Locations of $\mathbf{r}$ and $\mathbf{r}_s$ for (14-36).

The incident wave with unit amplitude is given by

$$\phi_i(r) = e^{ikz}. \tag{14-33}$$

Let us calculate the coherent field $\langle\psi\rangle$ inside the slab, which satisfies the Foldy–Twersky integral equation

$$\langle\psi^a\rangle = \phi_i^{\ a} + \int u_s^{\ a}\langle\psi^s\rangle\rho(\mathbf{r}_s)\, d\mathbf{r}_s. \tag{14-34}$$

As was discussed in Section 14-1, $u_s^{\ a}$ is an operator and $u_s^{\ a}\langle\psi^s\rangle$ indicates the field at $\mathbf{r}_a$ due to the scatterer at $\mathbf{r}_s$ when $\langle\psi^s\rangle$ is incident. Since the geometry of the slab and the incident wave are independent of the x and y coordinates, the coherent field $\langle\psi\rangle$ is also independent of x and y, and thus $\langle\psi\rangle$ should behave as a plane wave propagating in the $+z$ direction.

Let us consider $u_s{}^a\langle\psi^s\rangle$. This is the scattering characteristic of a single scatterer when the wave $\langle\psi\rangle$ is incident upon it. When $\mathbf{r}_a$ is in the far zone of the particle at $\mathbf{r}_s$, we can approximate it by

$$u_s{}^a\langle\psi^s\rangle = f(\hat{\mathbf{0}}, \hat{\mathbf{i}})\frac{\exp(ik|\mathbf{r}_a - \mathbf{r}_s|)}{|\mathbf{r}_a - \mathbf{r}_s|}\langle\psi^s\rangle, \tag{14-35}$$

where $\hat{\mathbf{i}}$ is a unit vector in the direction of propagation of $\langle\psi^s\rangle$ (and thus $\hat{\mathbf{i}}$ is $\hat{\mathbf{i}}_z$ in this case), and $\hat{\mathbf{0}}$ is a unit vector in the direction of $\mathbf{r}_a - \mathbf{r}_s$.

Using (14-33) and (14-35), we write the integral equation (14-34) as follows (see Fig. 14-10):

$$\langle\psi(z)\rangle = e^{ikz} + \int_0^d dz_s \int_{-\infty}^{\infty} dx_s \int_{-\infty}^{\infty} dy_s f(\hat{\mathbf{0}}, \hat{\mathbf{i}})\frac{\exp(ik|\mathbf{r} - \mathbf{r}_s|)}{|\mathbf{r} - \mathbf{r}_s|}\langle\psi(z_s)\rangle. \tag{14-36}$$

First we consider the coherent field within the slab $0 < z < d$. We perform the integration with respect to x_s and y_s using the method of stationary phase (see Appendix 14B, example I). The stationary phase point is given by

$$\frac{\partial}{\partial x_s}|r - r_s| = 0 \qquad \text{and} \qquad \frac{\partial}{\partial y_s}|r - r_s| = 0,$$

which yields $x_{s0} = x$ and $y_{s0} = y$. Thus we obtain

$$\begin{aligned}\int_{-\infty}^{\infty} dx_s \int_{-\infty}^{\infty} dy_s f(\hat{\mathbf{0}}, \hat{\mathbf{i}})\frac{\exp(ik|\mathbf{r} - \mathbf{r}_s|)}{|\mathbf{r} - \mathbf{r}_s|} &= \frac{2\pi i}{k}\exp[ik(z - z_s)]f(\hat{\mathbf{i}}, \hat{\mathbf{i}}) \quad \text{if} \quad z_s < z \\ &= \frac{2\pi i}{k}\exp[ik(z_s - z)]f(-\hat{\mathbf{i}}, \hat{\mathbf{i}}) \quad \text{if} \quad z_s > z.\end{aligned} \tag{14-37}$$

Using (14-37), we write (14-36) as

$$\begin{aligned}\langle\psi(z)\rangle = e^{ikz} &+ \int_0^z dz_s \frac{2\pi i}{k}\exp[ik(z - z_s)]f(\hat{\mathbf{i}}, \hat{\mathbf{i}})\rho\langle\psi(z_s)\rangle \\ &+ \int_z^d dz_s \frac{2\pi i}{k}\exp[ik(z_s - z)]f(-\hat{\mathbf{i}}, \hat{\mathbf{i}})\rho\langle\psi(z_s)\rangle.\end{aligned} \tag{14-38}$$

It is known that in general, for a large tenuous scatterer of size a, the magnitude of the ratio of the forward to backward scattering is $f(\hat{\mathbf{i}}, \hat{\mathbf{i}})/f(-\hat{\mathbf{i}}, \hat{\mathbf{i}}) \sim (ka)^4$, and for a large perfect conductor, $f(\hat{\mathbf{i}}, \hat{\mathbf{i}})/f(-\hat{\mathbf{i}}, \hat{\mathbf{i}}) \sim$

$(ka)^2$. Therefore we neglect the second integral in (14-38). Assuming that the density ρ is constant, (14–38) becomes

$$\langle \psi(z) \rangle = e^{ikz} \left[1 + \frac{2\pi i}{k} f(\hat{\mathbf{i}}, \hat{\mathbf{i}}) \rho \int_0^z \exp(-ikz_s) \langle \psi(z_s) \rangle \, dz_s \right]. \tag{14-39}$$

This integral equation can be solved exactly by substituting the following form into (14-39):

$$\langle \psi(z) \rangle = A e^{iKz}. \tag{14-40}$$

From this, we obtain

$$A = 1, \qquad K = k + \frac{2\pi f(\hat{\mathbf{i}}, \hat{\mathbf{i}})}{k} \rho. \tag{14-41}$$

This solution, (14-40) together with (14-41), means that the average field $\langle \psi \rangle$ for a plane wave propagates in a slab with the propagation constant K. More generally, the average field $\langle \psi \rangle$ for any wave in a medium containing random particles can be well represented by assuming that it satisfies the wave equation

$$(\nabla^2 + K^2) \langle \psi(\mathbf{r}) \rangle = 0, \tag{14-42}$$

where $K = k + [2\pi f(\hat{\mathbf{i}}, \hat{\mathbf{i}})/k]\rho$. We note that $f(\hat{\mathbf{i}}, \hat{\mathbf{i}})$ is in general complex even for a lossless scatterer, and so the coherent field $\langle \psi(\mathbf{r}) \rangle$ attenuates as it propagates through particles. This is of course due to scattering and is related to scattering cross section. To investigate this point, let us take the coherent intensity for an incident plane wave. We have

$$|\langle \psi(z) \rangle|^2 = \exp\left\{ - \left[\frac{4\pi\rho}{k} \operatorname{Im} f(\hat{\mathbf{i}}, \hat{\mathbf{i}}) \right] z \right\}. \tag{14-43}$$

We note that according to the "forward scattering theorem" (see Appendix 14C)

$$\frac{4\pi}{k} \operatorname{Im} f(\hat{\mathbf{i}}, \hat{\mathbf{i}}) = \sigma_s + \sigma_a, \tag{14-44}$$

where σ_s is the scattering cross section and σ_a is the absorption cross section. Therefore, (14-43) becomes

$$|\langle \psi(z) \rangle|^2 = \exp[-\rho(\sigma_s + \sigma_a)z], \qquad 0 < z < d. \tag{14-45}$$

Outside the slab, $z > d$, we substitute (14-40) into (14-39) and replace the upper limit of the integral by d. This gives the results

$$\begin{aligned} \langle \psi(z) \rangle &= \exp[iKd + ik(z-d)], \\ |\langle \psi(z) \rangle|^2 &= \exp[-\rho(\sigma_s + \sigma_a)d], \qquad z > d. \end{aligned} \tag{14-46}$$

This shows that the coherent intensity attenuates exponentially, and that the attenuation constant is proportional to the density ρ and to the total cross section $\sigma_s + \sigma_a$.

It should be noted that even though the analysis in this section is devoted to the problem of a plane wave incident on a slab, its generalization, given by (14-42), is considered to be a valid approximation in many practical situations.

14-6 PLANE WAVE INCIDENCE ON A SLAB OF SCATTERERS—"TOTAL INTENSITY"

In this section we examine the total intensity for a plane wave incident on a slab. A solution of this problem is by no means simple. In transport theory, a corresponding solution to this problem has been discussed in detail in Chapter 11 using Gauss's quadrature technique. Detailed complete solutions for Twersky's integral equations (14-28) and (14-29) have not appeared in the literature. However, Twersky gave an approximate solution to this problem which has been found to agree reasonably well with experimental data. Here we present this approximate solution. It should be emphasized, however, that although the solution to (14-42) gives a reasonable representation of the coherent field in most practical situations, there are no equations or solutions which can describe the total intensity in a similarly simple representation.

Let us consider the total intensity $\langle|\psi^a|^2\rangle$ defined in (14-20). This should satisfy Twersky's integral equations (14-28) and (14-29):

$$\langle|\psi^a|^2\rangle = |\langle\psi^a\rangle|^2 + \int |v_s{}^a|^2\langle|\psi^s|^2\rangle\rho(\mathbf{r}_s)\, d\mathbf{r}_s \tag{14-47}$$

and

$$v_s{}^a = u_s{}^a + \int u_t{}^a v_s{}^t \rho(\mathbf{r}_t)\, d\mathbf{r}_t\,. \tag{14-48}$$

Even though these are considered to be fundamental equations, their actual numerical solutions are difficult to obtain without some approximations.

First, we note that $v_s{}^a$ is an operator, but we can obtain an approximate representation in the following manner: We assume that in (14-48) the density $\rho(\mathbf{r}_t)$ is constant and $u_t{}^a$ is given by its far field approximation

$$u_t{}^a = f(\hat{\mathbf{i}}_{at}, \hat{\mathbf{i}}_{ts})\frac{\exp(ik|\mathbf{r}_a - \mathbf{r}_t|)}{|\mathbf{r}_a - \mathbf{r}_t|} \tag{14-49}$$

FIG. 14-11 Locations of $\mathbf{r}_a$, $\mathbf{r}_t$ and $\mathbf{r}_s$ for (14-49).

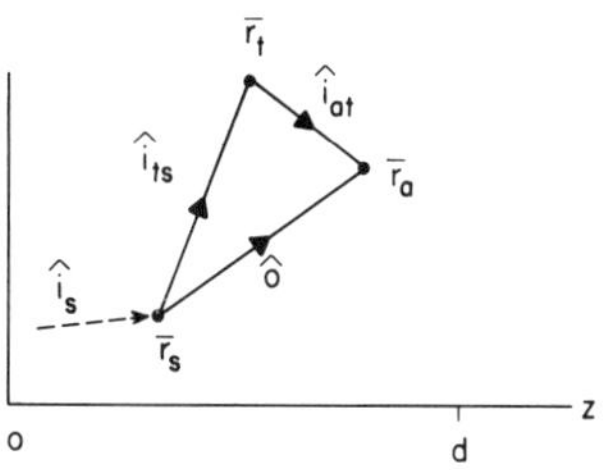

where $\hat{\mathbf{i}}_{at}$ is a unit vector in the direction of $\mathbf{r}_a - \mathbf{r}_t$, and $\hat{\mathbf{i}}_{ts}$ is a unit vector in the direction of $\mathbf{r}_t - \mathbf{r}_s$ (see Fig. 14-11). Similarly, $u_s{}^a$ is given by

$$u_s{}^a = f(\hat{\mathbf{0}}, \hat{\mathbf{i}}_s)\frac{\exp(ik|\mathbf{r}_a - \mathbf{r}_s|)}{|\mathbf{r}_a - \mathbf{r}_s|}, \tag{14-50}$$

where $\hat{\mathbf{0}}$ is a unit vector in the direction of $\mathbf{r}_a - \mathbf{r}_s$, and $\hat{\mathbf{i}}_s$ is a unit vector in the direction† of propagation of the total intensity $\langle |\psi^s|^2 \rangle$.

We now write $v_s{}^t$ in (14-48) as

$$v_s{}^t = u_s{}^t \xi_{ts}, \tag{14-51}$$

where ξ_{ts} is an unknown function. Then we obtain

$$u_s{}^a \xi_{as} = u_s{}^a + \int_0^d dz_t \int_{-\infty}^{\infty} dx_t \int_{-\infty}^{\infty} dy_t\, u_t{}^a u_s{}^t \xi_{ts}\, \rho. \tag{14-52}$$

The integral in (14-52) can be evaluated by the method of stationary phase (Appendix 14B, example II):

$$\begin{aligned}
&\int_{-\infty}^{\infty} dx_t \int_{-\infty}^{\infty} dy_t\, u_t{}^a u_s{}^t \xi_{ts} \\
&= \int_{-\infty}^{\infty} dx_t \int_{-\infty}^{\infty} dy_t\, f(\hat{\mathbf{i}}_{at}, \hat{\mathbf{i}}_{ts}) f(\hat{\mathbf{i}}_{ts}, \hat{\mathbf{i}}_s) \frac{\exp[ik(r_1 + r_2)]}{r_1 r_2} \xi_{ts} \\
&\simeq f(\hat{\mathbf{0}}, \hat{\mathbf{0}}) f(\hat{\mathbf{0}}, \hat{\mathbf{i}}_s) \exp(ik|\mathbf{r}_a - \mathbf{r}_s|) \frac{2\pi i}{k|\mathbf{r}_a - \mathbf{r}_s|(\hat{\mathbf{0}} \cdot \hat{\mathbf{i}}_z)} \xi_{ts} \\
&\qquad \text{when} \quad z_s < z_t < z_a
\end{aligned} \tag{14-53}$$

where $r_1 = |\mathbf{r}_t - \mathbf{r}_s|$, $r_2 = |\mathbf{r}_a - \mathbf{r}_t|$, and ξ_{ts} is evaluated at the stationary phase point. We neglect the contributions from the regions $z_t < z_s$ and $z_a < z_t$. Then (14-52) becomes

$$\xi_{as} = 1 + \frac{2\pi i f(\hat{\mathbf{0}}, \hat{\mathbf{0}})\rho}{k(\hat{\mathbf{0}} \cdot \hat{\mathbf{i}}_z)} \int_{z_s}^{z_a} \xi_{ts}\, dz_t. \tag{14-54}$$

† A more complete description of $\hat{\mathbf{i}}_s$ is given in the following section.

It is clear from this that ξ_{ts} is a function of z_t and z_s, and we obtain the solution

$$\xi_{as} = \exp[i\,\Delta_{as}(z_a - z_s)], \tag{14-55}$$

where

$$\Delta_{as} = \frac{2\pi\rho f(\hat{\mathbf{0}}, \hat{\mathbf{0}})}{k \cos\theta_{as}} \qquad \text{with} \quad \cos\theta_{as} = \hat{\mathbf{0}} \cdot \hat{\mathbf{i}}_z\,.$$

From (14-51) and (14-55), we obtain

$$v_s{}^a = f(\hat{\mathbf{0}}, \hat{\mathbf{i}}_s)\frac{\exp(iK|\mathbf{r}_a - \mathbf{r}_s|)}{|\mathbf{r}_a - \mathbf{r}_s|}, \tag{14-56}$$

where $K = k + [2\pi\rho f(\hat{\mathbf{0}}, \hat{\mathbf{0}})]/k$. Equation (14-56) indicates that $v_s{}^a$ has essentially the same characteristics as $u_s{}^a$ except that the propagation constant is K rather than k. This is reasonable because $u_s{}^a$ represents the radiation from s to a through free space, whereas $v_s{}^a$ represents the radiation from s to a through multiple scattering.

Now, we use (14-47) to find the total intensity $\langle|\psi^a|^2\rangle$ within the slab $0 < z_a < d$. We first note that

$$|v_s{}^a|^2 = |f(\hat{\mathbf{0}}, \hat{\mathbf{i}}_s)|^2 \frac{\exp[-\rho(\sigma_s + \sigma_a)|\mathbf{r}_a - \mathbf{r}_s|]}{|\mathbf{r}_a - \mathbf{r}_s|^2} \tag{14-57}$$

and therefore (14-47) becomes

$$\langle|\psi^a|^2\rangle = |\langle\psi^a\rangle|^2 + \int_{-\infty}^{\infty} dx_s \int_{-\infty}^{\infty} dy_s \int_0^d dz_s |v_s{}^a|^2 \langle|\psi^a|^2\rangle\rho. \tag{14-58}$$

Noting that for $z_a > z_s$, $x_s = (z_a - z_s)\tan\theta_s\cos\phi_s$, $y_s = (z_a - z_s)\tan\theta_s\sin\phi_s$, and, using $dx_s\,dy_s = r^2\sec\theta_s\,d\Omega$, $d\Omega = \sin\theta_s\,d\theta_s\,d\phi_s$, we get

$$\begin{aligned}\int_{-\infty}^{\infty} dx_s \int_{-\infty}^{\infty} dy_s &= \int_0^{\pi/2} d\theta_s \int_0^{2\pi} d\phi_s\, r^2 \frac{\sin\theta_s}{\cos\theta_s} \\ &= \int_{2\pi} r^2 \frac{d\Omega_s}{\cos\theta_s} \qquad \text{for} \quad z_a > z_s,\end{aligned} \tag{14-59}$$

where $r = |\mathbf{r}_a - \mathbf{r}_s|$ (see Fig. 14-12). We neglect the portion of the integral for $z_a < z_s$ because the backscattering is assumed to be small. Then we obtain

$$\begin{aligned}\langle|\psi^a|^2\rangle = |\langle\psi^a\rangle|^2 + \rho \int_{2\pi} d\Omega_s \int_0^{z_a} dz_s |f(\hat{\mathbf{0}}, \hat{\mathbf{i}}_s)|^2 \langle|\psi^s|^2\rangle \sec\theta_s \\ \times \exp[-\rho(\sigma_s + \sigma_a)(z_a - z_s)\sec\theta_s].\end{aligned} \tag{14-60}$$

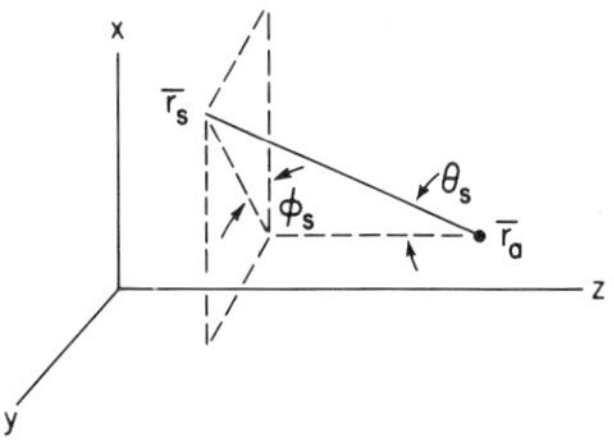

FIG. 14-12 Geometry showing θ_s and ϕ_s for (14-59).

We note that for the plane wave case, $\hat{\mathbf{i}}_s$, which is the direction of the wave propagation for $\langle |\psi^s|^2 \rangle$, is in the z direction†:

$$\hat{\mathbf{i}}_s = \hat{\mathbf{i}}_z. \tag{14-61}$$

Furthermore, we make the reasonable assumption that the scattering is mostly concentrated in the forward direction and thus

$$\theta_s \simeq \text{small} \qquad \text{and} \qquad \sec\theta_s \simeq 1. \tag{14-62}$$

We also note that since the backscattering is small,

$$\int_{2\pi} d\Omega_s |f(\hat{\mathbf{0}}, \hat{\mathbf{i}}_z)|^2 \simeq \int_{4\pi} d\Omega_s |f(\hat{\mathbf{0}}, \hat{\mathbf{i}}_z)|^2 = \sigma_s. \tag{14-63}$$

Under these conditions, we get

$$\langle |\psi^a|^2 \rangle = |\langle \psi^a \rangle|^2 + \rho\sigma_s \int_0^{z_a} \langle |\psi^s|^2 \rangle \exp[-\rho(\sigma_s + \sigma_a)(z_a - z_s)]\, dz_s, \tag{14-64}$$

where $|\langle \psi^a \rangle|^2$ is the coherent intensity given by

$$\exp[-\rho(\sigma_s + \sigma_a) z_a]. \tag{14-65}$$

Equation (14-64) can be solved exactly to yield

$$\langle |\psi^a|^2 \rangle = \exp(-\rho\sigma_a z_a) \qquad \text{for} \quad 0 < z_a < d. \tag{14-66}$$

This is the total intensity inside the slab, which depends only on the absorption cross section σ_a and not on the scattering cross section σ_s. In particular, if the scatterer is lossless, σ_a is zero and the total intensity is constant throughout the slab region. This is to be expected because the power should be conserved in a lossless medium if backscattering is neglected.

Let us next calculate the total intensity outside the slab. This is given by (14-58) when the total intensity given in (14-66) is inserted into the integral.

† This is only approximate. See the following section for a more rigorous discussion.

First, we evaluate $v_s{}^a = u_s{}^a \xi_{as}$ of (14-51) for $\mathbf{r}_a$ outside the slab. We substitute (14-55) into the integral on the right-hand side of (14-54) and get

$$\xi_{as} = 1 + \frac{2\pi i f(\hat{\mathbf{0}}, \hat{\mathbf{0}})\rho}{k(\hat{\mathbf{0}}, \hat{\mathbf{i}}_z)} \int_{z_s}^{d} \exp[i\Delta_{ts}(z_t - z_s)]\, dz_t. \tag{14-67}$$

Noting that $\mathbf{r}_t$ is the stationary phase point, and, therefore, $\Delta_{ts} = \Delta_{as} = 2\pi\rho f(\hat{\mathbf{0}}, \hat{\mathbf{0}})/(k \cos\theta_{as})$, we get $\xi_{as} = \exp[i\Delta_{as}(d - z_s)]$. Thus, for $\mathbf{r}_a$ outside the slab, we have

$$v_s{}^a = u_s{}^a \xi_{as} = f(\hat{\mathbf{0}}, \hat{\mathbf{i}}_s) \frac{\exp(ik|\mathbf{r}_a - \mathbf{r}_s|)}{|\mathbf{r}_a - \mathbf{r}_s|} \exp[i\Delta_{as}(d - z_s)], \tag{14-68}$$

and, therefore,

$$|v_s{}^a|^2 = |f(\hat{\mathbf{0}}, \hat{\mathbf{i}}_s)|^2 \frac{\exp[-\rho(\sigma_s + \sigma_a)(d - z_s)\sec\theta_{as}]}{|\mathbf{r}_a - \mathbf{r}_s|^2}. \tag{14-69}$$

Substituting (14-69) into (14-58) and using (14-66), we get for $\mathbf{r}_a$ outside the slab,

$$\begin{aligned}\langle|\psi^a|^2\rangle = {} & \exp[-\rho(\sigma_a + \sigma_s)d] + \rho \int_{\Omega_r} d\Omega_s \int_0^d dz_s \sec\theta_{as} |f(\hat{\mathbf{0}}, \hat{\mathbf{i}}_s)|^2 \\ & \times \exp[-\rho(\sigma_s + \sigma_a)(d - z_s)\sec\theta_{as} - \rho\sigma_a z_s]\end{aligned} \tag{14-70}$$

where θ_{as} is the angle between $\hat{\mathbf{0}}$ and $\hat{\mathbf{i}}_s = \hat{\mathbf{i}}_z$ (see Fig. 14-13), and Ω_r is the solid angle representing the receiving pattern of the receiver.

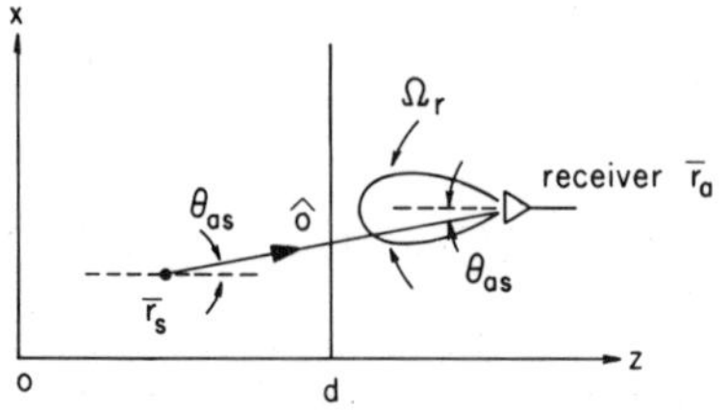

FIG. 14-13 Receiving angle θ_{as}, and Ω_r for (14-70).

Integrating with respect to z_s, we get

$$\begin{aligned}\langle|\psi^a|^2\rangle = {} & \exp[-\rho(\sigma_a + \sigma_s)d] + \int_{\Omega_r} d\Omega_s \sec\theta_{as} |f(\hat{\mathbf{0}}, \hat{\mathbf{i}}_s)|^2 \\ & \times \frac{\exp(-\rho\sigma_a d) - \exp[-\rho(\sigma_a + \sigma_s)d\sec\theta_{as}]}{(\sigma_s + \sigma_a)\sec\theta_{as} - \sigma_a}.\end{aligned} \tag{14-71}$$

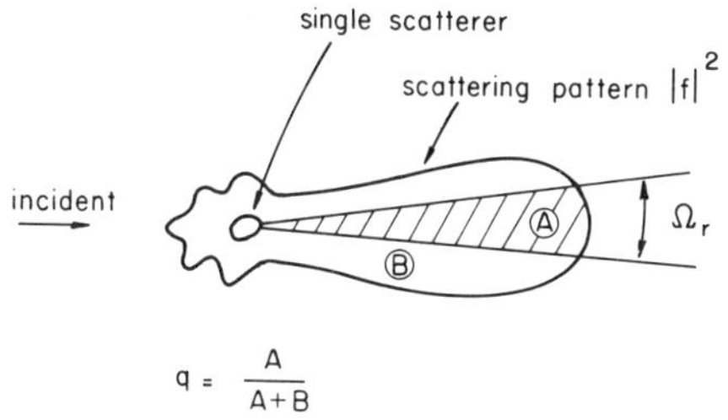

FIG. 14-14 Fraction of the scattered powers given by q in (14-73).

If θ_{as} is small, we can use $\sec\theta_{as} \simeq 1$. In this case,

$$\langle|\psi^a|^2\rangle = \exp[-\rho(\sigma_z+\sigma_s)d] + \int_{\Omega_r} d\Omega_s\,|f(\hat{\mathbf{0}},\hat{\mathbf{i}}_s)|^2\,\frac{\exp(-\rho\sigma_a d)(1-\exp(-\rho\sigma_s d))}{\sigma_s}. \qquad (14\text{-}72)$$

Noting that $\sigma_s = \int_{4\pi} d\Omega_s\,|f(\hat{\mathbf{0}},\hat{\mathbf{i}}_s)|^2$, we obtain the solution

$$\langle|\psi^a|^2\rangle = \exp(-\rho\sigma_a d)[\exp(-\rho\sigma_s d) + q(1-\exp(-\rho\sigma_s d))], \qquad (14\text{-}73)$$

where

$$q = \frac{\int_{\Omega_r} d\Omega_s\,|f(\hat{0},\hat{\mathbf{i}}_s)|^2}{\int_{4\pi} d\Omega_s\,|f(\hat{\mathbf{0}},\hat{\mathbf{i}}_s)|^2}$$

represents the fraction of the total scattered power within the receiving angle Ω_r (Fig. 14-14).

Equation (14-73) is a basic equation which has been used by many investigators. A typical plot of (14-73) is shown in Fig. 14-15. For small ρd ($\rho d < \sigma_s^{-1}\ln q^{-1}$), the coherent intensity dominates:

$$\ln\langle|\psi^a|^2\rangle \simeq -(\sigma_a+\sigma_s)\rho d. \qquad (14\text{-}74)$$

FIG. 14-15 Plot of the logarithm of the total intensity $T = \langle|\psi^a|^2\rangle$ given in (14-73).

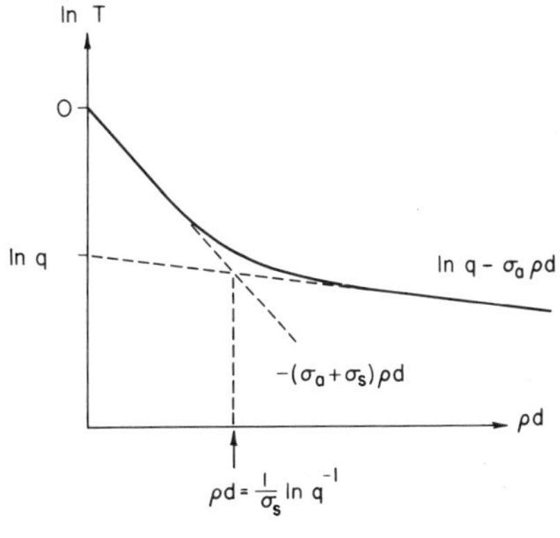

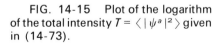

For large ρd, the incoherent intensity modified by q dominates:

$$\ln\langle|\psi^a|^2\rangle \simeq \ln q - \sigma_a \rho d. \tag{14-75}$$

Note that when Ω_r is 2π, the receiver collects almost all the scattered power, and thus within the approximation

$$\int_{4\pi} d\Omega_s |f(\hat{\mathbf{0}}, \hat{\mathbf{i}}_s)|^2 \simeq \int_{2\pi} d\Omega_s |f(\hat{\mathbf{0}}, \hat{\mathbf{i}}_s)|^2, \tag{14-76}$$

$q = 1$ and, therefore,

$$\langle|\psi^a|^2\rangle = \exp(-\rho\sigma_a d). \tag{14-77}$$

Equation (14-73) is applicable only in the region $\theta_{as} \simeq 0$. More generally, the angular dependence of the total intensity must be calculated by going back to (14-71).

14-7 RELATIONSHIP BETWEEN MULTIPLE SCATTERING THEORY AND TRANSPORT THEORY

In Section 14-6 we gave two integral equations (14-47) and (14-48) for the intensity. In Section 7-5 we derived integral equations for the intensities from the transport theory. These two formulations deal with the same problem of random scatterers and, therefore, we expect that they are closely related to each other. In this section, we show this relationship by deriving the transport equation from Twersky's integral equation under certain assumptions (Ishimaru, 1975):

We start with the two equations (14-47) and (14-48):

$$\langle|\psi^a|^2\rangle = |\langle\psi^a\rangle|^2 + \int |v_s{}^a|^2 \langle|\psi^s|^2\rangle \rho(\mathbf{r}_s)\, d\mathbf{r}_s, \tag{14-78}$$

$$v_s{}^a = u_s{}^a + \int u_t{}^a v_s{}^t \rho(\mathbf{r}_t)\, d\mathbf{r}_t. \tag{14-79}$$

First, we note that $v_s{}^a$ is an operator whose approximate form is given by (14-56):

$$v_s{}^a = f(\hat{\mathbf{0}}, \hat{\mathbf{i}}) \frac{\exp(iK|\mathbf{r}_a - \mathbf{r}_s|)}{|\mathbf{r}_a - \mathbf{r}_s|}, \tag{14-80}$$

where $K = k + [2\pi\rho f(\hat{\mathbf{0}}, \hat{\mathbf{0}})/k]$, $\operatorname{Im} K = \rho\sigma_t/2$, and $\hat{\mathbf{i}}$ is a unit vector in the direction of the incident wave. Since (14-80) is an operator, $|v_s{}^a|^2\langle|\psi^s|^2\rangle$ in the integral in (14-78) is not a product of $|v_s{}^a|^2$ and $\langle|\psi^s|^2\rangle$, but $v_s{}^a$ should represent the scattered field in the direction $\hat{\mathbf{0}}$ as given by (14-80) when a spatial spectrum of the power $\langle|\psi^s|^2\rangle$ is pointed in the direction $\hat{\mathbf{i}}$, and $|v_s{}^a|^2\langle|\psi^s|^2\rangle$ should represent the contribution from all incident directions $\hat{\mathbf{i}}$.

To express this concept in mathematical form, we express the correlation function in terms of the specific intensity $I(\mathbf{r}, \hat{\mathbf{s}})$:

$$\langle\psi(\mathbf{r}_a)\psi^*(\mathbf{r}_b)\rangle = \Gamma(\mathbf{r}_a, \mathbf{r}_b) = \Gamma(\mathbf{r}, \mathbf{r}_\mathrm{d}) \simeq \int I(\mathbf{r}, \hat{\mathbf{s}}) \exp(iK_\mathrm{r}\hat{\mathbf{s}} \cdot \mathbf{r}_\mathrm{d})\, d\omega, \quad (14\text{-}81)$$

where $\mathbf{r} = \frac{1}{2}(\mathbf{r}_a + \mathbf{r}_b)$ and $\mathbf{r}_\mathrm{d} = \mathbf{r}_a - \mathbf{r}_b$, and K_r is the real part of K given in (14-80). Equation (14-81) is valid when the correlation function is a slowly varying function of $\mathbf{r}$ (Barabanenkov, 1969).

The intensity $\langle|\psi(\mathbf{r})|^2\rangle$ is now given by

$$\langle|\psi(\mathbf{r})|^2\rangle = \Gamma(\mathbf{r}, 0) = \int I(\mathbf{r}, \hat{\mathbf{s}})\, d\omega. \quad (14\text{-}82)$$

We note that $\langle|\psi^s|^2\rangle$ in the integral of (14-78) should be decomposed as in (14-82) and each component $I(\mathbf{r}, \hat{\mathbf{s}})$ should produce scattering according to (14-80). Thus, we write

$$|v_s{}^a|^2\langle|\psi^s|^2\rangle = \int |f(\hat{\mathbf{s}}, \hat{\mathbf{s}}')|^2 \frac{\exp(-\rho\sigma_\mathrm{t}|\mathbf{r}_a - \mathbf{r}_s|)}{|\mathbf{r}_a - \mathbf{r}_s|^2} I(\mathbf{r}_s, \hat{\mathbf{s}}')\, d\omega', \quad (14\text{-}83)$$

where we have replaced $\hat{\mathbf{0}}$ and $\hat{\mathbf{i}}$ by $\hat{\mathbf{s}}$ and $\hat{\mathbf{s}}'$. Using the phase function [see Eq. (7-22)] $p(\hat{\mathbf{s}}, \hat{\mathbf{s}}')$ defined by

$$p(\hat{\mathbf{s}}, \hat{\mathbf{s}}') = \frac{4\pi}{\sigma_\mathrm{t}} |f(\hat{\mathbf{s}}, \hat{\mathbf{s}}')|^2 \quad (14\text{-}84)$$

and the average intensity [see Eq. (7-7)] $U(\mathbf{r})$

$$U(\mathbf{r}) = \frac{1}{4\pi}\int_{4\pi} I(\mathbf{r}, \hat{\mathbf{s}})\, d\omega, \quad (14\text{-}85)$$

and recognizing that the coherent intensity $|\langle\psi^a\rangle|^2$ attenuates in the same manner as the reduced incident intensity, we can write (14-78) in the form

$$U(\mathbf{r}_a) = U_\mathrm{ri}(\mathbf{r}_a) + \int d\mathbf{r}_s \frac{\exp(-\rho\sigma_\mathrm{t}|\mathbf{r}_a - \mathbf{r}_s|)}{4\pi|\mathbf{r}_a - \mathbf{r}_s|^2} \int \frac{\rho\sigma_\mathrm{t}}{4\pi} p(\hat{\mathbf{s}}, \hat{\mathbf{s}}') I(\mathbf{r}_s, \hat{\mathbf{s}}')\, d\omega'. \quad (14\text{-}86)$$

We recognize that this is identical to (7-41) derived from the transport theory. We note that the core of this equivalence is the approximation (14-80) and the approximation (14-81).

Equation (14-81) relates the specific intensity $I(\mathbf{r}, \hat{\mathbf{s}})$ to the mutual coherence function $\Gamma(\mathbf{r}_a, \mathbf{r}_b) = \langle\psi(\mathbf{r}_a)\psi^*(\mathbf{r}_b)\rangle$. Note that the concept of the specific intensity was heuristically derived in the radiative transfer theory and it was intended to show the amount and direction of the propagation of power and not the field quantities. However, (14-81) shows that the specific intensity does include information concerning the field quantities in the form of mutual coherence function. The relationship (14-81) therefore provides an important link between the radiative transfer theory and the multiple scattering theory. We note here that (14-81) is only approximate and that strictly speaking (14-81) cannot be consistent with the wave equation (for further work on this relationship, see Barabanenkov, 1969; Furutsu, 1975; Wolf, 1976).

14-8 APPROXIMATE INTEGRAL AND DIFFERENTIAL EQUATIONS FOR THE CORRELATION FUNCTION

In Section 14-4, we gave two integral equations (14-28) and (14-29) for the correlation function $\langle\psi^a\psi^{b*}\rangle$. We now use the assumptions of the preceding section to obtain an approximate representation of these integral equations. Making use of (14-80) and (14-81), we immediately write

$$\langle\psi^a\psi^{b*}\rangle = \langle\psi^a\rangle\langle\psi^{b*}\rangle + \Gamma_{fab}, \tag{14-87a}$$

where

$$\Gamma_{fab} = \int \rho\, d\mathbf{r}_s \frac{\exp(iK|\mathbf{r}_a - \mathbf{r}_s| - iK^*|\mathbf{r}_b - \mathbf{r}_s|)}{|\mathbf{r}_a - \mathbf{r}_s|\,|\mathbf{r}_b - \mathbf{r}_s|} \int d\omega' \times f(\hat{\mathbf{s}}_a, \hat{\mathbf{s}}')f^*(\hat{\mathbf{s}}_b, \hat{\mathbf{s}}')I(\mathbf{r}_s, \hat{\mathbf{s}}') \tag{14-87b}$$

and $\hat{\mathbf{s}}_a$ and $\hat{\mathbf{s}}_b$ are the unit vectors in the directions of $\mathbf{r}_a - \mathbf{r}_s$ and $\mathbf{r}_b - \mathbf{r}_s$, respectively (see Fig. 14-16). The specific intensity $I(\mathbf{r}_s, \hat{\mathbf{s}}')$ is related to the total intensity $\langle|\psi^s|^2\rangle$ through

$$\langle|\psi^s|^2\rangle = \int I(\mathbf{r}_s, \hat{\mathbf{s}}')\, d\omega'. \tag{14-88}$$

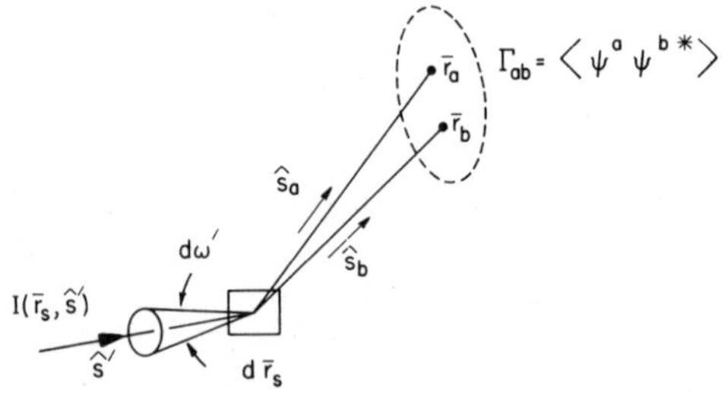

FIG. 14-16 Geometry showing the quantities used in (14-87b).

Equation (14-87a) constitutes the basic integral equation for the correlation function $\langle \psi^a \psi^{b*} \rangle$ and is equivalent to the first order smoothing approximation to the Bethe–Salpeter equation (Frisch, 1968; Ishimaru, 1975).

We can convert the integral equation (14-87a) into an approximate differential equation and show that this differential equation is identical to the equation of transfer (7-24) in the following manner. Using the center coordinate $\mathbf{r} = \frac{1}{2}(\mathbf{r}_a + \mathbf{r}_b)$ and the difference coordinate $\mathbf{r}_d = \mathbf{r}_a - \mathbf{r}_b$, we write [see (10-81)]

$$\langle \psi^a \psi^{b*} \rangle = \Gamma_{ab}(\mathbf{r}, \mathbf{r}_d) = \int I(\mathbf{r}, \hat{\mathbf{s}}) \exp(iK_r \hat{\mathbf{s}} \cdot \mathbf{r}_d)\, d\omega, \tag{14-89}$$

where K_r is the real part of K. We also define the coherent intensity I_c

$$\langle \psi^a \rangle \langle \psi^{b*} \rangle = \int I_c(\mathbf{r}, \hat{\mathbf{s}}) \exp(iK_r \hat{\mathbf{s}} \cdot \mathbf{r}_d)\, d\omega, \tag{14-90}$$

We assume that in (14-87b), the phase and amplitude may be approximated by

$$\begin{aligned} K|\mathbf{r}_a - \mathbf{r}_s| &\approx K|\mathbf{r} - \mathbf{r}_s| + (K/2)\mathbf{r}_d \cdot \hat{\mathbf{s}} \\ K|\mathbf{r}_b - \mathbf{r}_s| &\approx K|\mathbf{r} - \mathbf{r}_s| - (K/2)\mathbf{r}_d \cdot \hat{\mathbf{s}} \end{aligned} \tag{14-91a}$$

$$\frac{1}{|\mathbf{r}_a - \mathbf{r}_s|\,|\mathbf{r}_b - \mathbf{r}_s|} \approx \frac{1}{|\mathbf{r} - \mathbf{r}_s|^2} \tag{14-91b}$$

and $\hat{\mathbf{s}}_a \approx \hat{\mathbf{s}}$ and $\hat{\mathbf{s}}_b \approx \hat{\mathbf{s}}$ in f, where

$$\hat{\mathbf{s}} = \tfrac{1}{2}(\hat{\mathbf{s}}_a + \hat{\mathbf{s}}_b). \tag{14-91c}$$

These assumptions are valid if $|\mathbf{r}_d|$ is much smaller than the distance $|\mathbf{r}_a - \mathbf{r}_s|$ and $|\mathbf{r}_b - \mathbf{r}_s|$. Since the magnitude of $|\mathbf{r}_d|$ is generally limited within the correlation distance of the field, these assumptions are justified in almost all parts of the medium except a small volume near the observation points.

Using the assumption (14-91) we get

$$\begin{aligned} \Gamma_{fab} = \int \rho\, d\mathbf{r}_s \frac{\exp(iK_r \hat{\mathbf{s}} \cdot \mathbf{r}_d - \rho\sigma_t |\mathbf{r} - \mathbf{r}_s|)}{|\mathbf{r} - \mathbf{r}_s|^2} \\ \times \int d\omega' \,|f(\hat{\mathbf{s}}, \hat{\mathbf{s}}')|^2 I(\mathbf{r}_s, \hat{\mathbf{s}}') \end{aligned} \tag{14-92}$$

where $K_r = \frac{1}{2}(K + K^*)$ and $\rho\sigma_t = 2 \operatorname{Im} K = i^{-1}(K - K^*)$.

We can convert this into a differential equation by using $d\mathbf{r}_s =$

$|\mathbf{r} - \mathbf{r}_s|^2\, d\omega\, ds$ where ds is the elementary distance at $\mathbf{r}_s$ in the direction of $\mathbf{r} - \mathbf{r}_s$ [see (7-41)]:

$$I(\mathbf{r}, \hat{\mathbf{s}}) \exp(iK_r\hat{\mathbf{s}} \cdot \mathbf{r}_d) = I_c(\mathbf{r}, \hat{\mathbf{s}}) \exp(iK_r\hat{\mathbf{s}} \cdot \mathbf{r}_d) + \exp(iK_r\hat{\mathbf{s}} \cdot \mathbf{r}_d) \int_{\mathbf{r}_0}^{\mathbf{r}} \rho\, ds \exp(-\rho\sigma_t|\mathbf{r} - \mathbf{r}_s|) \times \int d\omega' |f(\hat{\mathbf{s}}, \hat{\mathbf{s}}')|^2 I(\mathbf{r}_s, \hat{\mathbf{s}}') \tag{14-93}$$

where $\mathbf{r}_0$ and ds are the point on the boundary and the differential distance at $\mathbf{r}_s$, respectively (see Fig. 14-17). We now remove the exponents $iK_r\hat{\mathbf{s}} \cdot \mathbf{r}_d$ in (14-93) and get the equation of transfer:

$$\frac{d}{ds} I(\mathbf{r}, \hat{\mathbf{s}}) = -\rho\sigma_t I(\mathbf{r}, \hat{\mathbf{s}}) + \frac{\rho\sigma_t}{4\pi} \int p(\hat{\mathbf{s}}, \hat{\mathbf{s}}') I(\mathbf{r}, \hat{\mathbf{s}}')\, d\omega' \tag{14-94a}$$

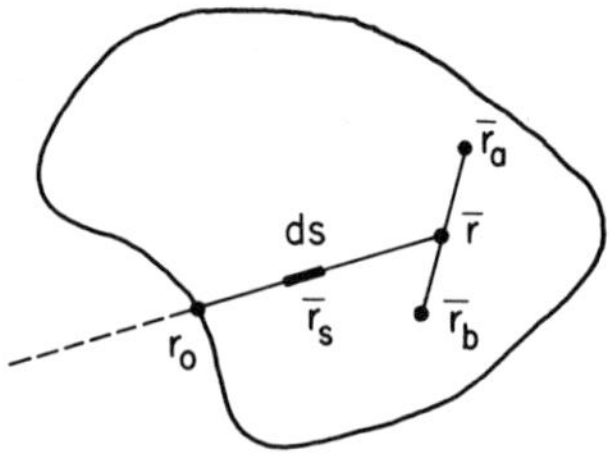

FIG. 14-17 Geometry showing the quantities in (14-93).

where $(\sigma_t/4\pi)p(\hat{\mathbf{s}}, \hat{\mathbf{s}}') = |f(\hat{\mathbf{s}}, \hat{\mathbf{s}}')|^2$ and $I_c(\mathbf{r}, \hat{\mathbf{s}})$ is identical to the reduced incident intensity and satisfies the equation

$$\frac{d}{ds} I_c(\mathbf{r}, \hat{\mathbf{s}}) = -\rho\sigma_t I_c(\mathbf{r}, \hat{\mathbf{s}}). \tag{14-94b}$$

We have shown in this section that Twersky's equations for multiple scattering are equivalent to the equation of transfer under certain assumptions. We note, therefore, that if one solves the equation of transfer for specific intensity $I(\mathbf{r}, \hat{\mathbf{s}})$ in a given problem, then the correlation function $\Gamma(\mathbf{r}, \mathbf{r}_d)$ can be obtained by taking the transform

$$\Gamma(\mathbf{r}, \mathbf{r}_d) = \langle \psi^a \psi^{b*} \rangle = \int I(\mathbf{r}, \hat{\mathbf{s}}) \exp(iK_r\hat{\mathbf{s}} \cdot \mathbf{r}_d)\, d\omega. \tag{14-95}$$

We also note that the total intensity $U_t(\mathbf{r})$ at $\mathbf{r}$ is the sum of the coherent intensity $U_c(\mathbf{r})$ and the incoherent intensity $U_i(\mathbf{r})$, and they are related to the specific intensities through the following:

$$U_t(\mathbf{r}) = \langle |\psi(\mathbf{r})|^2 \rangle = \int_{4\pi} I(\mathbf{r}, \hat{\mathbf{s}})\, d\omega$$

$$U_c(\mathbf{r}) = |\langle \psi(\mathbf{r}) \rangle|^2 = \int_{4\pi} I_c(\mathbf{r}, \hat{\mathbf{s}})\, d\omega \tag{14-96}$$

$$U_i(\mathbf{r}) = \langle |\psi_f(\mathbf{r})|^2 \rangle = \int_{4\pi} I_d(\mathbf{r}, \hat{\mathbf{s}})\, d\omega.$$

These quantities are equal to 4π times the average intensity, coherent intensity, and diffuse intensity, respectively, defined in Section 7-4.

14-9 FUNDAMENTAL EQUATIONS FOR MOVING PARTICLES

In the preceding chapters we have assumed that all the particles are stationary in time, and, therefore, the field quantity $\psi(\mathbf{r})$ is a function of position only. In this section, we introduce the motion of particles and consider its effects (Ishimaru, 1975). The field quantity ψ should then become a function of position and time. We develop a fundamental integral equation and an approximation to this equation that is of the same form as the equation of transfer.

To analyze the effects of the particle motion, we need to reexamine the statistical average discussed in Section 14-2. Let us consider two random functions $f_s(\mathbf{r}_s', t_s')$ and $g_s(\mathbf{r}_s'', t_s'')$ which depend on a particle moving with a velocity $\mathbf{V}$ which is located at $\mathbf{r}_s''$ at a time t_s'', and r_s' at a later time t_s', respectively (Fig. 14-18).

Following the procedure in Section 14-2, the average of $f_s g_s$ is then given by

$$\langle f_s(\mathbf{r}_s', t_s') g_s(\mathbf{r}_s'', t_s'') \rangle = \int f_s(\mathbf{r}_s', t_s') g_s(\mathbf{r}_s'', t_s'') \frac{\rho}{N}\, d\mathbf{r}_s \tag{14-97}$$

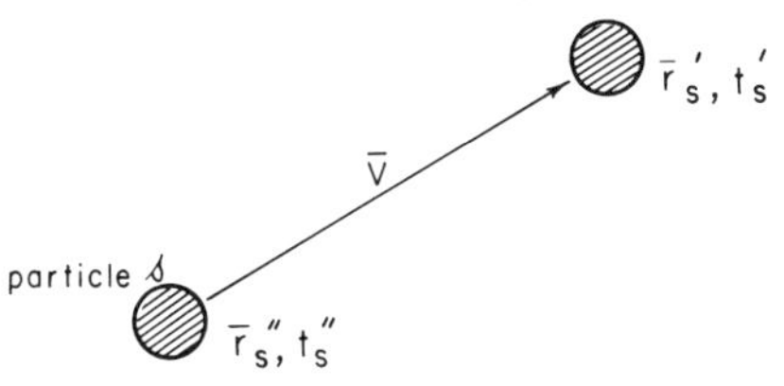

FIG. 14-18 Particle moves from $\mathbf{r}_s''$ at t_s'' to $\mathbf{r}_s'$ at t_s' with velocity V.

provided $\mathbf{r}_s'$ and $\mathbf{r}_s''$ are related through

$$\mathbf{r}_s' - \mathbf{r}_s'' = \mathbf{V}(t_s' - t_s'') \tag{14-98}$$

and $\mathbf{r}_s = \frac{1}{2}(\mathbf{r}_s' + \mathbf{r}_s'')$. Here we have assumed that the particle velocity $\mathbf{V}$ is constant during the time $t_s' - t_s''$.

We can express (14-97) and (14-98) in the more convenient form

$$\left\langle \sum_{s=1}^{N} f_s(\mathbf{r}_s', t_s') \sum_{s=1}^{N} g_s(\mathbf{r}_s'', t_s'') \right\rangle = \left\langle \sum_{s=1}^{N} f_s(\mathbf{r}_s', t_s') g_s(\mathbf{r}_s'', t_s'') \right\rangle$$

$$= \iint f_s(\mathbf{r}_s', t_s') g_s(\mathbf{r}_s'', t_s'') \rho(\mathbf{r}_s, t_s)\, \delta(\mathbf{r}_{sd} - \mathbf{V} t_{sd})\, d\mathbf{r}_s\, d\mathbf{r}_{sd}, \tag{14-99}$$

where $\mathbf{r}_{sd} = \mathbf{r}_s' - \mathbf{r}_s''$, $t_{sd} = t_s' - t_s''$, and $\rho(\mathbf{r}_s, t_s)$ means that the number density ρ must be evaluated at $\mathbf{r}_s = \frac{1}{2}(\mathbf{r}_s' + \mathbf{r}_s'')$ and $t_s = \frac{1}{2}(t_s' + t_s'')$. In (14-99) the average of the product of two series is reduced to that of one series because it is the sum of the average over the particle s and there are only N particles involved and not N^2 particles. Note that (14-99) is equivalent to writing

$$\begin{aligned} \sum_{s=1}^{N} f_s(\mathbf{r}_s', t_s') &= \int f_s(\mathbf{r}_s', t_s') \rho(\mathbf{r}_s', t_s')\, d\mathbf{r}_s' \\ \sum_{s=1}^{N} g_s(\mathbf{r}_s'', t_s'') &= \int g_s(\mathbf{r}_s'', t_s'') \rho(\mathbf{r}_s'', t_s'')\, d\mathbf{r}_s'' \end{aligned} \tag{14-100a}$$

and using the following:

$$\langle \rho(\mathbf{r}_s', t_s') \rho(\mathbf{r}_s'', t_s'') \rangle = \rho(\mathbf{r}_s, t_s)\, \delta(\mathbf{r}_{sd} - \mathbf{V} t_{sd}). \tag{14-100b}$$

Making use of this we write Twersky's integral equation (14-28) as follows:

$$\langle \psi^a \psi^{b*} \rangle = \langle \psi^a \rangle \langle \psi^{b*} \rangle + \iint V_{s'}^{a} V_{s''}^{b*} \langle |\psi^s|^2 \rangle \rho(\mathbf{r}_s, t_s)\, \delta(\mathbf{r}_{sd} - \mathbf{V} t_{sd})\, d\mathbf{r}_s\, d\mathbf{r}_{sd} \tag{14-101}$$

where $\psi^a = \psi(\mathbf{r}_a, t_a)$ is the field at $\mathbf{r}_a$ and t_a and $\psi^b = \psi(\mathbf{r}_b, t_b)$. The subscripts s' and s'' denote two points $\mathbf{r}_s'$ and $\mathbf{r}_s''$ which a single moving particle occupies at t_s' and t_s''. These two times t_s' and t_s'' are the retarded times due to the wave propagation and are given by

$$t_s' = t_a - \frac{|\mathbf{r}_a - \mathbf{r}_s'|}{v_p}, \qquad t_s'' = t_b - \frac{|\mathbf{r}_b - \mathbf{r}_s''|}{v_p}. \tag{14-102}$$

where v_p is the phase velocity of the wave in the medium and is given by ω/K_r.

We write (14-101) in the more explicit form

$$\langle\psi^a\psi^{b*}\rangle = \langle\psi^a\rangle\langle\psi^{b*}\rangle + \Gamma_{fab}, \tag{14-103a}$$

$$\Gamma_{fab} = \int \rho\, d\mathbf{r}_s\, d\mathbf{r}_{sd}\, \delta(r_{sd} - \mathbf{V}t_{sd}) \frac{\exp(iK|\mathbf{r}_a - \mathbf{r}_s'| - iK^*|\mathbf{r}_b - \mathbf{r}_s''|)}{|\mathbf{r}_a - \mathbf{r}_s'|\,|\mathbf{r}_b - \mathbf{r}_s''|}$$

$$\times \int d\omega' f(\hat{\mathbf{s}}_a, \hat{\mathbf{s}}')f^*(\hat{\mathbf{s}}_b, \hat{\mathbf{s}}')I(\mathbf{r}_s, \hat{\mathbf{s}}', t_s, t_{sd}) \exp(iK_r\hat{\mathbf{s}}' \cdot \mathbf{r}_{sd}) \tag{14-103b}$$

where we have used the notations $\mathbf{r}_d = \mathbf{r}_a - \mathbf{r}_b$, $t_d = t_a - t_b$, and $t = \frac{1}{2}(t_a + t_b)$ (see Fig. 14-19).

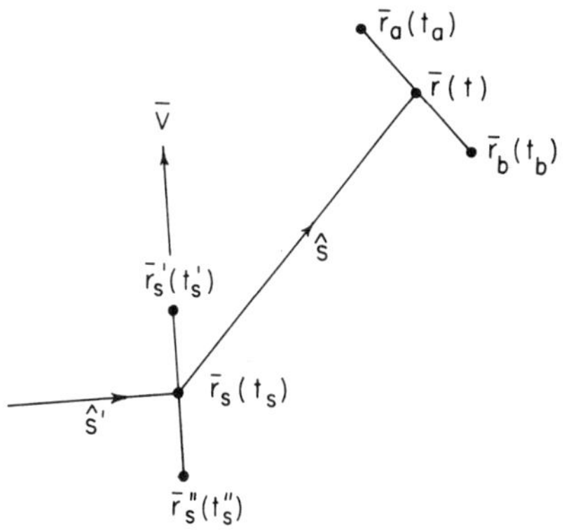

FIG. 14-19 Notations used in (14-103b).

The specific intensity in (14-103b) is time varying and is related to the correlation function through

$$\langle\psi^a\psi^{b*}\rangle = \langle\psi(\mathbf{r}_a, t_a)\psi(\mathbf{r}_b, t_b)^*\rangle$$

$$= \int I(\mathbf{r}, \hat{\mathbf{s}}, t, t_d) \exp(iK_r\hat{\mathbf{s}} \cdot \mathbf{r}_d)\, d\omega$$

$$\langle\psi^a\rangle\langle\psi^{b*}\rangle = \int I_c(\mathbf{r}, \hat{\mathbf{s}}, t, t_d) \exp(iK_r\hat{\mathbf{s}} \cdot \mathbf{r}_d)\, d\omega. \tag{14-104a}$$

The correlation function $\langle\psi^a\psi^{b*}\rangle$ is also called the mutual coherence function

$$\Gamma(\mathbf{r}_a, t_a; \mathbf{r}_b, t_b) = \langle\psi(\mathbf{r}_a, t_a)\psi(\mathbf{r}_b, t_b)^*\rangle. \tag{14-104b}$$

The time t, the retarded time t_s, the time difference t_d at the observation point, and the time difference t_{sd} as the particle s moves through the distance $\mathbf{r}_{sd}$ are related by

$$t_s = t - \frac{|\mathbf{r} - \mathbf{r}_s|}{v_p}, \qquad \mathbf{r}_{sd} = \mathbf{V} t_{sd}, \tag{14-105a}$$

$$t_{sd} = t_d - \frac{1}{v_p}\frac{\mathbf{r}_d}{2} \cdot \hat{\mathbf{s}} + \frac{1}{v_p}(\mathbf{V} \cdot \hat{\mathbf{s}}) t_{sd}. \tag{14-105b}$$

Equations (14-103a) and (14-103b) are the fundamental integral equations for moving particles.

Equation (14-103b) can be further simplified by using the approximations

$$|\mathbf{r}_a - \mathbf{r}_s'| \approx |\mathbf{r} - \mathbf{r}_s| + \tfrac{1}{2}(\mathbf{r}_d - \mathbf{r}_{sd}) \cdot \hat{\mathbf{s}} \tag{14-106a}$$

$$|\mathbf{r}_b - \mathbf{r}_s''| \approx |\mathbf{r} - \mathbf{r}_s| - \tfrac{1}{2}(\mathbf{r}_d - \mathbf{r}_{sd}) \cdot \hat{\mathbf{s}} \tag{14-106b}$$

$$t_{sd} \approx t_d. \tag{14-106c}$$

Equations (14-106a) and (14-106b) are identical to (14-91), and (14-106c) is valid as long as the velocity of the particle $|\mathbf{V}|$ is much smaller than the velocity of the wave propagation v_p because t_d is on the order of the correlation time and $|\mathbf{r}_d|$ is on the order of $|\mathbf{V}|$ times the correlation time [see (14-105b)].

Equation (14-103b) then becomes

$$\Gamma_{fab} = \int \rho \, d\mathbf{r}_s \frac{\exp(iK_r \hat{\mathbf{s}} \cdot \mathbf{r}_d - iK_r(\hat{\mathbf{s}} - \hat{\mathbf{s}}') \cdot \mathbf{V} t_{sd} - \rho\sigma_t |\mathbf{r} - \mathbf{r}_s|)}{|r - r_s|^2}$$

$$\times \int d\omega' |f(\hat{\mathbf{s}}, \hat{\mathbf{s}}')|^2 I(\mathbf{r}_s, \hat{\mathbf{s}}', t_s, t_{sd}). \tag{14-107}$$

Employing the same procedure used to obtain (14-94), we obtain the following equation of transfer for the time-varying specific intensity $I(\mathbf{r}, \hat{\mathbf{s}}, t, t_d)$:

$$\frac{dI(\mathbf{r}, \hat{\mathbf{s}}, t, t_d)}{ds} = -\rho\sigma_t I(\mathbf{r}, \hat{\mathbf{s}}, t, t_d) + \int \rho F(\hat{\mathbf{s}}, \hat{\mathbf{s}}', t_d) I(\mathbf{r}, \hat{\mathbf{s}}', t, t_d)\, d\omega' \tag{14-108a}$$

where

$$F(\hat{\mathbf{s}}, \hat{\mathbf{s}}', t_d) = |f(\hat{\mathbf{s}}, \hat{\mathbf{s}}')|^2 \exp[-iK_r(\hat{\mathbf{s}} - \hat{\mathbf{s}}') \cdot \mathbf{V} t_d]. \tag{14-108b}$$

Equations (14-108a) and (14-108b) constitute the equation of transfer when the particles are moving with the velocity $\mathbf{V}$. Note that the scattering function $|f(\hat{\mathbf{s}}, \hat{\mathbf{s}}')|^2$ for the time-invariant particles is replaced by (14-108b).

If the velocity $\mathbf{V}$ consists of the average velocity $\mathbf{U}$ and the velocity fluctuation $\mathbf{V}_f$, we need to take the average over $\mathbf{V}_f$ in (14-108b). It is reasonable to assume that the specific intensity I depends on contributions from all parts of the random medium and therefore it is almost independent of the velocity fluctuation $\mathbf{V}_f$ at the same location. Under this assumption, letting $\mathbf{V} = \mathbf{U} + \mathbf{V}_f$, we have

$$F(\hat{\mathbf{s}}, \hat{\mathbf{s}}', t_d) = |f(\hat{\mathbf{s}}, \hat{\mathbf{s}}')|^2 \chi[-K_r(\hat{\mathbf{s}} - \hat{\mathbf{s}}')t_d] \exp[-iK_r(\hat{\mathbf{s}} - \hat{\mathbf{s}}') \cdot \mathbf{U}t_d], \quad (14\text{-}108c)$$

where χ is the characteristic function (see Davenport and Root, 1958) given by

$$\chi[-K_r(\hat{\mathbf{s}} - \hat{\mathbf{s}}')t_d] = \langle \exp[-iK_r(\hat{\mathbf{s}} - \hat{\mathbf{s}}') \cdot \mathbf{V}_f t_d] \rangle. \quad (14\text{-}108d)$$

Equations (14-108a) and (14-108c) constitute the basic equation of transfer when the motions of the particles consist of the average velocity and the velocity fluctuation.

As an example, if the velocity fluctuation $\mathbf{V}_f$ is assumed to be Gaussian with the probability density

$$p(V_f) = (2\pi\sigma_v^2)^{-3/2} \exp(-V_f^2/2\sigma_v^2) \quad (14\text{-}109a)$$

and the variance σ_v^2, then the characteristic function becomes

$$\chi(-K_r(\hat{\mathbf{s}} - \hat{\mathbf{s}}')t_d) = \exp[-\tfrac{1}{2}(K_r^2 |\hat{\mathbf{s}} - \hat{\mathbf{s}}'|^2 \sigma_v^2 t_d^2)]. \quad (14\text{-}109b)$$

Up to this point, we have discussed the equations for the time-varying mutual coherence function $\langle \psi(\mathbf{r}_a, t_a)\psi^*(\mathbf{r}_b, t_b) \rangle$ and the time-varying specific intensity $I(\mathbf{r}, \hat{\mathbf{s}}, t, t_d)$. They represent the correlation in time difference t_d. If we take the Fourier transform in t_d, we get the temporal spectrum. We therefore define the temporal frequency spectrum $W(\mathbf{r}, t, \Omega)$ and the temporal frequency spectrum of the specific intensity $W_I(\mathbf{r}, \hat{\mathbf{s}}, t, \Omega)$ as follows:

$$W(\mathbf{r}, t, \Omega) = 2 \int_{-\infty}^{\infty} \langle \psi(\mathbf{r}, t_a)\psi(\mathbf{r}, t_b)^* \rangle \exp(i\Omega t_d)\, dt_d \quad (14\text{-}110a)$$

$$W_I(\mathbf{r}, \hat{\mathbf{s}}, t, \Omega) = 2 \int_{-\infty}^{\infty} I(\mathbf{r}, \hat{\mathbf{s}}, t, t_d) \exp(i\Omega t_d)\, dt_d. \quad (14\text{-}110b)$$

where Ω is the temporal frequency and W and W_I are related by the integral

$$W(\mathbf{r}, t, \Omega) = \int_{4\pi} W_I(\mathbf{r}, \hat{\mathbf{s}}, t, \Omega)\, d\omega. \quad (14\text{-}111)$$

We note here that even though W and W_I are functions of time t in general, the field fluctuations are statistically stationary in most practical cases, and therefore W and W_I are independent of time t.

It is possible to convert the equation of transfer (14-108a) into the equation for the temporal frequency spectrum. We have

$$\frac{d}{ds} W_I(\mathbf{r}, \hat{\mathbf{s}}, \Omega) = -\rho\sigma_t W_I(\mathbf{r}, \hat{\mathbf{s}}, \Omega) + \frac{1}{4\pi}\int_{-\infty}^{\infty} d\Omega' \times \left[\int \rho G(\hat{\mathbf{s}}, \hat{\mathbf{s}}', \Omega - \Omega') W_I(\mathbf{r}, \hat{\mathbf{s}}', \Omega')\, d\omega'\right], \tag{14-112a}$$

where

$$G(\hat{\mathbf{s}}, \hat{\mathbf{s}}', \Omega) = 2\int_{-\infty}^{\infty} F(\hat{\mathbf{s}}, \hat{\mathbf{s}}', t_d) \exp(+i\Omega t_d)\, dt_d. \tag{14-112b}$$

If the velocity is uniform and has no fluctuation, we let $\mathbf{V} = \mathbf{U}$ in (14-108b) and substituting into (14-112a) and (14-112b) we get

$$\frac{d}{ds} W_I(\mathbf{r}, \hat{\mathbf{s}}, \Omega) = -\rho\sigma_t W_I(\mathbf{r}, \hat{\mathbf{s}}, \Omega) + \int \rho |f(\hat{\mathbf{s}}, \hat{\mathbf{s}}')|^2 W_I(\mathbf{r}, \hat{\mathbf{s}}', \Omega')\, d\omega', \tag{14-113}$$

where $\Omega' = \Omega - K_r(\hat{\mathbf{s}} - \hat{\mathbf{s}}') \cdot \mathbf{U}$. This clearly shows the Doppler shift of frequency from Ω' to Ω. The wave with frequency Ω' is incident on the particle in the direction $\hat{\mathbf{s}}'$ and the scattered wave in the direction $\hat{\mathbf{s}}$ has the frequency Ω. The scattering amplitude $f(\hat{\mathbf{s}}, \hat{\mathbf{s}}')$ should be considered as the scattered wave in the direction $\hat{\mathbf{s}}$ with the frequency Ω when the incident wave is incident in the direction $\hat{\mathbf{s}}'$ with the frequency Ω', and, therefore, more strictly we should write $f(\hat{\mathbf{s}}, \Omega, \hat{\mathbf{s}}', \Omega')$.

We can also obtain a similar equation when the velocity $\mathbf{V}$ consists of the average $\mathbf{U}$ and the fluctuation $\mathbf{V}_f$. In this case we have

$$G(\hat{\mathbf{s}}, \hat{\mathbf{s}}', \Omega) = \frac{4\pi |f(\hat{\mathbf{s}}, \hat{\mathbf{s}}')|^2}{[2K_r^2 |\hat{\mathbf{s}} - \hat{\mathbf{s}}'|^2 \sigma_v^2]^{1/2}} \exp\left\{-\frac{[\Omega - K_r(\hat{\mathbf{s}} - \hat{\mathbf{s}}') \cdot \mathbf{U}]^2}{2K_r^2 |\hat{\mathbf{s}} - \hat{\mathbf{s}}'|^2 \sigma_v^2}\right\}, \tag{14-114}$$

showing the Doppler shift due to $\mathbf{U}$ and the spectrum broadening due to the velocity fluctuation.

We note that even though Eq. (14-112a) is available for W_I, it is often more convenient first to find the equation for the time-varying specific intensity or mutual coherence function and then obtain the temporal spectrum by taking the Fourier transform (14-110a) or (14-110b).

In this section, we have shown that when the particles are in motion, the equation of transfer may be modified by using the time-correlated scattering amplitude (14-108c) in the equation of transfer (14-108a). We note that the time-correlated scattering amplitude (14-108c) is identical to (4-48) in Section 4-5 with the correspondence $\hat{\mathbf{s}} = \hat{\mathbf{0}}$, $\hat{\mathbf{s}}' = \hat{\mathbf{i}}$, $t_d = \tau$, and $K_r = k$. This is to be expected since both deal with the same moving scatterers. In fact, the equation of transfer (14-108a) can be derived simply by recognizing this correspondence. However, the derivation given in this section is more fundamental and shows the approximation needed in the derivation.

14-10 FLUCTUATIONS DUE TO THE SIZE DISTRIBUTION

In the preceding section we assumed that all particles in the medium have the same size. In practice, however, the particles are often not one size, but their sizes are distributed over a certain range. We define the probability density function $W(D)$ for finding the particle size between D and $D + dD$:

$$W(D) = \frac{n(D)}{\rho} \qquad \text{where} \quad \int_0^\infty W(D)\, dD = 1 \tag{14-115}$$

and $n(D)\, dD$ is the number of particles per unit volume having a dimension between D and $D + dD$. ρ is the total number of particles per unit volume and is called number density or simply density.

$$\rho = \int_0^\infty n(D)\, dD. \tag{14-116}$$

Making use of this, we can generalize the specific intensity obtained for particles with a certain density ρ and size D to include the size distribution. The average of a quantity $f(D)$ which depends on the size D is obtained by

$$\langle f(D) \rangle_s = \int_0^\infty f(D) W(D)\, dD. \tag{14-117}$$

For example, consider the attenuation of the reduced incident intensity

$$I_{ri} = I_0 \exp(-\rho\sigma_t L) \tag{14-118}$$

obtained when the particles have the same size D and the density ρ. The average of the attenuation constant $\rho\sigma_t$ when the size is distributed is then given by

$$\langle \rho\sigma_t \rangle_s = \int_0^\infty \rho\sigma_t W(D)\, dD = \int_0^\infty \sigma_t(D) n(D)\, dD. \tag{14-119}$$

The variance of the quantity $f(D)$ is given by

$$\sigma_f{}^2 = \langle [f(D) - \langle f(D) \rangle]^2 \rangle_s = \int_0^\infty [f(D) - \langle f(D) \rangle]^2 W(D)\, dD. \tag{14-120}$$

If $f(D)$ is a slowly varying monotonic function of D and the spread of the size distribution is small, then we can obtain an approximate expression of the variance σ_f^2 in the following manner. We expand $f(D)$ in Taylor's series about the average size $D_0 = \langle D \rangle$ and keep the first two terms:

$$f(D) = f(D_0) + (D - D_0)\,\partial f/\partial D\,|_{D_0} + \cdots \tag{14-121}$$

where $D_0 = \langle D \rangle = \int_0^\infty D W(D)\,dD$ and $f(D_0) = \langle f(D) \rangle$. Substituting (14-121) into (14-120), we get an approximation

$$\sigma_f^2 \approx (\partial f/\partial D)_{D_0}^2 \sigma_D^2 \tag{14-122}$$

where $\sigma_D^2 = \int_0^\infty (D - D_0)^2 W(D)\,dD$ is the variance of the size distribution.

For example, the variance of $f = \rho\sigma_t$ is given approximately by

$$\sigma_f^2 \approx \rho^2\,|\partial\sigma_t/\partial D\,|_{D_0}^2 \sigma_D^2. \tag{14-123}$$

APPENDIX 14A EXAMPLE OF TWERSKY'S SCATTERING PROCESS WHEN $N = 3$

The scattering process for the incoherent intensity contained in Twersky's integral equation when only three scatterers are present is shown in the following. Altogether, 159 different processes are represented by Twersky's equation [taken from Twersky, 1964, Eq. (65)].

(a) *Single Scattering Process* There are three cases ($s = 1, 2, 3$) involved in this process (Fig. 14A-1).

(b) *Double Scattering Process* There are five distinct cases for each s and t as shown below. There are six combinations of s and t, and thus, there are $5 \times 6 = 30$ processes altogether (Fig. 14A-2):

s	1 1	2 2	3 3
t	2 3	1 3	1 2

(c) *Triple Scattering Process* There are six combinations of s, t, and m:

s	1 1	2 2	3 3
t	2 3	1 3	1 2
m	3 2	3 1	2 1

For each s, t, and m, there are 21 cases shown. Altogether, there are 126 cases (Fig. 14A-3).

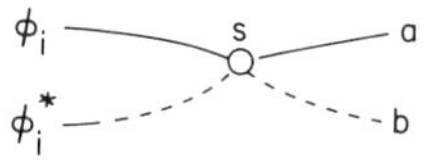

FIG. 14A-1 Single scattering.

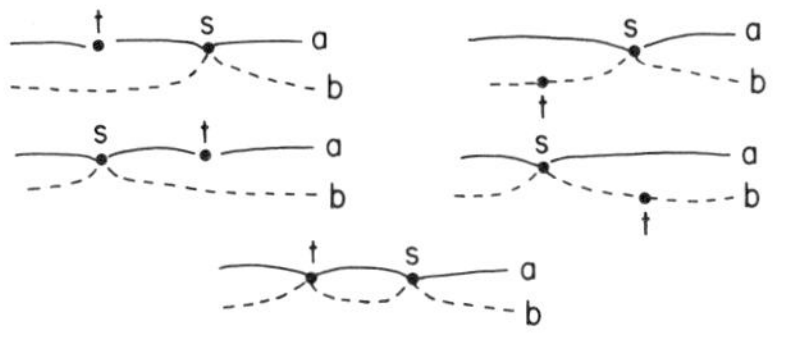

FIG. 14A-2 Double scattering.

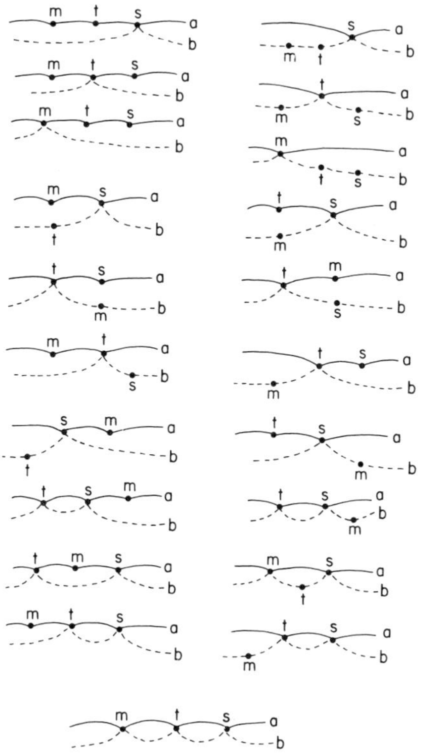

FIG. 14A-3 Triple scattering.

APPENDIX 14B STATIONARY PHASE EVALUATION OF A MULTIPLE INTEGRAL I

$$I = \int_{-\infty}^{\infty} dx_1 \int_{-\infty}^{\infty} dx_2 \cdots \int_{-\infty}^{\infty} dx_N A(x_1 \cdots x_N) \exp[if(x_1, x_2, \ldots, x_N)]. \tag{14B-1}$$

First we find a stationary phase point $(x_{10}, x_{20}, x_{30}, \ldots, x_{NO})$ by satisfying N equations:

$$\partial f/\partial x_1 = \partial f/\partial x_2 = \partial f/\partial x_3 = \cdots = \partial f/\partial x_N = 0. \tag{14B-2}$$

Then, we expand f about this stationary phase point:

$$\begin{aligned} f(x_1, x_2, \ldots, x_N) = f(x_{10}, x_{20}, \ldots, x_{NO}) &\\ + \frac{1}{2!}\bigg[(x_1 - x_{10})\frac{\partial}{\partial x_1} + (x_2 - x_{20})\frac{\partial}{\partial x_2} &\\ + \cdots + (x_N - x_{NO})\frac{\partial}{\partial x_N}\bigg]^2 f\bigg|_{x_{10}, x_{20}, \ldots} &\\ + \text{higher order terms.} & \end{aligned}$$

These higher order terms contribute little to the integral unless the second derivatives of f are small. Furthermore, we assume that the amplitude $A(x_1, \ldots, x_N)$ is a slowly varying function of $x_1, \ldots, x_N$, and thus we approximate

$$A(x_1, \ldots, x_N) \simeq A(x_{10}, x_{20}, \ldots, x_{NO}).$$

Then we write, letting $x_1 - x_{10} = x_1'$, $x_2 - x_{20} = x_2'$, ...,

$$\begin{aligned} I = A(x_{10}, x_{20}, \ldots, x_{NO}) \exp[if(x_{10}, x_{20}, \ldots, x_N)] &\\ \times \int_{-\infty}^{\infty} dx_1' \int_{-\infty}^{\infty} dx_2' \cdots \int_{-\infty}^{\infty} dx_N' \exp(i\tfrac{1}{2}[T]), & \end{aligned} \tag{14B-3}$$

where

$$[T] = \left(x_1' \frac{\partial}{\partial x_1'} + x_2' \frac{\partial}{\partial x_2'} + \cdots + x_N' \frac{\partial}{\partial x_N'}\right)^2 f.$$

We write $[T]$ in the matrix form

$$[T] = \tilde{x} F x$$

$$x = \begin{bmatrix} x_1' \\ \vdots \\ x_N' \end{bmatrix}, \qquad F = \begin{bmatrix} f_{11} & f_{12} & \cdots & f_{1N} \\ f_{21} & & & \\ \vdots & & & \\ f_{N1} & \cdots & \cdots & f_{NN} \end{bmatrix} \tag{14B-4}$$

where $\tilde{x}$ is a transpose of the matrix x, and

$$f_{ij} = \frac{\partial^2}{\partial x_i' \, \partial x_j'} f \Bigg|_{x_1'=0,\, x_2'=0,\, \dots} .$$

Next, we note that by the orthogonal transformation of X to Y†, $X = PY$, and

$$Y = \begin{bmatrix} y_1 \\ y_2 \\ \vdots \\ y_N \end{bmatrix}$$

we can convert $[T]$ into the diagonal form

$$[T] = \tilde{X} F X = \tilde{Y} \tilde{P} F P Y = \tilde{Y} \alpha Y \tag{14B-5}$$

where $\tilde{X}$ is a transpose of X and

$$\alpha = \begin{bmatrix} \alpha_1{}^2 & 0 & 0 & 0 & 0 \\ 0 & \alpha_2{}^2 & 0 & 0 & 0 \\ 0 & 0 & \alpha_3{}^2 & & 0 \\ \vdots & \vdots & \vdots & \ddots & \vdots \\ 0 & 0 & 0 & & \alpha_N{}^2 \end{bmatrix}$$

Thus we obtain

$$[T] = \alpha_1{}^2 y_1{}^2 + \alpha_2{}^2 y_2{}^2 + \cdots + \alpha_N{}^2 y_N{}^2. \tag{14B-6}$$

Also, note that the Jacobian of $x_1', \dots, x_N'$ with respect to $y_1, y_2, \dots, y_N$ is one:

$$dx_1' \, dx_2' \cdots dx_N' = \frac{\partial(x_1', \dots, x_N')}{\partial(y_1, \dots, y_N)} dy_1 \cdots dy_N$$

$$\text{Jacobian} = \frac{\partial(x_1' \cdots x_N')}{\partial(y_1 \cdots y_N)}$$

$$= \begin{vmatrix} \dfrac{\partial x_1'}{\partial y_1} & \cdots & \dfrac{\partial x_N'}{\partial y_1} \\ \vdots & & \\ \dfrac{\partial x_1'}{\partial y_N} & \cdots & \dfrac{\partial x_N'}{\partial y_N} \end{vmatrix} = |P| = 1.$$

† For two variables, this corresponds to a rotation of the coordinate system about the origin to make the axes coincide with the major and minor axes.

Therefore, we obtain

$$\int_{-\infty}^{\infty} dx_1' \int_{-\infty}^{\infty} dx_2' \cdots \int_{-\infty}^{\infty} dx_N' \exp(i\tfrac{1}{2}[T])$$

$$= \int_{-\infty}^{\infty} dy_1 \int_{-\infty}^{\infty} dy_2 \cdots \int_{-\infty}^{\infty} dy_N \exp(i[\alpha_1{}^2 y_2{}^2 + \alpha_2{}^2 y_2{}^2 + \cdots + \alpha_N{}^2 y_N{}^2])$$

$$= \frac{(2\pi)^{n/2} e^{in\pi/4}}{\sqrt{\alpha_1{}^2 \alpha_2{}^2 \cdots \alpha_N{}^2}}. \tag{14B-7}$$

But, since

$$\alpha_1{}^2 \alpha_2{}^2 \cdots \alpha_N{}^2 = |\alpha| = |\tilde{P}FP| = |\tilde{P}|\,|F|\,|P| = |F|,$$

we obtain

$$I = A(x_{10}, x_{20}, \ldots, x_{N0}) \exp[if(x_{10}, x_{20}, \ldots, x_{N0})] \frac{(2\pi)^{n/2} e^{in\pi/4}}{\sqrt{\Delta}}, \tag{14B-8}$$

where

$$\Delta = \text{the determinant of } F = |F|,$$

and is called Hesse's determinant. For $N = 2$, we have

$$I = \int_{-\infty}^{\infty} dx_1 \int_{-\infty}^{\infty} dx_2\, A(x_1 x_2) \exp[if(x_1 x_2)]$$

$$= A(x_{10} x_{20}) \exp[if(x_{10} x_{20})] \frac{(2\pi) e^{i\pi/2}}{\sqrt{f_{11} f_{22} - f_{12}^2}}, \tag{14B-9}$$

where x_{10}, x_{20} are given by $\partial f/\partial x_1 = 0$ and $\partial f/\partial x_2 = 0$.

Example I

$$I = \int_{-\infty}^{\infty} dx_s \int_{-\infty}^{\infty} dy_s A(x_s, y_s) \frac{\exp(ik|r_s - r_a|)}{|r_s - r_a|}. \tag{14B-10}$$

In this case,

$$f = k\sqrt{(x_s - x_a)^2 + (y_s - y_a)^2 + (z_s - z_a)^2} = kr$$

and thus the stationary phase point (x_{s0}, y_{s0}) is given by

$$\frac{\partial f}{\partial x_s} = k \frac{\partial r}{\partial x_s} = k \frac{(x_s - x_a)}{r} = 0, \qquad \frac{\partial f}{\partial y_s} = k \frac{(y_s - y_a)}{r} = 0. \tag{14B-11}$$

Therefore, $x_{s0} = x_a$ and $y_{s0} = y_a$. Furthermore,

$$\frac{\partial^2 f}{\partial x_s{}^2} = k\left(\frac{1}{r} - \frac{(x_s - x_a)^2}{r^3}\right), \qquad \frac{\partial^2 f}{\partial y_s{}^2} = k\left(\frac{1}{r} - \frac{(y_s - y_a)^2}{r^3}\right)$$

$$\frac{\partial^2 f}{\partial x_s\, \partial y_s} = k(x_s - x_a)\left(-\frac{1}{r^2}\right)\frac{(y_s - y_a)}{r},$$

and therefore

$$I \simeq A(x_{s0}, y_{s0})\frac{2\pi i}{k}\exp(ik\,|z_s - z_a|). \tag{14B-12}$$

Example II

$$I = \int_{-\infty}^{\infty} dx \int_{-\infty}^{\infty} dy A(x, y)\frac{\exp[ik(r_1 + r_2)]}{r_1 r_2}, \tag{14B-13}$$

where $r_1{}^2 = x^2 + y^2 + z^2$ and $r_2{}^2 = (X - x)^2 + (Y - y)^2 + (Z - z)^2$. The stationary phase point (x_s, y_s, z) is given by

$$f = k(r_1 + r_2), \qquad \frac{\partial f}{\partial x} = k\left(\frac{x}{r_1} - \frac{(X - x)}{r_2}\right) = 0,$$

$$\frac{\partial f}{\partial y} = k\left(\frac{y}{r_1} - \frac{(Y - y)}{r_2}\right) = 0,$$

yielding (see Fig. 14B-1)

$$\frac{x_s}{X} = \frac{y_s}{Y} = \frac{r_{1s}}{r_{1s} + r_{2s}}.$$

We note that

$$\frac{\partial^2 f}{\partial x^2} = k\left[\frac{1}{r_1}\left(1 - \frac{x^2}{r_1{}^2}\right) + \frac{1}{r_2}\left(1 - \frac{(x - X)^2}{r_2{}^2}\right)\right]$$

$$\frac{\partial^2 f}{\partial y^2} = k\left[\frac{1}{r_1}\left(1 - \frac{y^2}{r_1{}^2}\right) + \frac{1}{r_2}\left(1 - \frac{(y - Y)^2}{r_2{}^2}\right)\right]$$

$$\frac{\partial^2 f}{\partial x\, \partial y} = k\left[-\frac{xy}{r_1{}^3} - \frac{(x - X)(y - Y)}{r_2{}^3}\right].$$

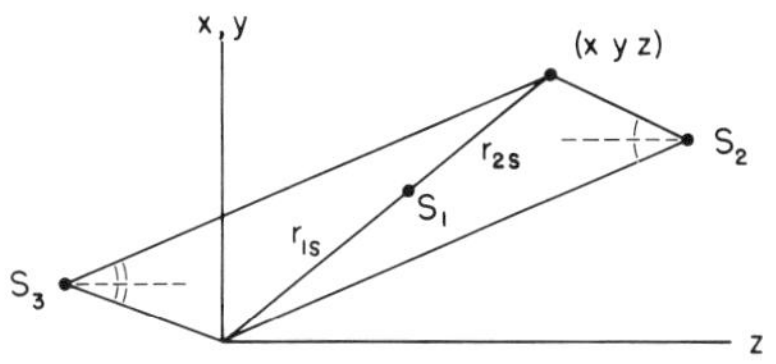

FIG. 14B-1 Stationary phase points S_1, S_2, and S_3 for (14B-13).

At the stationary phase point, we obtain

$$\frac{\partial^2 f}{\partial x^2} = k\left(\frac{1}{r_{1s}} + \frac{1}{r_{2s}}\right)\left(1 - \frac{x_s^{\,2}}{r_{1s}^2}\right)$$

$$\frac{\partial^2 f}{\partial y^2} = k\left(\frac{1}{r_{1s}} + \frac{1}{r_{2s}}\right)\left(1 - \frac{y_s^{\,2}}{r_{1s}^2}\right)$$

$$\frac{\partial^2 f}{\partial x\,\partial y} = k\left(\frac{1}{r_{1s}} + \frac{1}{r_{2s}}\right)\left(1 - \frac{x_s y_s}{r_{1s}^2}\right)$$

$$\begin{aligned}\frac{\partial^2 f}{\partial x^2}\frac{\partial^2 f}{\partial y^2} - \left(\frac{\partial^2 f}{\partial x\,\partial y}\right)^2 &= k^2\left(\frac{1}{r_{1s}} + \frac{1}{r_{2s}}\right)^2\left(1 - \frac{x_s^{\,2} + y_s^{\,2}}{r_1^{\,2}}\right)\\ &= k^2\left(\frac{1}{r_{1s}} + \frac{1}{r_{2s}}\right)^2\left(1 - \frac{X^2 + Y^2}{(r_{1s} + r_{2s})^2}\right)\\ &= k^2\frac{Z_s^{\,2}}{(r_{1s}r_{2s})^2}, \qquad Z_s^{\,2} = (r_{1s} + r_{2s})^2 - (X^2 + Y^2).\end{aligned}$$

Therefore, we get

$$\begin{aligned} I &\simeq A(x_s y_s)\frac{\exp[ik(r_{1s} + r_{2s})]}{r_{1s}r_{2s}}\frac{2\pi i}{(kZ_s/r_{1s}r_{2s})}\\ &= A(x_s y_s)\exp[ik(r_{1s} + r_{2s})]\frac{2\pi i}{kZ_s}. \end{aligned} \tag{14B-14}$$

For $0 < z < Z$, $Z_s = Z$; for $z < 0$, $Z_s = Z + 2|z|$; for $z > Z$, $Z_s = 2z - Z$.

APPENDIX 14C FORWARD SCATTERING THEOREM

Let us consider a plane wave of unit amplitude ψ_i incident on a scatterer with relative dielectric constant $\varepsilon(r)$. In general, $\varepsilon(r)$ is complex (lossy). The incident wave is given by

$$\psi_i(r) = \exp(ik\hat{\mathbf{i}}\cdot\mathbf{r}) \tag{14C-1}$$

where $\hat{\mathbf{i}}$ is a unit vector in the direction of propagation and $k = 2\pi/\lambda$, the free space wave number.

The total field $\psi(r)$ satisfies the wave equation

$$[\nabla^2 + k^2\varepsilon(r)]\psi(\mathbf{r}) = 0. \tag{14C-2}$$

First, we write this as

$$[\nabla^2 + k^2]\psi = -k^2(\varepsilon - 1)\psi$$

and obtain the integral equation for ψ:

$$\psi(\mathbf{r}) = \psi_i(\mathbf{r}) + \int_{V'}\frac{\exp(ik|\mathbf{r} - \mathbf{r}'|)}{4\pi|\mathbf{r} - \mathbf{r}'|}k^2(\varepsilon - 1)\psi(\mathbf{r}')\,dV'. \tag{14C-3}$$

The scattered field $u(\mathbf{r})$ is then

$$u(\mathbf{r}) = \int_{V'} \frac{\exp(ik|\mathbf{r} - \mathbf{r}'|)}{4\pi|\mathbf{r} - \mathbf{r}'|} k^2(\varepsilon - 1)\psi(\mathbf{r}')\, dV',$$

which at a large distance from the scatterer becomes

$$u_s{}^a = \frac{\exp(ik|\mathbf{r}_a - \mathbf{r}_s|)}{|\mathbf{r}_a - \mathbf{r}_s|} f(\hat{\mathbf{0}}, \hat{\mathbf{i}}), \tag{14C-4}$$

where the "scattering amplitude" $f(\hat{\mathbf{0}}, \hat{\mathbf{i}})$ is given by

$$f(\hat{\mathbf{0}}, \hat{\mathbf{i}}) = \int_{V'} \frac{k^2(\varepsilon - 1)}{4\pi} \exp(-ik\hat{\mathbf{0}} \cdot \mathbf{r}')\psi(\mathbf{r}')\, dV'.$$

First, we get an expression for the forward scattering amplitude $f(\hat{\mathbf{i}}, \hat{\mathbf{i}})$. To do this, we consider Green's theorem:

$$\int_V (v_1 \nabla^2 v_2 - v_2 \nabla^2 v_1)\, dV = \int_S \left(v_1 \frac{\partial v_2}{\partial n} - v_2 \frac{\partial v_1}{\partial n}\right) dS. \tag{14C-5}$$

We take the surface S so as to enclose the scatterer. Now we let $v_1 = \psi_i{}^*$ and $v_2 = \psi$. Then, we get

$$-\int_{V'} k^2(\varepsilon - 1)\psi_i{}^*\psi\, dV' = \int_S \left(\psi_i{}^* \frac{\partial \psi}{\partial n} - \psi \frac{\partial \psi_i{}^*}{\partial n}\right) dS. \tag{14C-6}$$

But since $\psi_i{}^* = \exp(-ik\hat{\mathbf{i}} \cdot \mathbf{r})$, the left-hand side of (14C-6) becomes, using (14C-4),

$$L_1 H_1 = 4\pi f(\hat{\mathbf{i}}, \hat{\mathbf{i}}). \tag{14C-7}$$

Therefore†

$$-4\pi \operatorname{Im} f(\hat{\mathbf{i}}, \hat{\mathbf{i}}) = \int_S \operatorname{Im}\left[\psi_i{}^* \frac{\partial u}{\partial n} - u \frac{\partial \psi_i{}^*}{\partial n}\right] dS. \tag{14C-8}$$

Now we define the absorption cross section σ_a. σ_a is the total absorbed power within the scatterer per uni' incident intensity. We define the power flux $\mathbf{P}$:

$$\mathbf{P} = \operatorname{Re}(\psi^* \nabla\psi / ik) = \operatorname{Im}(\psi^* \nabla\psi / k). \tag{14C-9}$$

† Note that $\psi = \psi_i + u$ and letting $v_1 = \psi_i{}^*$ and $v_2 = \psi_i$ in (14C-5),

$$\int_S \left(\psi_i{}^* \frac{\partial \psi_i}{\partial n} - \psi_i \frac{\partial \psi_i{}^*}{\partial n}\right) = 0.$$

Then the absorption cross section is given by

$$\sigma_a = -\int_S \operatorname{Im}\left(\frac{\psi^* \, \partial\psi/\partial n}{k}\right) dS, \tag{14C-10}$$

where $\partial/\partial n$ is the outward normal derivative and S is a surface enclosing the scatterer. S need not be the surface of the scatterer and can be taken at a large distance from it.

On the other hand, the scattering cross section σ_s is given by

$$\sigma_s = \int_S \operatorname{Im}\left(\frac{u^* \, \partial u/\partial n}{k}\right) dS, \tag{14C-11a}$$

which, using (14C-4), can be written as

$$\sigma_s = \int_{4\pi} |f(\hat{\mathbf{o}}, \hat{\mathbf{i}})|^2 \, d\Omega, \tag{14C-11b}$$

where Ω is the solid angle ($d\Omega = \sin\theta \, d\theta \, d\phi$). Now, we note from (14C-10) that

$$\sigma_a = -\frac{1}{2ki}\int_S \left(\psi^* \frac{\partial\psi}{\partial n} - \psi \frac{\partial\psi^*}{\partial n}\right) dS. \tag{14C-12a}$$

We note that

$$\psi^* \frac{\partial\psi}{\partial n} - \psi \frac{\partial\psi^*}{\partial n} = (\psi_i^* + u^*)\frac{\partial}{\partial n}(\psi_i + u) - (\psi_i + u)\frac{\partial}{\partial n}(\psi_i^* + u^*)$$

and, therefore,

$$\begin{aligned}\sigma_a = &-\frac{1}{2ki}\int_S \left[\left(\psi_i^* \frac{\partial u}{\partial n} - u \frac{\partial \psi_i^*}{\partial n}\right)\right. \\ &\left. - \left(\psi_i \frac{\partial u^*}{\partial n} - u^* \frac{\partial \psi_i}{\partial n}\right) + u^* \frac{\partial u}{\partial n} - u \frac{\partial u^*}{\partial n}\right] dS \\ = &-\frac{1}{k}\int_S \left\{\operatorname{Im}\left[\psi_i^* \frac{\partial u}{\partial n} - u \frac{\partial \psi_i^*}{\partial n}\right] + \operatorname{Im}\left(u^* \frac{\partial u}{\partial n}\right)\right\} dS.\end{aligned} \tag{14C-12b}$$

Substituting (14C-8) and (14C-11a) into (14C-12b), we obtain

$$\sigma_a = \frac{4\pi}{k} \operatorname{Im} f(\hat{\mathbf{i}}, \hat{\mathbf{i}}) - \sigma_s,$$

which completes the proof of the forward scattering theorem:

$$\frac{4\pi}{k} \operatorname{Im} f(\hat{\mathbf{i}}, \hat{\mathbf{i}}) = \sigma_s + \sigma_a. \tag{14C-13}$$

CHAPTER 15 □ MULTIPLE SCATTERING THEORY OF WAVE FLUCTUATIONS AND PULSE PROPAGATION IN RANDOMLY DISTRIBUTED SCATTERERS

In Chapters 4–6, we discussed the characteristics of wave fluctuations and pulse propagation for the special case of "tenuous" distribution of particles. Since particle density was assumed to be low, it was possible to make use of the single scattering approximation and its slight extension (first order multiple scattering). In Chapters 7–13, we dealt with the more general case in which the single scattering approximation is no longer valid, and we made use of the transport theory.

In Chapter 14, we presented general discussions on multiple scattering theory and derived fundamental equations for correlation functions in stationary as well as moving scatterers. Based on the multiple scattering theory presented in Chapter 14, we can now discuss in more detail fluctuation characteristics of waves and pulse propagation for the strong fluctuation case.

In this chapter we first summarize the fundamental equations. We then discuss the correlation function, angular spectrum, and frequency spectrum for the case in which particle sizes are comparable or greater than a wavelength. We show the general solutions and an example of plane wave incidence. The limitation on image resolution imposed by randomly distributed particles is then discussed. Backscattering and pulse propagation in strong fluctuation regions are analyzed and useful universal forms of pulse propagation characteristics are presented.

15-1 FUNDAMENTAL EQUATIONS FOR MOVING SCATTERERS

Let us restate the fundamental equations for the correlation function Γ of the field $\psi(\mathbf{r}, t)$ in randomly distributed scatterers. As shown in (14-104) and

(14-108), the correlation function (mutual coherence function) is related to the specific intensity I through a Fourier transform:

$$\Gamma(\mathbf{r}_a, t_a; \mathbf{r}_b, t_b) = \langle \psi(\mathbf{r}_a, t_a)\psi^*(\mathbf{r}_b, t_b)\rangle$$

$$= \int I(\mathbf{r}, \hat{\mathbf{s}}, t, t_d) \exp(iK_r \hat{\mathbf{s}} \cdot \mathbf{r}_d)\, d\omega. \tag{15-1}$$

The time-varying specific intensity satisfies

$$\frac{dI(\mathbf{r}, \hat{\mathbf{s}}, t, t_d)}{ds} = -\rho\sigma_t I(\mathbf{r}, \hat{\mathbf{s}}, t, t_d)$$

$$+ \int \rho F(\hat{\mathbf{s}}, \hat{\mathbf{s}}', t_d) I(\mathbf{r}, \hat{\mathbf{s}}', t, t_d)\, d\omega' \tag{15-2}$$

$$F(\hat{\mathbf{s}}, \hat{\mathbf{s}}, t_d) = |f(\hat{\mathbf{s}}, \hat{\mathbf{s}}')|^2 \chi[-K_r(\hat{\mathbf{s}} - \hat{\mathbf{s}}')t_d]$$

$$\times \exp[-iK_r(\hat{\mathbf{s}} - \hat{\mathbf{s}}') \cdot \mathbf{U} t_d] \tag{15-3}$$

where $\psi(\mathbf{r}, t)$ is the field at $\mathbf{r}$ and t, Γ the mutual coherence function, I the time-varying specific intensity, K_r the real part of the complex wave number $K = k + [2\pi f(\hat{\mathbf{s}}, \hat{\mathbf{s}})\rho]/k$, ρ the number density, $f(\hat{\mathbf{s}}, \hat{\mathbf{s}}')$ the scattering amplitude, χ the characteristic function of the fluctuation of the particle velocity, and $\mathbf{U}$ the average velocity of the particle. Equation (15-2) is identical to the equation of transfer except the time-correlated scattering amplitude (15-3), and therefore the solutions discussed in the preceding chapters can be used directly to obtain the time-varying specific intensity $I(\mathbf{r}, \hat{\mathbf{s}}, t, t_d)$. Once the specific intensity is obtained, the correlation function (mutual coherence function) $\Gamma(\mathbf{r}_a, t_a; \mathbf{r}_b, t_b)$ can be obtained by the Fourier transform (15-1). In the following sections we show some examples of the solution.

15-2 CORRELATION FUNCTION, ANGULAR SPECTRUM, AND FREQUENCY SPECTRUM IN THE SMALL ANGLE APPROXIMATION

If the particle sizes are comparable to or large compared with a wavelength, the wave scattered by the particles is largely confined within a small angle in the forward direction and it is possible to simplify the equation of transfer and get an exact general solution. Under this approximation, a variety of useful quantities can be calculated. In this section, we present general solutions for correlation function, angular spectrum, and frequency spectrum of the fluctuation of the field in randomly distributed scatterers.

The small angle approximation was discussed in Chapter 13. Following the derivation given in Section 13-1, and noting (13-5) the equation of transfer is approximated by

$$\frac{\partial}{\partial z} I(z, \boldsymbol{\rho}, \mathbf{s}, t_d) + \mathbf{s} \cdot \nabla_t I(z, \boldsymbol{\rho}, \mathbf{s}, t_d)$$

$$= -\rho_n \sigma_t I(z, \boldsymbol{\rho}, \mathbf{s}, t_d) + \frac{\rho_n \sigma_t}{4\pi} \iint_{-\infty}^{\infty} p(\mathbf{s} - \mathbf{s}', t_d) I(z, \boldsymbol{\rho}, \mathbf{s}, t_d)\, d\mathbf{s} \quad (15\text{-}4)$$

where the number density is denoted by ρ_n to avoid confusion with the radial vector $\boldsymbol{\rho}$ and

$$\mathbf{s} = l\hat{\mathbf{x}} + m\hat{\mathbf{y}}, \qquad \mathbf{r} = \boldsymbol{\rho} + z\hat{\mathbf{z}}, \qquad F(\hat{\mathbf{s}}, \hat{\mathbf{s}}', t_d) = \frac{\sigma_t}{4\pi} p(\mathbf{s} - \mathbf{s}', t_d).$$

The general solution of (15-4) was already obtained in (13-18):

$$I(z, \boldsymbol{\rho}, \mathbf{s}, t_d) = \frac{1}{(2\pi)^4} \int d\boldsymbol{\kappa} \int d\mathbf{q} \exp(-i\boldsymbol{\kappa} \cdot \boldsymbol{\rho} - i\mathbf{s} \cdot \mathbf{q}) F_0(\boldsymbol{\kappa}, \mathbf{q} + \boldsymbol{\kappa} z)$$
$$\times K(z, \boldsymbol{\kappa}, \mathbf{q}, t_d) \quad (15\text{-}5)$$

where

$$F_0(\boldsymbol{\kappa}, \mathbf{q}) = \iint I_0(\boldsymbol{\rho}, \mathbf{s}) \exp(-i\boldsymbol{\kappa} \cdot \boldsymbol{\rho} + i\mathbf{s} \cdot \mathbf{q})\, d\boldsymbol{\rho}\, d\mathbf{s}$$

$$K(z, \boldsymbol{\kappa}, \mathbf{q}, t_d) = \exp\left[-\int_0^z \rho_n \sigma_t \left\{1 - \frac{1}{4\pi} P(\mathbf{q} + \boldsymbol{\kappa}(z - z'), t_d)\right\} dz'\right]$$

$$P(\mathbf{q}, t_d) = \iint_{-\infty}^{\infty} p(\mathbf{s}, t_d)\, e^{i\mathbf{s} \cdot \mathbf{q}}\, d\mathbf{s}$$

where $I_0(\boldsymbol{\rho}, \mathbf{s})$ is the specific intensity at $z = 0$.

The correlation function at $(z, \boldsymbol{\rho})$ is then given by

$$\Gamma(z, \boldsymbol{\rho}, \boldsymbol{\rho}_d, t_d) = \langle \psi(z, \boldsymbol{\rho}_1, t_1) \psi^*(z, \boldsymbol{\rho}_2, t_2) \rangle$$
$$= \int I(z, \boldsymbol{\rho}, \mathbf{s}, t_d) \exp(iK_r \mathbf{s} \cdot \boldsymbol{\rho}_d)\, d\mathbf{s} \quad (15\text{-}6a)$$

where $\boldsymbol{\rho} = \frac{1}{2}(\boldsymbol{\rho}_1 + \boldsymbol{\rho}_2)$, $\boldsymbol{\rho}_d = \boldsymbol{\rho}_1 - \boldsymbol{\rho}_2$, and $t_d = t_1 - t_2$.

We can restate the formal solution (15-5) in terms of the mutual coherence function (15-6a). Using the mutual coherence function $\Gamma_0(\boldsymbol{\rho}, \boldsymbol{\rho}_d)$ at $z = 0$,

$$\Gamma_0(\boldsymbol{\rho}, \boldsymbol{\rho}_d) = \int I_0(\boldsymbol{\rho}, \mathbf{s}) \exp(iK_r \mathbf{s} \cdot \boldsymbol{\rho}_d)\, d\mathbf{s}, \quad (15\text{-}6b)$$

we get

$$\Gamma(z, \boldsymbol{\rho}, \boldsymbol{\rho}_d, t_d)$$
$$= \left(\frac{K_r}{2\pi z}\right)^2 \int d\boldsymbol{\rho}' \int d\boldsymbol{\rho}_d' \, \Gamma_0(\boldsymbol{\rho}', \boldsymbol{\rho}_d') \exp\left[i \frac{K_r}{z} (\boldsymbol{\rho}_d - \boldsymbol{\rho}_d') \cdot (\boldsymbol{\rho} - \boldsymbol{\rho}') - H\right]$$

$$H = \int_0^z \rho_n \sigma_t \left\{1 - \frac{1}{4\pi} P\left[\mathbf{K}_r \boldsymbol{\rho}_d' + K_r(\boldsymbol{\rho}_d - \boldsymbol{\rho}_d')\frac{z'}{z}, t_d\right]\right\} dz'. \tag{15-6c}$$

The angular spectrum at $(z, \boldsymbol{\rho})$ is given by the specific intensity

$$I(z, \boldsymbol{\rho}, \mathbf{s}, t_d = 0) = I(z, \boldsymbol{\rho}, l\hat{\mathbf{x}} + m\hat{\mathbf{y}}, t_d = 0),$$
$$l = \sin\theta \cos\phi, \qquad m = \sin\theta \sin\phi \tag{15-7}$$

where θ and ϕ are defined in spherical coordinates.

The frequency spectrum W at $(z, \boldsymbol{\rho})$ is given by

$$W(z, \boldsymbol{\rho}, \boldsymbol{\rho}_d = 0, \omega) = 2 \int_{-\infty}^{\infty} \Gamma(z, \boldsymbol{\rho}, \boldsymbol{\rho}_d = 0, t_d) \exp(i\omega t_d)\, dt_d$$
$$= 4 \int_0^{\infty} \Gamma(z, \boldsymbol{\rho}, \boldsymbol{\rho}_d = 0, t_d) \cos(\omega t_d)\, dt_d. \tag{15-8a}$$

The integral of the frequency spectrum over all frequencies is then equal to the total intensity P_t:

$$P_t(z, \boldsymbol{\rho}) = \Gamma(z, \boldsymbol{\rho}, \boldsymbol{\rho}_d = 0, t_d = 0)$$
$$= \frac{1}{4\pi} \int_{-\infty}^{\infty} W(z, \boldsymbol{\rho}, \boldsymbol{\rho}_d = 0, \omega)\, d\omega$$
$$= \int_0^{\infty} W(z, \boldsymbol{\rho}, \boldsymbol{\rho}_d = 0, \omega)\, df \tag{15-8b}$$

where $\omega = 2\pi f$.

In (15-7) we identified the specific intensity as the angular spectrum. Note that the angular spectrum is a two-dimensional Fourier transform (15-6a) of the mutual coherence function and the temporal spectrum (15-8a) is a one-dimensional Fourier transform in time.

15-3 PLANE WAVE SOLUTION

As an example, let us consider a plane wave incident upon randomly distributed scatterers. In this case, the specific intensity at $z = 0$ is given by [see (13-22)]

$$I_0(\boldsymbol{\rho}, \mathbf{s}) = I_0\, \delta(\mathbf{s}), \qquad F_0(\boldsymbol{\kappa}, \mathbf{q}) = (2\pi)^2 I_0\, \delta(\boldsymbol{\kappa}). \tag{15-9}$$

Let us assume that the particles are moving with the average velocity $\mathbf{U} = U\hat{\mathbf{x}}$ and the velocity fluctuation with the variance σ_v^2 [see (14-108) and (14-109)]. We also approximate the scattering pattern by a Gaussian function [see (6-20), (6-21), (6-129), and (13-24)]. We then have the following expression for F:

$$F(\mathbf{s}, t_d) = \frac{\alpha_p \sigma_s}{\pi} \exp\left[-\alpha_p s^2 - iK_r \mathbf{s} \cdot \mathbf{U} t_d - \frac{K_r^2 s^2 \sigma_v^2 t_d^2}{2}\right] \tag{15-10}$$

where α_p is approximately proportional to $(D/\lambda)^2$ and becomes equal to $2.66(D/\lambda)^2$ as the particle diameter D becomes large compared with a wavelength. We also get from (15-5)

$$P(\mathbf{q}, t_d) = \frac{4\pi W_0}{(1 + A)} \exp\left[-\frac{|\mathbf{q} - K_r \mathbf{U} t_d|^2}{4\alpha_p(1 + A)}\right] \tag{15-11}$$

where $A = K_r^2 \sigma_v^2 t_d^2 / 2\alpha_p$. Substituting (15-9) and (15-11) into (15-5), we get

$$I(z, \mathbf{s}, t_d) = \frac{I_0}{(2\pi)^2} \int d\mathbf{q} e^{-i\mathbf{s} \cdot \mathbf{q}} K(z, \mathbf{q}, t_d)$$

$$\begin{aligned} \Gamma(z, \mathbf{q}, t_d) &= I_0 K(z, \mathbf{q}, t_d) \\ &= I_0 \exp\left\{-\rho_n \sigma_t z \left[1 - \frac{W_0}{1 + A} \exp\left(-\frac{|\mathbf{q} - K_r \mathbf{U} t_d|^2}{4\alpha_p(1 + A)}\right)\right]\right\}. \end{aligned} \tag{15-12}$$

Note that since this is a plane wave solution, I and Γ are independent of $\boldsymbol{\rho}$. We also note that since Γ and I are related through (15-6), the function Γ in (15-12) is the mutual coherence function and $\mathbf{q}$ is equal to $K_r \boldsymbol{\rho}_d$:

$$\Gamma(z, \boldsymbol{\rho}_d, t_d) = \Gamma(z, \mathbf{q}, t_d)\Big|_{\mathbf{q} = K_r \boldsymbol{\rho}_d}. \tag{15-13}$$

Let us consider the angular spectrum (15-7). Letting $\mathbf{q} = q(\cos\phi'\hat{\mathbf{x}} + \sin\phi'\hat{\mathbf{y}})$ and $\mathbf{s} = \sin\theta(\cos\phi\hat{\mathbf{x}} + \sin\phi\hat{\mathbf{y}})$ in (5-12) and integrating with respect to ϕ', we obtain

$$I(z, \theta) = \frac{I_0}{2\pi} \int_0^\infty q\, dq J_0(q \sin\theta) \exp\left\{-\rho_n \sigma_t z \left[1 - W_0 \exp\left(-\frac{q^2}{4\alpha_p}\right)\right]\right\}. \tag{15-14}$$

We can evaluate (15-14) approximately for small and large optical distances $\tau = \rho_n \sigma_t z$. For small optical distance $\tau < 1$, we write approximately

$$\begin{aligned} \Gamma(z, \mathbf{q}, t_d = 0) &= I_0 \exp\{-\tau[1 - W_0 \exp(-q^2/4\alpha_p)]\} \\ &\approx I_0 \exp(-\tau)[1 + \tau W_0 \exp(-q^2/4\alpha_p)]. \end{aligned} \tag{15-15}$$

Substituting (15-15) into (15-14), we get

$$I(z, \theta) = I_0 e^{-\tau}[\delta(\theta) + (\alpha_p W_0 \tau/\pi) \exp(-\alpha_p \sin^2 \theta)]. \qquad (15\text{-}16)$$

The first term is the coherent intensity and is the same as the incident wave (15-9) except for the attenuation $\exp(-\tau)$. The second term represents the incoherent intensity; its magnitude increases with the distance τ, and its angular spread is identical to that of the scattering characteristics (15-10) of a particle. Note also that the flux F is given by

$$F(z) = \int I(z, \theta)\, d\mathbf{s} = I_0 e^{-\tau}[1 + W_0 \tau]. \qquad (15\text{-}17)$$

This is the first term of the series expansion of the total flux:

$$F(z) = I_0 \exp(-\rho_n \sigma_a z) = I_0 \exp[-\tau(1 - W_0)]. \qquad (15\text{-}18)$$

Next, let us consider the large optical distance $\tau \gg 1$. In this case we can approximate

$$\Gamma(z, \mathbf{q}, t_d = 0) \approx I_0\{\exp(-\tau) + \exp[-\tau(1 - W_0 + W_0 q^2/4\alpha_p)]\}. \quad (15\text{-}19)$$

Note that in this approximation, the first term gives rise to the coherent intensity and the second term gives the incoherent intensity. It should be noted that as $q \to \infty$, (15-14) reduces to the coherent intensity, and the approximation (15-19) is consistent with this behavior since the second term disappears as $q \to \infty$.

Substituting (15-19) into (15-14), we get

$$I(z, \theta) = I_0 e^{-\tau}\, \delta(\theta) + I_0(\alpha_p/\pi W_0 \tau) \exp\{-\tau(1 - W_0) - [(\alpha_p \sin^2 \theta)/W_0 \tau]\}. \qquad (15\text{-}20)$$

This shows the broadening of the angular spectrum of the incoherent intensity as $z^{1/2}$ [see also (13-27)].

Let us next consider the frequency spectrum (15-8). For a plane wave case, we have from (15-12),

$$W(z, \omega) = 2 \int_{-\infty}^{\infty} \Gamma(z, t_d) \exp(i\omega t_d)\, dt_d$$

$$\Gamma(z, t_d) = I_0 \exp\left[-\tau\left\{1 + \frac{W_0}{1 + A} \exp\left[-\frac{(K_r U t_d)^2}{4\alpha_p(1 + A)}\right]\right\}\right] \qquad (15\text{-}21)$$

where U is the component of the particle velocity transverse to the z axis and A is given in (15-11).

Let us consider an example in which the particle velocity is constant and the velocity fluctuation is zero ($A = 0$). Then we have

$$\Gamma(z, t_d) = I_0 \exp[-\tau\{1 + W_0 \exp[-(K_r U t_d)^2/4\alpha_p]\}]. \qquad (15\text{-}22)$$

Comparing with (15-15), we note that if the particle velocity is constant and its transverse component is U, then the spatial correlation function $\Gamma(z, \mathbf{q}) = \Gamma(z, K_r \boldsymbol{\rho}_d)$ given in (15-15) is identical to the temporal correlation function (15-22) with the replacement $\boldsymbol{\rho}_d = \mathbf{U}t_d$:

$$\Gamma(z, \boldsymbol{\rho}_d) = \Gamma(z, \mathbf{U}t_d). \tag{15-23}$$

The temporal spectrum $W(z, \omega)$ in (15-21) with $A = 0$ can be calculated for small and large optical distances. For small optical distance using (15-15), we get

$$W(z, \omega) = I_0 \exp(-\tau)\left[4\pi\,\delta(\omega) + \frac{4\tau W_0 \pi^{1/2}}{\omega_c} \exp\left(-\frac{\omega^2}{\omega_c{}^2}\right)\right] \tag{15-24}$$

where $\omega_c = K_r U/\alpha_p^{1/2}$. For a large optical distance $\tau \gg 1$, we use (15-19) and get

$$W(z, \omega) = I_0\{4\pi\,\delta(\omega)\exp(-\tau) + (4\pi^{1/2}/\omega_c)\exp[-\tau(1 - W_0) - (\omega^2/\omega_c{}^2)]\} \tag{15-25}$$

where $\omega_c = (K_r U)[(\tau W_0)^{1/2}/\alpha_p^{1/2}]$. The first term in (15-24) and (15-25) represents the coherent field and therefore there is no frequency spread. Note that the delta function $\delta(\omega)$ means that there is no frequency spread from the carrier frequency. The second terms in (15-24) and (15-25) give the frequency spread due to the incoherent field.

15-4 LIMITATION ON IMAGE RESOLUTION IMPOSED BY RANDOMLY DISTRIBUTED SCATTERERS

Suppose that a plane wave propagated through randomly distributed scatterers is observed by an image-forming receiver such as a lens or a parabolic antenna. If the scatterers were absent, the Airy disk would be formed at a focal plane. In this section, we discuss the effect of the scatterers on this image.

Let us consider a circular aperture lens with diameter $2a$ which focuses the incoming plane wave at a focal distance f (see Fig. 15-1). We let $\psi(z, \boldsymbol{\rho}')$

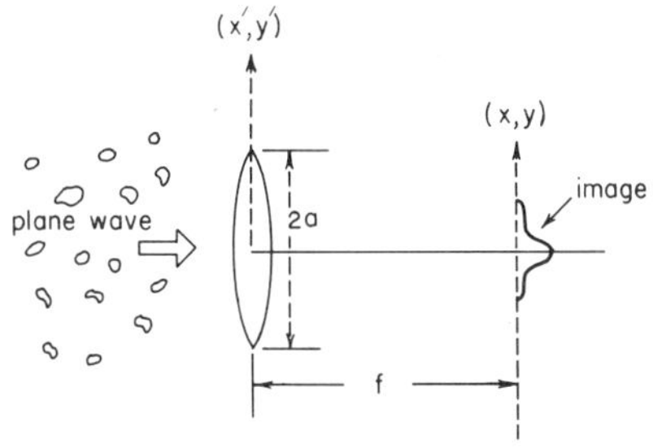

FIG. 15-1 Plane wave propagated through randomly distributed scatterers is incident on a focusing lens with diameter $2a$ and focal length f.

be the field incident on the lens. The field $\psi_f(\boldsymbol{\rho})$ at the focal plane is given by the Kirchhoff formula:

$$\psi_f(\boldsymbol{\rho}) = \frac{k}{2\pi i} \int_s \frac{\exp(ikr + i\phi)}{r} \psi(z, \boldsymbol{\rho}')\, d\boldsymbol{\rho}' \tag{15-26}$$

where s is the area of the circular aperture of the lens, and r the distance between the point on the aperture and the point on the focal plane:

$$r = [(x' - x)^2 + (y' - y)^2 + f^2]^{1/2}. \tag{15-27}$$

The phase ϕ is the additional phase introduced by the lens to focus the wave at the focal point.

We use the Fresnel approximation

$$\frac{e^{ikr}}{r} \approx \frac{1}{f} \exp ik \left[f + \frac{(x' - x)^2 + (y' - y)^2}{2f} \right],$$

$$\phi \approx -\frac{k}{2f}(x'^2 + y'^2). \tag{15-28}$$

We then get

$$\psi_f(\boldsymbol{\rho}) = \frac{k}{2\pi f i} \exp\left(ikf + i\frac{k\rho^2}{2f}\right) \int_s \exp\left(i\frac{k}{f}\boldsymbol{\rho} \cdot \boldsymbol{\rho}'\right) \psi(z, \boldsymbol{\rho}')\, d\boldsymbol{\rho}' \tag{15-29}$$

where $\boldsymbol{\rho} = x\hat{\mathbf{x}} + y\hat{\mathbf{y}}$ and $\boldsymbol{\rho}' = x'\hat{\mathbf{x}} + y'\hat{\mathbf{y}}$.

The intensity $P_f(\boldsymbol{\rho})$ at the focal plane is then given by

$$\begin{aligned} P_f(\boldsymbol{\rho}) &= \langle |\psi_f(\boldsymbol{\rho})|^2 \rangle \\ &= \frac{k^2}{(2\pi f)^2} \iint_s \exp\left(i\frac{k}{f}\boldsymbol{\rho} \cdot \boldsymbol{\rho}_d'\right) \Gamma(z, \boldsymbol{\rho}_d')\, d\boldsymbol{\rho}_d'\, d\boldsymbol{\rho}_c' \end{aligned} \tag{15-30}$$

where $\boldsymbol{\rho}_d' = \boldsymbol{\rho}_1' - \boldsymbol{\rho}_2'$, $\boldsymbol{\rho}_c' = \frac{1}{2}(\boldsymbol{\rho}_1' + \boldsymbol{\rho}_2')$, $d\boldsymbol{\rho}_d'\, d\boldsymbol{\rho}_c' = d\boldsymbol{\rho}_1'\, d\boldsymbol{\rho}_2'$, and from (15-12)

$$\begin{aligned} \Gamma(z, \boldsymbol{\rho}_d') &= \langle \psi(z, \boldsymbol{\rho}_1') \psi^*(z, \boldsymbol{\rho}_2') \rangle \\ &= I_0 \exp\left\{ -\tau \left[1 - W_0 \exp\left(-\frac{K_r^2 \rho_d'^2}{4\alpha_p} \right) \right] \right\}. \end{aligned} \tag{15-31}$$

The integration in (15-30) is over the circular aperture of radius a. We can rewrite (15-30) by using a function $F(\boldsymbol{\rho})$ defined by

$$F(\boldsymbol{\rho}) = \begin{cases} 1 & \text{when} \quad |\boldsymbol{\rho}| < a \\ 0 & \text{when} \quad |\boldsymbol{\rho}| > a. \end{cases} \tag{15-32}$$

We can then extend the limit of integration and get

$$P_f(\boldsymbol{\rho}) = \frac{k^2}{(2\pi f)^2} \iint_{-\infty}^{\infty} \exp\left(i\frac{k}{f}\boldsymbol{\rho}\cdot\boldsymbol{\rho}_d'\right)\Gamma(z, \boldsymbol{\rho}_d')F(\boldsymbol{\rho}_1')F(\boldsymbol{\rho}_2')\, d\boldsymbol{\rho}_d'\, d\boldsymbol{\rho}_c'. \quad (15\text{-}33)$$

The integral

$$K(\boldsymbol{\rho}_d') = \iint_{-\infty}^{\infty} F\left(\boldsymbol{\rho}_c' + \frac{\boldsymbol{\rho}_d'}{2}\right)F\left(\boldsymbol{\rho}_c' - \frac{\boldsymbol{\rho}_d'}{2}\right) d\boldsymbol{\rho}_c' \quad (15\text{-}34)$$

is the area of the intersection of two circles of radius a whose centers are separated by $\boldsymbol{\rho}_d'$, and this can be calculated to give

$$K(\boldsymbol{\rho}_d') = K(\rho_d') = \begin{cases} 2a^2\left[\cos^{-1}\left(\dfrac{\rho_d'}{2a}\right) - \dfrac{\rho_d'}{2a}\left(1 - \dfrac{\rho_d'^2}{4a^2}\right)^{1/2}\right] & \text{for } \rho_d' < 2a \\ 0 & \text{for } \rho_d' > 2a. \end{cases} \quad (15\text{-}35)$$

This function $K(\rho_d')$ is equal to πa^2 for $\rho_d' = 0$ and reduces to zero as $\rho_d' \to 2a$.

Substituting (15-35) into (15-33), we get the intensity distribution in the focal plane†:

$$\begin{aligned} P_f(\boldsymbol{\rho}) &= P_f(\rho) \\ &= \frac{k^2}{2\pi f^2}\int_0^{2a} \rho_d'\, d\rho_d'\, J_0\left(\frac{k\rho}{f}\rho_d'\right)\Gamma(z, \rho_d')K(\rho_d') \end{aligned} \quad (15\text{-}36)$$

where $\Gamma(z, \rho_d')$ is given in (15-31), $K(\rho_d')$ is given in (15-35), and J_0 is the Bessel function of order zero.

Equation (15-36) gives the general formula for the intensity distribution at the focal plane and can be calculated numerically. Also note that if the scatterers were absent, $\Gamma = I_0$ in (15-30) and we should get

$$\begin{aligned} P_a(\rho) &= \frac{k^2 I_0}{(2\pi f)^2}\left[2\pi\int_0^a J_0\left(\frac{k}{f}\rho\rho'\right)\rho'\, d\rho'\right]^2 \\ &= \frac{k^2 I_0}{2\pi f^2}\int_0^{2a} \rho_d'\, d\rho_d'\, J_0\left(\frac{k\rho}{f}\rho_d'\right)K(\rho_d') = I_0\frac{a^2}{\rho^2}J_1^{\,2}\left(\frac{k\rho a}{f}\right). \end{aligned} \quad (15\text{-}37)$$

This is the Airy disk pattern of a circular aperture. The first zero occurs at the radius $\rho_a = (3.832f)/ka$.

† This is also called the point spread function, and its Fourier transform is the modulation transfer function (MTF) (see Section 20-20).

In order to examine the general characteristics of the pattern (15-36), let us consider the case of large optical distance ($\tau \gg 1$). In this case we use (15-19). We then obtain

$$P_f(\rho) = P_c(\rho) + P_i(\rho),$$

$$P_c(\rho) = I_0 e^{-\tau}(a/\rho)^2 J_1{}^2(k\rho a/f) \tag{15-38}$$

$$P_i(\rho) = \frac{k^2}{2\pi f^2}\int_0^{2a} \rho_d{}'\, d\rho_d{}'\, J_0\left(\frac{k\rho}{f}\rho_d{}'\right)\Gamma_i(z, \rho_d{}')K(\rho_d{}')$$

where $\Gamma_i(z, \rho_d{}') = I_0 \exp\{-\tau[1 - W_0 + (W_0 K_r{}^2 \rho_d'^2/4\alpha_p)]\}$.

In these equations, $P_c(\rho)$ is the coherent intensity and has the same Airy disk as in free space except for the attenuation $\exp(-\tau)$. $P_i(\rho)$ is the incoherent intensity representing multiple scattering.

The correlation distance of the wave incident upon the lens is given by

$$\rho_0 = (4\alpha_p/\tau W_0 K_r{}^2)^{1/2}. \tag{15-39}$$

Since $\alpha_p = 2.66(D/\lambda)^2$ when the particle diameter $D > \lambda$, and $K_r \approx k$, the correlation distance ρ_0 is approximately given by

$$\rho_0 = 0.52D/(\tau W_0)^{1/2} = 0.52D/(\rho_n \sigma_s z)^{1/2}. \tag{15-40}$$

In many practical cases, the correlation distance ρ_0 is much smaller than the aperture size a. In this case, we can approximate $K(\rho_d{}') \approx K(0) = \pi a^2$ and extend the limit of integration in (15-38) to infinity. Then we get

$$P_i(\rho) = \frac{k^2}{2\pi f^2}\int_0^{\infty} \rho_d{}'\, d\rho_d{}'\, J_0\left(\frac{k\rho}{f}\rho_d{}'\right)\Gamma_i(z, \rho_d{}')\pi a^2$$

$$= I_0(a/\rho_i)^2 \exp[-\tau(1 - W_0) - (\rho^2/\rho_i{}^2)] \tag{15-41}$$

where $\rho_i{}^2 = (K_r f/k)^2(\tau W_0/\alpha_p) = (f/k)^2(4/\rho_0{}^2) = 0.2724\rho_a{}^2(a/\rho_0)^2$. This shows that the incoherent intensity $P_i(\rho)$ is spread out at the focal plane and this spread ρ_i compared with the Airy disk size ρ_a is proportional to the ratio of the aperture size a to the correlation distance ρ_0.

If the correlation distance of the incoherent field is much greater than the aperture size a, then we can approximate Γ_i by $I_0 \exp[-\tau(1 - W_0)]$ and obtain a pattern similar to the Airy disk:

$$P_i(\rho) = I_0 \exp[-\tau(1 - W_0)](a/\rho)^2 J_1{}^2(k\rho a/f). \tag{15-42}$$

This is to be expected because the plane wave with the correlation distance ρ_0 large compared with the aperture size a should give an image not much different from that of a plane wave with $\rho_0 \to \infty$.

Let us examine the physical meaning of (15-38)–(15-40). The coherent intensity $P_c(\rho)$ has the same pattern as in free space and therefore this has the same resolution characteristics as in free space. The incoherent intensity $P_i(\rho)$, on the other hand, spreads out considerably (see Fig. 15-2). As long as

FIG. 15-2 The intensity as a function of the distance ρ in the focal plane. The total intensity consists of the coherent intensity $P_c(\rho)$ and the incoherent intensity $P_i(\rho)$. As the propagation distance increases, the incoherent intensity P_i increases and finally exceeds the coherent intensity P_c. Then the image cannot be seen.

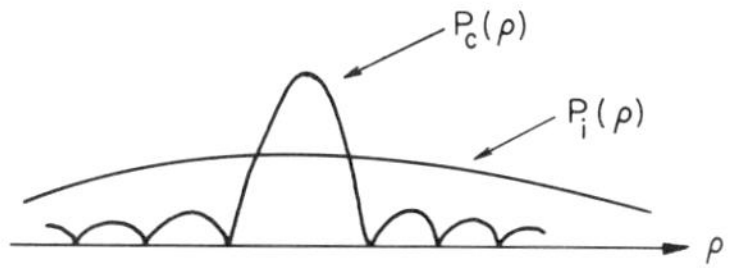

the coherent intensity is greater than the incoherent intensity in the neighborhood of the focal point $\rho = 0$, it should be possible to form a reasonably good image of a distant source. However, if P_i becomes comparable to P_c, then it is not possible to form a good image. The condition $P_i = P_c$ at $\rho = 0$ is given by

$$(K_r a)^2/4\alpha_p = \exp(\tau W_0)/\tau W_0 . \tag{15-43}$$

Since α_p is proportional to $(D/\lambda)^2$ where D is the particle diameter, the left-hand side of (15-43) is proportional to $(a/D)^2$, which may be very large. Therefore (15-43) is satisfied only at a large optical distance. This optical distance is approximately given by

$$\tau W_0 = \rho_n \sigma_t z W_0 = \rho_n \sigma_s z = \ln B + \ln(\ln B) \tag{15-44}$$

where $B = (K_r a)^2/4\alpha_p$. If D/λ is large, α_p is $2.66(D/\lambda)^2$ and also $K_r \approx k$. Therefore B is given by

$$B = 3.71(a/D)^2 \tag{15-45}$$

where a is the aperture radius and D the particle diameter.

In underwater photography (Duntley, 1974, p. 135; Mertens, 1970), it has been reported that clear photographs can be obtained through water with large optical distance. In the analysis just shown, if a large aperture is used to photograph a distant object in water, it is possible to get a reasonably clear image within the optical distance given in (15-44) and (15-45). The contrast is not good because of the incoherent intensity in the background.

This analysis was made on the basis of the point spread function (15-36). This can also be analyzed using the modulation transfer function (MTF). Since the MTF is the Fourier transform of $P_f(\boldsymbol{\rho})$, we get from (15-33) and (15-36) that the MTF is proportional to $\Gamma(z, \rho_d')K(\rho_d')$. Care should be taken to analyze the MTF, because although $\Gamma(z, \rho_d')$ decreases as ρ_d' increases, $\Gamma(z, \rho_d')$ approaches a constant value $I_0 \exp(-\tau)$ as $\rho_d' \to \infty$, and this constant value corresponds to the coherent intensity. At a large optical distance τ, $I_0 \exp(-\tau)$ may be small compared with the incoherent intensity, but the incoherent intensity spreads out in the focal plane whereas the coherent intensity is concentrated within the Airy disk, and therefore the coherent intensity should not be neglected. If we analyze the MTF only for small ρ_d' (corresponding to small spatial frequency), we obtain the behavior of the incoherent intensity, but this cannot yield complete information about the image resolution. This explains the apparent contradiction (Duntley, 1974) that at a large optical distance (15–20) in water containing scatterers, the MTF is found to roll off at a small spatial frequency both from theoretical study and experimental measurements, yet clear photographs of objects can be obtained. Beyond the distance given by (15-43), the coherent intensity is negligible, and the image resolution is determined by ρ_i in (15-41), and the angular resolution is given by $\rho_i/f \approx \lambda/\rho_0$.

15-5 OUTPUT FROM RECEIVER IN RANDOMLY DISTRIBUTED SCATTERERS

Suppose that a plane wave propagated through randomly distributed scatterers is incident on a lens or a parabolic antenna. The output from the receiver is fluctuating in time. In this section, we discuss this fluctuation chararacteristic of the output.

We let $A_r(\theta, \phi)$ be the receiving cross section of the receiver. Then the correlation function $B_v(t_d)$ of the output voltage $V(t)$ is given by

$$B_v(t_d) = \langle V(t_1)V^*(t_2)\rangle = \int A_r(\mathbf{s})I(z, \mathbf{s}, t_d)\, d\mathbf{s} \tag{15-46}$$

where $t_d = t_1 - t_2$, $\mathbf{s} = \sin\theta\cos\phi\hat{\mathbf{x}} + \sin\theta\sin\phi\hat{\mathbf{y}}$, $|\mathbf{s}| = \sin\theta$, and $I(z, \mathbf{s}, t_d)$ is given in (15-12). In (15-46), we normalized the voltage $V(t)$ so that $B_v(t_d = 0)$ is the received power P_r:

$$P_r = \langle |V|^2\rangle = \int A_r(\mathbf{s})I(z, \mathbf{s}, t_d = 0)\, d\mathbf{s}. \tag{15-47}$$

We also used $\mathbf{s}$ with the range of integration $0 \le |\mathbf{s}| \le \infty$ consistent with the small angle approximation (15-12).

As an example, assume that the receiving cross section $A_r(\theta, \phi)$ is approximated by a Gaussian function

$$A_r(\theta, \phi) = A_{r0}\exp(-\alpha_r s^2) \tag{15-48}$$

where $\alpha_r = (4\ln 2)/\theta_b^2$ and θ_b is the half-power beamwidth [see (4-14)]. We then get

$$B_v(t_d) = \frac{A_{r0}}{4\pi\alpha_r}\int d\mathbf{q}\exp\left(-\frac{q^2}{4\alpha_r}\right)\Gamma(z, \mathbf{q}, t_d) \tag{15-49}$$

where $\Gamma(z, \mathbf{q}, t_d)$ is given in (15-12).

The temporal frequency spectrum $W_v(\omega)$ of the output is given by

$$W_v(\omega) = \frac{A_{r0}}{4\pi\alpha_r} \int d\mathbf{q} \exp\left(-\frac{q^2}{4\alpha_r}\right) W(z, \mathbf{q}, \omega) \tag{15-50}$$

where $W(z, \mathbf{q}, \omega) = 2 \int_{-\infty}^{\infty} \Gamma(z, \mathbf{q}, t_d) \exp(i\omega t_d)\, dt_d$.

Let us consider the received power P_r. From (15-49), we obtain

$$P_r = \frac{A_{r0} I_0}{2\alpha_r} \int_0^{\infty} q\, dq \exp\left\{-\frac{q^2}{4\alpha_r} - \rho_n \sigma_t z \left[1 - W_0 \exp\left(-\frac{q^2}{4\alpha_p}\right)\right]\right\}. \tag{15-51}$$

For large optical distance $\tau = \rho_n \sigma_t z \gg 1$, we can use the approximation (15-19) and obtain

$$\begin{aligned} P_r &= P_{rc} + P_{ri}, \qquad P_{rc} = (A_{r0} I_0) \exp(-\tau), \\ P_{ri} &= (A_{r0} I_0) \frac{\exp[-\tau(1 - W_0)]}{1 + (\tau W_0 \alpha_r/\alpha_p)}. \end{aligned} \tag{15-52}$$

The first term P_{rc} is the coherent intensity and the second term P_{ri} is the incoherent intensity. Note that α_r is approximately equal to $(D_r/\lambda)^2$ where D_r is the diameter of the receiver, and α_p is approximately equal to $2.66(D/\lambda)^2$ where D is the particle diameter. Therefore the incoherent intensity P_{ri} is given by

$$P_{ri} = (A_{r0} I_0) \frac{\exp(-\rho_n \sigma_a z)}{1 + c(\rho_n \sigma_s z)(D_r/D)^2} \tag{15-53}$$

where c is a constant on the order of unity whose value depends on the aperture field distribution of the receiver and the particle characteristics. Note that for a point receiver ($D_r \to 0$), the incoherent intensity P_{ri} is much greater than the coherent intensity. However, for large D_r/D, the incoherent intensity P_{ri} can be considerably small. The distance z within which the coherent intensity is greater than the incoherent intensity is therefore approximately given by

$$\rho_n \sigma_s z < \ln[(D_r/D)^2 \ln(D_r/D)^2]. \tag{15-54}$$

This indicates that for a receiver with large aperture size, the output is dominated by the coherent intensity and the fluctuation is small over a large optical distance. Not surprisingly, this condition (15-54) is the same as (15-44) for image resolution. It should be noted that a detailed analysis has been reported on the limits imposed by atmospheric inhomogeneities on the directivity of large antennas (Shifrin, 1971, p. 282).

15-6 SPHERICAL WAVE IN RANDOMLY DISTRIBUTED PARTICLES

If a point source is located at the origin, it is convenient to use (15-6c) and the mutual coherence function Γ_0 at $z = 0$:

$$\Gamma_0(\boldsymbol{\rho}', \boldsymbol{\rho}_d') = (2\pi/K_r)^2\, \delta(\boldsymbol{\rho}')\, \delta(\boldsymbol{\rho}_d'). \tag{15-55}$$

Note that in free space ($H = 0$), (15-55) substituted in (15-6c) gives the correct mutual coherence function Γ at z:

$$\Gamma(z, \boldsymbol{\rho}, \boldsymbol{\rho}_d) = z^{-2} \exp[i(K_r/z)\boldsymbol{\rho} \cdot \boldsymbol{\rho}_d]. \tag{15-56}$$

The mutual coherence function in randomly distributed scatterers is obtained by substituting (15-55) into (15-6c). We have

$$\begin{aligned} \Gamma(z, \boldsymbol{\rho}, \boldsymbol{\rho}_d, t_d) &= z^{-2} \exp[i(K_r/z)\boldsymbol{\rho} \cdot \boldsymbol{\rho}_d - H] \\ H &= \int_0^z \rho_n \sigma_t \left\{1 - \frac{1}{4\pi} P\left(K_r \boldsymbol{\rho}_d \frac{z'}{z}, t_d\right)\right\} dz'. \end{aligned} \tag{15-57}$$

For the scattering characteristics given in (15-10), we have

$$P\left(K_r \boldsymbol{\rho}_d \frac{z'}{z}, t_d\right) = \frac{\sigma_s}{1 + A} \exp\left[-\frac{|(K_r \boldsymbol{\rho}_d z'/z) - K_r \mathbf{U} t_d|^2}{4\alpha_p(1 + A)}\right] \tag{15-58}$$

where A is given in (15-11).

Using (15-57), we can easily obtain the formulas for the angular spectrum and the frequency spectrum.

15-7 BACKSCATTERING FROM RANDOMLY DISTRIBUTED SCATTERERS

Consider a plane wave propagating in the z direction which is incident normally on a slab of randomly distributed scatterers. The specific intensity $I(\mathbf{r}, \hat{\mathbf{s}})$ satisfies the equation of transfer:

$$\frac{d}{ds} I(\mathbf{r}, \hat{\mathbf{s}}) = -\rho\sigma_t I(\mathbf{r}, \hat{\mathbf{s}}) + \frac{\rho\sigma_t}{4\pi} \int_{4\pi} p(\hat{\mathbf{s}}, \hat{\mathbf{s}}') I(\mathbf{r}, \hat{\mathbf{s}}')\, d\omega'. \tag{15-59}$$

In order to study the backscattering, it is convenient to divide the specific intensity into the forward specific intensity $I_+(\mathbf{r}, \hat{\mathbf{s}})$ and the backward specific intensity $I_-(\mathbf{r}, \hat{\mathbf{s}})$ (Ishimaru 1977a):

$$I(\mathbf{r}, \hat{\mathbf{s}}) = \begin{cases} I_+(\mathbf{r}, \hat{\mathbf{s}}) & \text{when} \quad \hat{\mathbf{s}} \cdot \hat{\mathbf{z}} > 0 \\ I_-(\mathbf{r}, \hat{\mathbf{s}}) & \text{when} \quad \hat{\mathbf{s}} \cdot \hat{\mathbf{z}} < 0. \end{cases} \tag{15-60}$$

Using (15-60), we write (15-59) in the following two equations. For $\hat{\mathbf{s}} \cdot \hat{\mathbf{z}} > 0$,

$$\frac{d}{ds} I_+(\mathbf{r}, \hat{\mathbf{s}}) = -\rho\sigma_t I_+(\mathbf{r}, \hat{\mathbf{s}}) + \frac{\rho\sigma_t}{4\pi} \int_{+2\pi} p(\hat{\mathbf{s}}, \hat{\mathbf{s}}') I_+(\mathbf{r}, \hat{\mathbf{s}}')\, d\omega' + \frac{\rho\sigma_t}{4\pi} \int_{-2\pi} p(\hat{\mathbf{s}}, \hat{\mathbf{s}}') I_-(\mathbf{r}, \hat{\mathbf{s}}')\, d\omega' \tag{15-61}$$

where the integral over $+2\pi$ means the integration with respect to $\hat{\mathbf{s}}'$ in the range $\hat{\mathbf{s}}' \cdot \hat{\mathbf{z}} > 0$ and the integral over -2π means the integration in the range $\hat{\mathbf{s}}' \cdot \hat{\mathbf{z}} < 0$. Similarly, for $\hat{\mathbf{s}} \cdot \hat{\mathbf{z}} < 0$, we have

$$\frac{d}{ds} I_-(\mathbf{r}, \hat{\mathbf{s}}) = -\rho\sigma_t I_-(\mathbf{r}, \hat{\mathbf{s}}) + \frac{\rho\sigma_t}{4\pi} \int_{-2\pi} p(\hat{\mathbf{s}}, \hat{\mathbf{s}}') I_-(\mathbf{r}, \hat{\mathbf{s}})\, d\omega' + \frac{\rho\sigma_t}{4\pi} \int_{+2\pi} p(\hat{\mathbf{s}}, \hat{\mathbf{s}}') I_+(\mathbf{r}, \hat{\mathbf{s}}')\, d\omega'. \tag{15-62}$$

Now we consider an iteration solution of (15-61) and (15-62): I_{+n} and I_{-n}. Noting that the nth iteration for I_+ is generated by the $(n-1)$th iteration solution for I_-, we write (15-61) in the form

$$\frac{d}{ds} I_{+n}(\mathbf{r}, \hat{\mathbf{s}}) = -\rho\sigma_t I_{+n}(\mathbf{r}, \hat{\mathbf{s}}) + \frac{\rho\sigma_t}{4\pi} \int_{+2\pi} p(\hat{\mathbf{s}}, \hat{\mathbf{s}}') I_{+n}(\mathbf{r}, \hat{\mathbf{s}}')\, d\omega' + \frac{\rho\sigma_t}{4\pi} \int_{-2\pi} p(\hat{\mathbf{s}}, \hat{\mathbf{s}}') I_{-(n-1)}(\mathbf{r}, \hat{\mathbf{s}}')\, d\omega'. \tag{15-63}$$

Similarly, (15-62) can be written as

$$\frac{d}{ds} I_{-n}(\mathbf{r}, \hat{\mathbf{s}}) = -\rho\sigma_t I_{-n}(\mathbf{r}, \hat{\mathbf{s}}) + \frac{\rho\sigma_t}{4\pi} \int_{-2\pi} p(\hat{\mathbf{s}}, \hat{\mathbf{s}}') I_{-n}(\mathbf{r}, \hat{\mathbf{s}})\, d\omega' + \frac{\rho\sigma_t}{4\pi} \int_{+2\pi} p(\hat{\mathbf{s}}, \hat{\mathbf{s}}') I_{+(n-1)}(\mathbf{r}, \hat{\mathbf{s}}')\, d\omega'. \tag{15-64}$$

As the first approximation, we let $I_{-0}(\mathbf{r}, \hat{\mathbf{s}}') = 0$ in (15-63) and obtain the equation of transfer for $I_+ = I_{+1}$:

$$\frac{d}{ds} I_+(\mathbf{r}, \hat{\mathbf{s}}) = -\rho\sigma_t I_+(\mathbf{r}, \hat{\mathbf{s}}) + \frac{\rho\sigma_t}{4\pi} \int_{+2\pi} p(\hat{\mathbf{s}}, \hat{\mathbf{s}}') I_+(\mathbf{r}, \hat{\mathbf{s}}')\, d\omega'. \tag{15-65}$$

The equation for the first approximation for the backward intensity $I_- = I_{-1}$ is obtained by substituting I_{+1} in (15-64):

$$\frac{d}{ds} I_-(\mathbf{r}, \hat{\mathbf{s}}) = -\rho\sigma_t I_-(\mathbf{r}, \hat{\mathbf{s}}) + \frac{\rho\sigma_t}{4\pi} \int_{-2\pi} p(\hat{\mathbf{s}}, \hat{\mathbf{s}}') I_-(\mathbf{r}, \hat{\mathbf{s}}')\, d\omega' + \frac{\rho\sigma_t}{4\pi} \int_{2\pi} p(\hat{\mathbf{s}}, \hat{\mathbf{s}}') I_+(\mathbf{r}, \hat{\mathbf{s}}')\, d\omega'. \tag{15-66}$$

Equations (15-65) and (15-66) constitute the basic equations for the first order forward and backward specific intensities.

Let us consider the small angle approximation of these two basic equations (15-65) and (15-66). Equation (15-65) becomes identical to (15-4) and its general solution is already given in (15-5a). For (15-66), noting $\hat{\mathbf{s}} \cdot \hat{\mathbf{z}} < 0$, we obtain

$$\frac{\partial}{\partial z} I_-(z, \boldsymbol{\rho}, \mathbf{s}) + \mathbf{s} \cdot \nabla_t I_-(z, \boldsymbol{\rho}, \mathbf{s}) = -\rho_n \sigma_t I_-(z, \boldsymbol{\rho}, \mathbf{s}) + \frac{\rho_n \sigma_t}{4\pi} \iint_{-\infty}^{\infty} p(\mathbf{s} - \mathbf{s}') I_-(z, \boldsymbol{\rho}, \mathbf{s})\, ds' + Q$$

$$Q = \frac{\rho_n \sigma_t}{4\pi} \int_{2n} p(\mathbf{s} - \mathbf{s}') I_+(z, \boldsymbol{\rho}, \mathbf{s}')\, ds' \tag{15-67}$$

where we used ρ_n for number density to avoid confusion with the radial position vector $\boldsymbol{\rho}$.

As an example, let us consider the problem of plane wave incidence on a slab. The forward intensity I_+ is already given in (15-12). We substitute this I_+ into (15-67) and calculate Q. The integral for Q must be performed with $\hat{\mathbf{s}}'$ pointed in the forward direction ($\hat{\mathbf{s}}' \cdot \hat{\mathbf{z}} > 0$) and $\hat{\mathbf{s}}$ pointed in the backward direction ($\hat{\mathbf{s}} \cdot \hat{\mathbf{z}} > 0$). In many practical situations, the particles have a certain size distribution and some absorption. In this case, the backscattering pattern is a slowly varying function of the angle. Therefore, it is reasonable to approximate $p(\mathbf{s} - \mathbf{s}')$ by a constant:

$$p(\mathbf{s} - \mathbf{s}') = \sigma_b / \sigma_t \tag{15-68}$$

where $\sigma_b = 4\pi |f(-\hat{\mathbf{i}}, \hat{\mathbf{i}})|^2$ is the backscattering cross section.

Substituting (15-68) into (15-67) and using I_+ in (15-12), we get for stationary scatterers:

$$Q = (\rho_n \sigma_b / 4\pi) I_0 K(z, \mathbf{q} = 0) = (\rho_n \sigma_b / 4\pi) I_0 \exp(-\rho_n \sigma_a z). \tag{15-69}$$

For moving scatterers, we get

$$Q = \frac{\rho_n \sigma_b}{4\pi} I_0 \exp\left[-\rho_n \sigma_t z \left\{ 1 - \frac{W_0}{1 + A} \exp\left[-\frac{(K_n U t_d)}{4\alpha_p (1 + A)} \right] \right\} \right]. \tag{15-70}$$

Let us consider the solution for stationary scatterers. Using (15-69), we can obtain the exact plane wave solution for (15-67). We note that in (15-67), $\nabla_t = 0$ for the plane wave case. Taking a Fourier transform

$$F(z, \mathbf{q}) = \int I_-(z, \mathbf{s})\, e^{-i\mathbf{s} \cdot \mathbf{q}}\, ds, \tag{15-71}$$

we get

$$-\frac{\partial}{\partial z}F + \rho_n\sigma_t\left(1 - \frac{P(\mathbf{q})}{4\pi}\right)F = -\frac{\rho_n\sigma_b}{4\pi}I_0\exp(-\rho_n\sigma_a z)(2\pi)^2\,\delta(\mathbf{q}). \quad (15\text{-}72)$$

This can be solved easily to yield

$$I_-(z,\mathbf{s}) = \frac{1}{(2\pi)^2}\int Fe^{i\mathbf{s}\cdot\mathbf{q}}\,d\mathbf{q} = \left(\frac{\rho_n\sigma_b I_0}{4\pi}\right)\int_0^d \exp(-2\rho_n\sigma_a z)\,dz$$
$$= \left(\frac{\rho_n\sigma_b I_0 d}{4\pi}\right)\left[\frac{1-\exp(-2\rho_n\sigma_a d)}{2\rho_n\sigma_a d}\right] \quad (15\text{-}73)$$

where d is the thickness of the slab (Fig. 15-3).

Equation (15-73) may be compared with the single scattering (first order multiple scattering) solution I_s

$$I_s = \frac{\rho_n\sigma_b I_0}{4\pi}\int_0^d \exp(-2\rho_n\sigma_t z)\,dz$$
$$= \left(\frac{\rho_n\sigma_b I_0 d}{4\pi}\right)\left[\frac{1-\exp(-2\rho_n\sigma_t d)}{2\rho_n\sigma_t d}\right]. \quad (15\text{-}74)$$

Note the significant difference between the multiple scattering solution (15-73) and the first order solution (15-74).

Equation (15-73) is valid in the range $\theta > (kD)^{-1}$ where D is the particle size and $|\mathbf{s}| = \sin\theta$ (see Fig. 15-3). In the range $\theta < (kD)^{-1}$, it is known† (Watson, 1969; de Wolf, 1971) that the backscattered intensity I_b is given by

$$I_b = 2I_- - I_s. \quad (15\text{-}75)$$

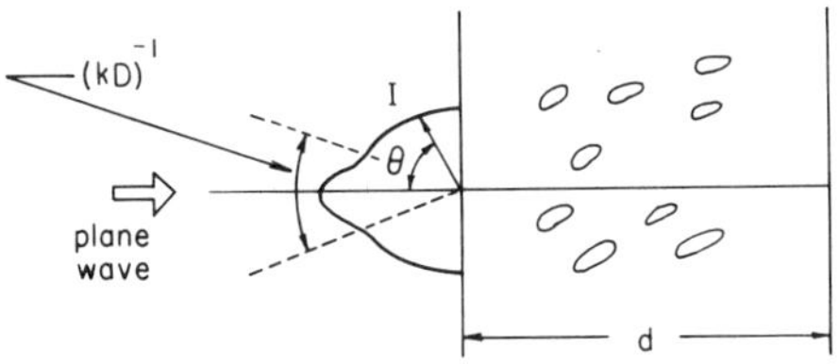

FIG. 15-3 Backscattering from randomly distributed scatterers. The intensity for $\theta > (kD)^{-1}$ is given by I_- in (15-73) and the intensity for $\theta < (kD)^{-1}$ is given by I_b in (15-76).

† More rigorously, the backscattered intensity requires consideration of the fourth order moment.

Physically, this means that the backscattered intensity I_m due to the multiple scattering must be counted twice because the same path is used twice to get the backscattering, but the backscattering due to the first order scattering I_s should be counted only once (see Fig. 15-4). Therefore we have for $\theta < (kD)^{-1}$

$$I_b = 2I_m + I_s = 2(I_m + I_s) - I_s = 2I_- - I_s$$
$$= \left(\frac{\rho_n \sigma_b I_0 d}{4\pi}\right)[2f_- - f_s] \tag{15-76}$$

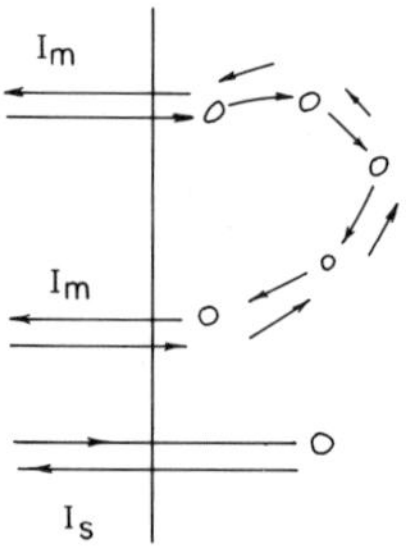

FIG. 15-4

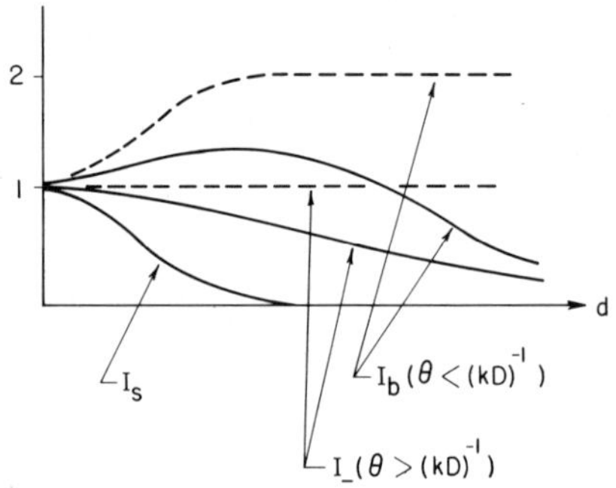

FIG. 15-5

FIG. 15-4 Multiple scattering I_m must be counted twice while the single scattering is counted once.

FIG. 15-5 Backscattered intensity normalized by $\rho \sigma_b I_0 d/4\pi$. The solid curves result when the absorption is not zero and the dashed curves result when the particles are lossless.

where

$$f_- = \frac{1 - \exp(-2\rho_n \sigma_a d)}{2\rho_n \sigma_a d}, \qquad f_s = \frac{1 - \exp(-2\rho_n \sigma_t d)}{2\rho_n \sigma_t d}.$$

When $\theta > (kD)^{-1}$, $I_b = I_-$ (see Fig. 15-5).

The backscattering cross section σ per unit area of the slab of scatterers with thickness d is therefore given by

$$\sigma = \frac{4\pi I_b}{I_0} = (\rho_n \sigma_b d)[2f_- - f_s]. \tag{15-77}$$

The lidar backscattering from clouds has been measured (Milton *et al.*, 1972; Anderson and Browell, 1973; Brinkworth, 1973; Cohen, 1975), and it was noted that the single scattering theory is inadequate to explain the data. However, comparisons between the multiple scattering theory in this section and experiments have not yet been established.

15-8 PULSE PROPAGATION IN RANDOMLY DISTRIBUTED SCATTERERS

In Sections 5-1 and 5-2, we discussed the general formulation of pulse propagation in a random medium. The correlation function B_u of the complex envelope is given by

$$B_u(t_1 t_2) = \int_{-\infty}^{\infty} d\omega_1 \int_{-\infty}^{\infty} d\omega_2\, U_i(\omega_1)U_i^*(\omega_2)\Gamma \exp(-i\omega_1 t_1 + i\omega_2 t_2) \tag{15-78}$$

where $U_i(\omega)$ is the spectrum of the complex envelope of the input pulse $u_i(t)$:

$$u_i(t) = \int_{-\infty}^{\infty} U_i(\omega)e^{-i\omega t}\, d\omega, \tag{15-79}$$

and Γ is the two-frequency correlation function.

The intensity $P(t)$ of the output pulse is given by

$$P(t) = \int_{-\infty}^{\infty} d\omega_1 \int_{-\infty}^{\infty} d\omega_2\, U_i(\omega_1)U_i^*(\omega_2)\Gamma_0 \exp[-i(\omega_1 - \omega_2)t] \tag{15-80}$$

where Γ_0 is Γ evaluated at $t_1 = t_2$.

In many practical problems, the two-frequency correlation function is a slowly varying function of $t = \frac{1}{2}(t_1 + t_2)$ and $\omega_c = \frac{1}{2}(\omega_1 + \omega_2)$ and therefore we can assume that Γ is a function of t_d and ω_d only. This is called the wide-sense stationary uncorrelated scattering (WSSUS) channel. In this case we can express (15-80) in a convolution integral of the input intensity $I_i(t)$:

$$I_i(t) = \int_{-\infty}^{\infty} d\omega_1 \int_{-\infty}^{\infty} d\omega_2\, U_i(\omega_1)U_i^*(\omega_2) \exp[-i(\omega_1 - \omega_2)t] \tag{15-81}$$

and the response $G(t)$ to the delta function input power.

In order to find the response $G(t)$ to the delta function, we first let $I_i(t) = \delta(t)$ in (15-81) and find

$$\int d\omega_c\, U_i(\omega_1)U_i^*(\omega_2) = \frac{1}{2\pi}. \tag{15-82}$$

Substituting this into (15-80), we obtain

$$G(t) = \frac{1}{2\pi}\int d\omega_d\, \Gamma_0(\omega_d) \exp(-i\omega_d t). \tag{15-83}$$

Now for a general $I_i(t)$, we get from (15-81)

$$\int d\omega_c\, U_i(\omega_1)U_i^*(\omega_2) = \frac{1}{2\pi}\int I_i(t) \exp(i\omega_d t)\, dt. \tag{15-84}$$

Substituting this into (15-80) and using (15-83), we get the convolution integral

$$I(t) = \int G(t - t')I_i(t')\,dt'. \tag{15-85}$$

This shows that the output for a general input pulse can be easily obtained once the impulse response $G(t)$ as given in (15-83) is found.

The total energy E_o of the output pulse is given by

$$E_o = \int_{-\infty}^{\infty} I(t)\,dt. \tag{15-86}$$

Using (15-83), we get

$$E_o = \Gamma_0(\omega_d = 0)E_i, \qquad E_i = \int_{-\infty}^{\infty} I_i(t)\,dt. \tag{15-87}$$

It is clear from (15-87) thqt Γ_0 at $\omega_d = 0$ gives the ratio of the total output energy to the total input energy. For example, for a plane wave $\Gamma_0(\omega_d = 0)$ must be unity if there is no absorption and backscattering is neglected.

15-9 INTEGRAL AND DIFFERENTIAL EQUATIONS FOR TWO-FREQUENCY MUTUAL COHERENCE FUNCTION IN RANDOMLY DISTRIBUTED SCATTERERS

In the preceding section we summarized the general formulation for a pulse propagation in a random medium. Central to this formulation is the knowledge of the two-frequency mutual coherence function Γ. The integral equation for Γ can be obtained following the derivations in Sections 14-8 and 14-9.

We express Γ as a sum of the coherent part Γ_c and the incoherent part Γ_i:

$$\Gamma = \Gamma_c + \Gamma_i. \tag{15-88}$$

The coherent part Γ_c is given by

$$\Gamma_c = \langle H(\omega_1, \mathbf{r}, t_1)\rangle\langle H(\omega_2, \mathbf{r}_2, t_2)^*\rangle \tag{15-89}$$

where $H(\omega, \mathbf{r}, t)$ is the field at $\mathbf{r}$ and t when the input is a time harmonic function $\exp(-i\omega t)$ (see Section 5-1). The coherent field $\langle H\rangle$ satisfies the wave equation

$$(\nabla^2 + L_l{}^2)\langle H_l\rangle = 0, \qquad l = 1, 2$$

$$K_l = k_l + [2\pi\rho f_l(\hat{\mathbf{i}}, \hat{\mathbf{i}})]/k_l, \qquad \langle H_l\rangle = \langle H(\omega_l, \mathbf{r}, t_l)\rangle. \tag{15-90}$$

In order to derive Γ_i, we first define the generalized specific intensity I:

$$\Gamma(\omega_1, \omega_2, \mathbf{r}_1, \mathbf{r}_2, t_1, t_2) = \int I(\omega_1, \omega_2, \mathbf{r}_c, \hat{\mathbf{s}}, t_1, t_2) \exp(iK_r \hat{\mathbf{s}} \cdot \mathbf{r}_d)\, d\Omega \qquad (15\text{-}91)$$

where $\mathbf{r}_c = \frac{1}{2}(\mathbf{r}_1 + \mathbf{r}_2)$, $\mathbf{r}_d = (\mathbf{r}_1 - \mathbf{r}_2)$, $K_r = \mathrm{Re}[\frac{1}{2}(K_1 + K_2^*)]$, and $d\Omega$ is an elementary solid angle in the direction defined by the unit vector $\hat{\mathbf{s}}$. We assume that the particle velocity $\mathbf{V}$ is much smaller than the velocity of wave propagation. Following the procedure in Section 14-8, we get (Ishimaru and Hong, 1975a; Hong and Ishimaru, 1976)

$$\Gamma_i = \int \rho\, d\mathbf{r}_s \{[\exp(iK_1 R_1 - iK_2^* R_2)]/R_1 R_2\} \times \int d\Omega' f_1 f_2 I \exp(iK_r \hat{\mathbf{s}}' \cdot \mathbf{V} t_d), \qquad (15\text{-}92)$$

where

$$R_1 = |\mathbf{r}_1 - \tfrac{1}{2}\mathbf{V}t_d - \mathbf{r}_s|, \qquad R_2 = |\mathbf{r}_2 + \tfrac{1}{2}\mathbf{V}t_d - \mathbf{r}_s|,$$

$$t_d = t_1 - t_2,$$

$$f_1 = f_1(\hat{\mathbf{s}}_1, \hat{\mathbf{s}}'), \qquad f_2 = f_2(\hat{\mathbf{s}}_2, \hat{\mathbf{s}}'), \qquad I = I(\omega_1, \omega_2, \mathbf{r}_s, \hat{\mathbf{s}}', t_1, t_2),$$

$d\Omega'$ is the elementary solid angle in the direction $\hat{\mathbf{s}}'$, and $\hat{\mathbf{s}}_1$ and $\hat{\mathbf{s}}_2$ are the unit vectors in the directions $\mathbf{r}_1 - \mathbf{r}_s$ and $\mathbf{r}_2 - \mathbf{r}_s$, respectively (Fig. 15-6).

Equations (15-88)–(15-92) constitute the basic integral equation for the two-frequency mutual coherence function Γ. General solutions of this integral equation are not available. However, we can obtain a simpler approximate differential equation when the particle sizes are comparable to or greater than a wavelength. In this case, the waves are scattered mostly in the

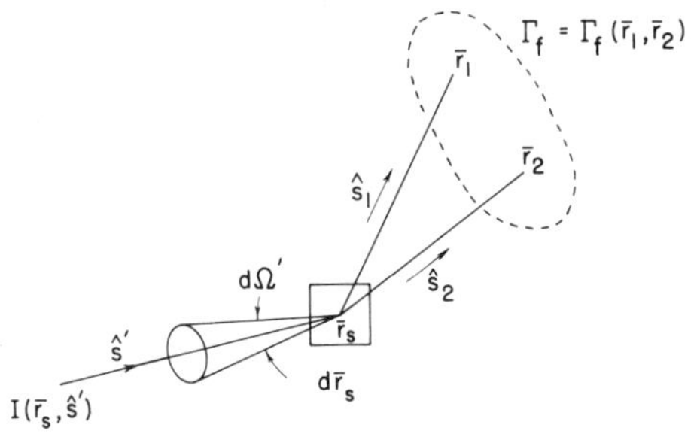

FIG. 15-6 Geometry defining the quantities in the integral equations (15-92).

forward direction and the small angle approximation may be employed (see Section 15-2).

The differential equation for the two-frequency mutual coherence function Γ at two points $(\boldsymbol{\rho}, z)$ and $(\boldsymbol{\rho}_2, z)$ in the same transverse plane $(z_1 = z_2 = z)$ at two different times t_1 and t_2 may be obtained using the parabolic approximation for $\exp(ik_1 R_1)/R_1$ and assuming that the scattering amplitude f is a function of $\mathbf{s} - \mathbf{s}'$:

$$\left\{\frac{\partial}{\partial z} - \frac{i}{2}\left[\frac{1}{K_{r1}}\nabla_1{}^2 - \frac{1}{K_{r2}}\nabla_2{}^2\right] - i(K_1 - K_2{}^*) - P(\boldsymbol{\rho}_d - \mathbf{V}t_d)\right\}\Gamma = 0 \tag{15-93}$$

where

$$P(\boldsymbol{\rho}_d) = \int \rho_n f_1(\mathbf{s}) f_2{}^*(\mathbf{s}) \exp(-iK_r \mathbf{s} \cdot \boldsymbol{\rho}_d)\, ds$$

$$K_1 = k_1 + \frac{2\pi\rho_n f_1(0)}{k_1}, \qquad K_2 = k_2 + \frac{2\pi\rho_n f_2(0)}{k_2}$$

$$K_{r1} = \mathrm{Re}(K_1), \qquad K_{r2} = \mathrm{Re}(K_2), \qquad \boldsymbol{\rho}_d = \boldsymbol{\rho}_1 - \boldsymbol{\rho}_2, \qquad t_d = t_1 - t_2.$$

$f_1(\mathbf{s})$ is the scattering amplitude of a single particle at ω_1 in the direction (θ, ϕ) when a wave is incident in the direction $(\theta = 0, \phi = 0)$, and $\mathbf{s} = l\hat{\mathbf{x}} + m\hat{\mathbf{y}} = \sin\theta\cos\phi\hat{\mathbf{x}} + \sin\theta\sin\phi\hat{\mathbf{y}}$. Similarly, $f_2(\mathbf{s})$ is the scattering amplitude at ω_2. $\nabla_1{}^2$ and $\nabla_2{}^2$ are two-dimensional Laplacian operators with respect to $\boldsymbol{\rho}_1$ and $\boldsymbol{\rho}_2$, respectively. The number density is designated by ρ_n to avoid confusion with the radial vector $\boldsymbol{\rho}$.

Equation (15-93) is the basic differential equation in the small angle approximation.

15-10 TWO-FREQUENCY MUTUAL COHERENCE FUNCTION FOR THE PLANE WAVE CASE

Equation (15-93) has been solved exactly when $\omega_1 = \omega_2$. However, there is no analytical solution available for $\omega_1 \neq \omega_2$. In this section, we analyze a special case in which a plane wave is propagating in the z direction.

For a plane wave case, Γ is independent of $\boldsymbol{\rho}_c = \frac{1}{2}(\boldsymbol{\rho}_1 + \boldsymbol{\rho}_2)$ and (15-93) becomes

$$\left[\frac{\partial}{\partial z} + i\frac{K_{ri} - K_{r2}}{2K_{r1}K_{r2}}\nabla_d{}^2 - i(K_1 - K_2{}^*) - P(\boldsymbol{\rho}_d - \mathbf{V}t_d)\right]\Gamma = 0 \tag{15-94}$$

where $\nabla_d{}^2$ is the Laplacian with respect to $\boldsymbol{\rho}_d$.

For most practical situations, the difference between K_r and k is very small. Also, the frequency difference $\omega_1 - \omega_2$ is much smaller than the carrier frequency ω_0. We can therefore use the approximations

$$K_{r1} \approx k_1, \qquad K_{r2} \approx k_2, \qquad K_{r1} K_{r2} \approx k_1 k_2 \approx k^2 \tag{15-95}$$

where $k = \omega_0 / c$.

We also note that the scattering amplitude f is approximately proportional to k when the particle size is large. In fact, as shown in (6-21), we have for $a_0 \gg \lambda$ (a_0 is the particle radius)

$$f(0) = \frac{ia_0}{\sin\theta} J_1(ka_0 \sin\theta)\Big|_{\theta=0} = \tfrac{1}{2}(ika_0{}^2). \tag{15-96}$$

Since f/k is almost independent of k, we can write

$$i(K_1 - K_2{}^*) = i(K_{r1} - K_{r2}) - \rho_n \sigma_t \approx i(k_1 - k_2) - \rho_n \sigma_t \tag{15-97}$$

where the total cross section σ_t is evaluated at the carrier frequency.

For the function $P(\boldsymbol{\rho}_d)$, we can approximate

$$f_1(\mathbf{s}) f_2(\mathbf{s})^* \approx (1 - k_d{}^2/4k^2) |f(\mathbf{s})|^2 \tag{15-98}$$

where $k_d = k_1 - k_2$ and $f(\mathbf{s})$ is evaluated at ω_0. And we can further approximate $|f|^2$ by a Gaussian function [see (6-20)]

$$|f(\mathbf{s})|^2 = \frac{\alpha_p}{\pi} \sigma_s \exp(-\alpha_p s^2) \tag{15-99}$$

where σ_s is the scattering cross section evaluated at ω_0, $\alpha_p = 2.77/\theta_{pb}^2$, and θ_{pb} is the half-power beamwidth of the scattering pattern of a single particle. From (15-93), we get

$$\begin{aligned} P(\boldsymbol{\rho}_d) &= \rho_n \sigma_s (1 - k_d{}^2/4k^2) \exp(-k^2 \rho_d{}^2/4\alpha_p) \\ &\approx \rho_n \sigma_s \exp(-k^2 \rho_d{}^2/4\alpha_p). \end{aligned} \tag{15-100}$$

We now substitute (15-95), (15-97), and (15-100) into (15-94). We also write

$$\Gamma = \Gamma_1 \exp(ik_d z - \rho_n \sigma_a z). \tag{15-101}$$

Then we get the following equation for Γ_1:

$$[\partial/\partial z + ia\nabla_d{}^2 + \rho_n \sigma_s - P(\boldsymbol{\rho}_d)]\Gamma_1 = 0 \tag{15-102}$$

where $a = k_d/2k^2$. The boundary condition when a plane wave is incident at $z = 0$ is

$$\Gamma_1 = 1 \quad \text{at} \quad z = 0. \tag{15-103}$$

Equations (15-101) and (15-102) together with (15-103) constitute the basic mathematical formulation for the two-frequency mutual coherence function.

The general solution of (15-102) can be obtained by a numerical technique. However, it may be more instructive to give an approximate analytical solution. This is discussed in the following sections.

15-11 WEAK FLUCTUATION SOLUTION OF A PLANE PULSE WAVE

Let us consider a wave at a relatively short distance where the coherent intensity is dominant over the incoherent intensity. In this range we write

$$\Gamma_1 = \exp(\psi). \tag{15-104}$$

Substituting this into (15-102), we get

$$\frac{\partial}{\partial z}\psi + ia(\nabla_d{}^2\psi + \nabla_d\psi \cdot \nabla_d\psi) + \rho_n\sigma_s - P(\boldsymbol{\rho}_d) = 0. \tag{15-105}$$

The function ψ is an even function of ρ_d and $\nabla_d{}^2\psi$ is generally maximum at $\rho_d = 0$, decaying to zero at $\rho_d \to \infty$, while $|\nabla_d\psi|^2$ is zero near $\rho_d = 0$ and as $\rho_d \to \infty$. At a short distance, therefore, it is expected (and verified by exact numerical calculations) that the nonlinear term $|\nabla_d\psi|^2$ is much smaller than $\nabla_d{}^2\psi$ and can be neglected. Thus we have approximately

$$\frac{\partial}{\partial z}\psi + ia\nabla_d{}^2\psi + \rho_n\sigma_s - P(\boldsymbol{\rho}_d) = 0. \tag{15-106}$$

The boundary condition for ψ is $\psi(z = 0) = 0$. Equation (15-106) can be solved exactly by using the Fourier transform:

$$\boldsymbol{\psi}(z, \mathbf{s}) = \frac{k^2}{(2\pi)^2}\int \psi(z, \boldsymbol{\rho}_d)\exp(ik\mathbf{s}\cdot\boldsymbol{\rho}_d)\,d\boldsymbol{\rho}_d \tag{15-107}$$

where we let $K_\gamma \approx k$ to be consistent with (15-95). We then obtain

$$\frac{\partial}{\partial z}\boldsymbol{\psi} - iak^2s^2\boldsymbol{\psi} + \rho_n\sigma_s k^2\,\delta(k\mathbf{s}) - \rho_n f_1(\mathbf{s})f_2{}^*(\mathbf{s}) = 0. \tag{15-108}$$

This can be solved easily, and we get

$$\psi(z, \boldsymbol{\rho}_d) = -\rho_n\sigma_s z + \int d\mathbf{s}\exp(-ik\mathbf{s}\cdot\boldsymbol{\rho}_d)$$

$$\times\,\rho_n f_1(\mathbf{s})f_2(\mathbf{s})\int_0^z dz'\exp(iak^2s^2z'). \tag{15-109}$$

The two-frequency mutual coherence function in (15-101) using (15-100), (15-104), and (15-109) is then given by (Ishimaru and Hong, 1975)

$$\begin{aligned} \Gamma(z, \boldsymbol{\rho}_d = 0, \omega_d) &= A(z, \omega_d) \exp[ik_d z + i\phi(z, \omega_d)] \\ A(z, \omega_d) &= \exp\{-\rho_n \sigma_t z[1 - (W_0/x) \tan^{-1} x]\} \\ \phi(z, \omega_d) &= (\rho_n \sigma_t z/2)[(W_0/x) \ln(1 + x^2)] \end{aligned} \qquad (15\text{-}110)$$

$$W_0 = \frac{\sigma_s}{\sigma_t} = \text{albedo of a single particle}$$

$$x = k_d z/2\alpha_p = \omega_d/\omega_r, \qquad \omega_r = 2\alpha_p c/z,$$

$$k_d = \omega_d/c.$$

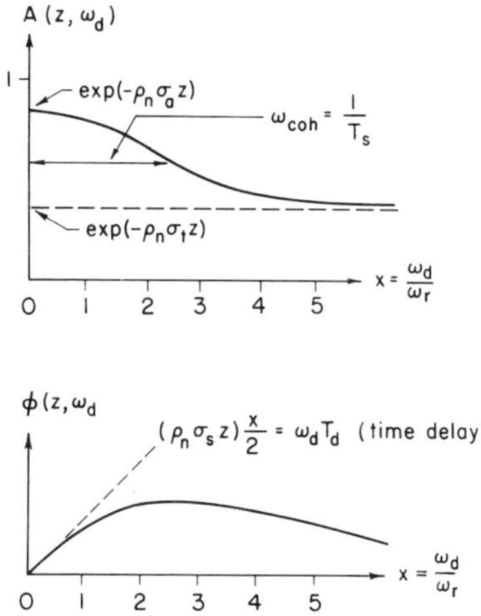

FIG. 15-7 Amplitude $A(z, \omega_d)$ and phase $\phi(z, \omega_d)$ of the two-frequency mutual coherence function in the weak fluctuation case.

The expression (15-110) should be applicable to the case in which the optical distance is less than about 3. When the optical distance is much greater than unity, we cannot ignore the nonlinear term in (15-105) and a different approach should be used. This is discussed in the following section.

Let us examine the solution (15-110). General shapes of the amplitude $A(z, \omega_d)$ and the phase $\phi(z, \omega_d)$ are shown in Fig. 15-7. Note that as $\omega_d \to \infty$, $A \to \exp(-\rho_n \sigma_t z)$ and $\phi \to 0$. This limiting value corresponds to the coherent part Γ_c. We write

$$\Gamma = \Gamma_c + \Gamma_i, \qquad \Gamma_c = \exp(-\rho_n \sigma_t z). \qquad (15\text{-}111)$$

Then $\Gamma_i \to 0$ as $\omega_d \to \infty$. The coherence bandwidth ω_{coh} (see Fig. 15-7) of the incoherent part Γ_i may be approximately given by letting $\tan^{-1} x = x - x^3/3 + \cdots$. We then obtain

$$\omega_{\text{coh}} \approx [3/\rho_n \sigma_s z]^{1/2}, \qquad \omega_r = [3/\rho_n \sigma_s]^{1/2}(2\alpha_p c/z^{3/2}) \tag{15-112}$$

when $\rho_n \sigma_s z$ is comparable to unity. When $\rho_n \sigma_s z$ is smaller than unity, the coherence bandwidth ω_{coh} is approximately given by

$$\omega_{\text{coh}} \approx 5\omega_r = 10\alpha_p c/z. \tag{15-113}$$

The inverse of the coherence bandwidth ω_{coh} represents the spread of the pulse shape

$$T_s = 1/\omega_{\text{coh}}. \tag{15-114}$$

Figure 15-7 also shows the phase as a function of $x = \omega_d/\omega_r$. For small x, the phase $\phi(z, \omega_d)$ is $(\rho_n \sigma_s z)(x/2)$. This linear portion of the phase represents a time delay T_d of the pulse:

$$T_d = \rho_n \sigma_s z/2\omega_r = \rho_n \sigma_s z^2/4\alpha_p c. \tag{15-115}$$

We also note that at $\omega_d = 0$, $\Gamma = \exp(-\rho_n \sigma_a z)$, consistent with the energy conservation relationship (15-87).

The output pulse $I(t)$ is given by the convolution integral (15-85), and the impulse response $G(t)$ is given by the Fourier transform (15-83):

$$G(t) = \frac{1}{2\pi} \int_{-\infty}^{\infty} A(z, \omega_d) \exp\left[i\phi(z, \omega_d) - i\omega_d\left(t - \frac{z}{c}\right)\right] d\omega_d. \tag{15-116}$$

The general shape of this impulse response is sketched in Fig. 15-8a. Note that the impulse response consists of the coherent part (delta function) and the incoherent part. The incoherent part is characterized by a time delay T_d and a spread T_s. If the input pulse is shorter than T_s, the pulse shape of the coherent part is almost identical to the input pulse but the incoherent part is spread out (see Fig. 15-8(b). If the input is broader than T_s, the output pulse shape is almost the same as the input pulse shape (see Fig. 15-8c).

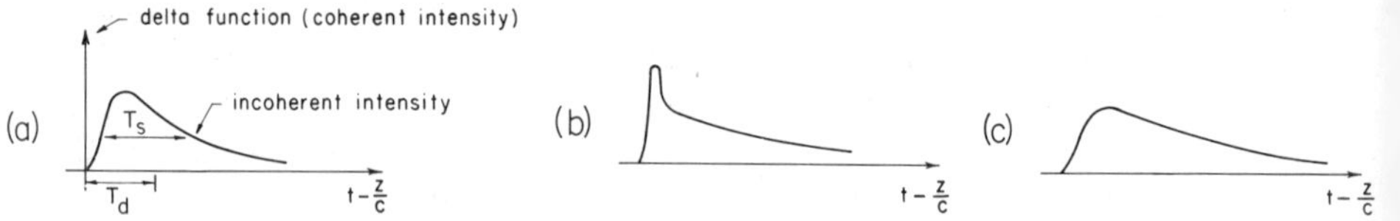

FIG. 15-8 (a) Impulse response consists of the coherent part (delta function) and the incoherent part. (b) When the input pulse is narrower than the inverse of the coherence bandwidth, the coherent intensity is still distinct from the incoherent intensity. (c) When the input pulse is broader than the inverse of the coherence bandwidth, the coherent and the incoherent intensities merge.

15-12 STRONG FLUCTUATION SOLUTION OF A PLANE PULSE WAVE

As indicated in the preceding section, if the optical distance $\rho_n \sigma_t z$ is much greater than unity, the nonlinear term in (15-105) cannot be ignored. We should then go back to (15-102). Let us use (15-100) and write (15-102) as

$$\partial/\partial z + ia\nabla_d{}^2 + \rho_n \sigma_s[1 - \exp(k^2 \rho_d{}^2/4\alpha_p)]\}\Gamma_1 = 0. \qquad (15\text{-}117)$$

It can be shown (Hong and Ishimaru, 1976; Shishov, 1974; Lee and Jokipii, 1975a,b) that if the optical distance $\rho_n \sigma_t z$ is much greater than unity, the incoherent part Γ_{1i} dominates over the coherent part Γ_{1c} and the incoherent part Γ_{1i} satisfies the following equation which is obtained by expanding the exponential in (15-117) and keeping the first two terms:

$$[\partial/\partial z + ia\nabla_d{}^2 + b\rho_d{}^2]\Gamma_{1i} = 0 \qquad (15\text{-}118)$$

where $a = k_d/2k^2$ and $b = \rho_n \sigma_s k^2/4\alpha_p$.

We can normalize (15-118) by letting

$$z/L = z', \qquad \rho_d/\rho_0 = \rho_d{}', \qquad (15\text{-}119)$$

where L is the propagation distance ($0 \le z' \le 1$). ρ_0 is chosen such that (15-118) takes the form

$$[\partial/\partial z' + i(\omega_d/\omega_{coh})\nabla^2_{d'} + \rho'^2_d]\Gamma_{1i} = 0. \qquad (15\text{-}120)$$

To obtain (15-120), we require

$$\rho_0 = (4\alpha_p/\rho_n \sigma_s z k^2)^{1/2}, \qquad (15\text{-}121)$$

$$\omega_{coh} = 8\alpha_p c/\rho_n \sigma_s z^2. \qquad (15\text{-}122)$$

The radius ρ_0 is identical to the correlation distance given in (15-39). We can solve (15-120) exactly in the following manner (Sreenivasiah *et al.*, 1976). We assume the following form for Γ_{1i}:

$$\Gamma_{1i} = [f(z')]^{-1} \exp[g(z')\rho'^2_d] \qquad (15\text{-}123)$$

where $f(z)$ and $g(z)$ are functions of z only. Substituting this into (15-120) we obtain

$$-\frac{1}{f}\frac{\partial f}{\partial z'} + i4\alpha g + \left[\frac{\partial g}{\partial z'} + i4\alpha g^2 + 1\right]\rho'^2_d = 0 \qquad (15\text{-}124)$$

where $\alpha = \omega_d/\omega_{coh}$.

Now we note that the first two terms are functions of z only regardless of ρ_d'. Therefore we should require

$$-\frac{1}{f}\frac{\partial f}{\partial z'} + i4\alpha g = 0 \tag{15-125}$$

$$\frac{\partial g}{\partial z'} + i4\alpha g^2 + 1 = 0. \tag{15-126}$$

The boundary conditions should be

$$g(z') = 0 \tag{15-127}$$

$$f(z') = 1 \tag{15-128}$$

at $z' = 0$.

Equation (15-126) is a Riccati equation and with (15-127) can be easily solved:

$$g(z') = -\frac{\tan[(i4\alpha)^{1/2}z']}{(i4\alpha)^{1/2}}. \tag{15-129}$$

Using this, f can be obtained from (15-125):

$$f(z') = \cos[(i4\alpha)^{1/2}z']. \tag{15-130}$$

The final solution when $z' = 1$ is then given by

$$\Gamma_{1i} = \frac{1}{\cos(i4\alpha)^{1/2}} \exp\left[-\frac{\tan(i4\alpha)^{1/2}}{(i4\alpha)^{1/2}}\rho_d'^2\right] \tag{15-131}$$

where $\alpha = \omega_d/\omega_{coh}$. This solution when $\rho_d = 0$ is shown in Fig. 15-9.

Equation (15-131) is the exact two-frequency mutual function satisfying (15-120). Let us next consider the pulse wave. The impulse response $G(t)$ is given by

$$G(t) = \frac{1}{2\pi}\int_{-\infty}^{\infty} \frac{\exp[-i\omega_d(t - z/c)]}{\cos(i4\alpha)^{1/2}}\, d\omega_d$$

$$= \frac{\omega_{coh}}{2\pi}\int_{-\infty}^{\infty} \frac{\exp(-i\alpha T)}{\cos(i4\alpha)^{1/2}}\, d\alpha \tag{15-132}$$

where $T = \omega_{coh}(t - z/c)$.

Consider the integrand in (15-132). Since a cosine function is an even function, there is no branch point in the α plane. The poles $\alpha = \alpha_n$ in the integrand are given by

$$(i4\alpha_n)^{1/2} = (2n + 1)(\pi/2), \qquad n = 0, \pm 1, \pm 2, \ldots. \tag{15-133}$$

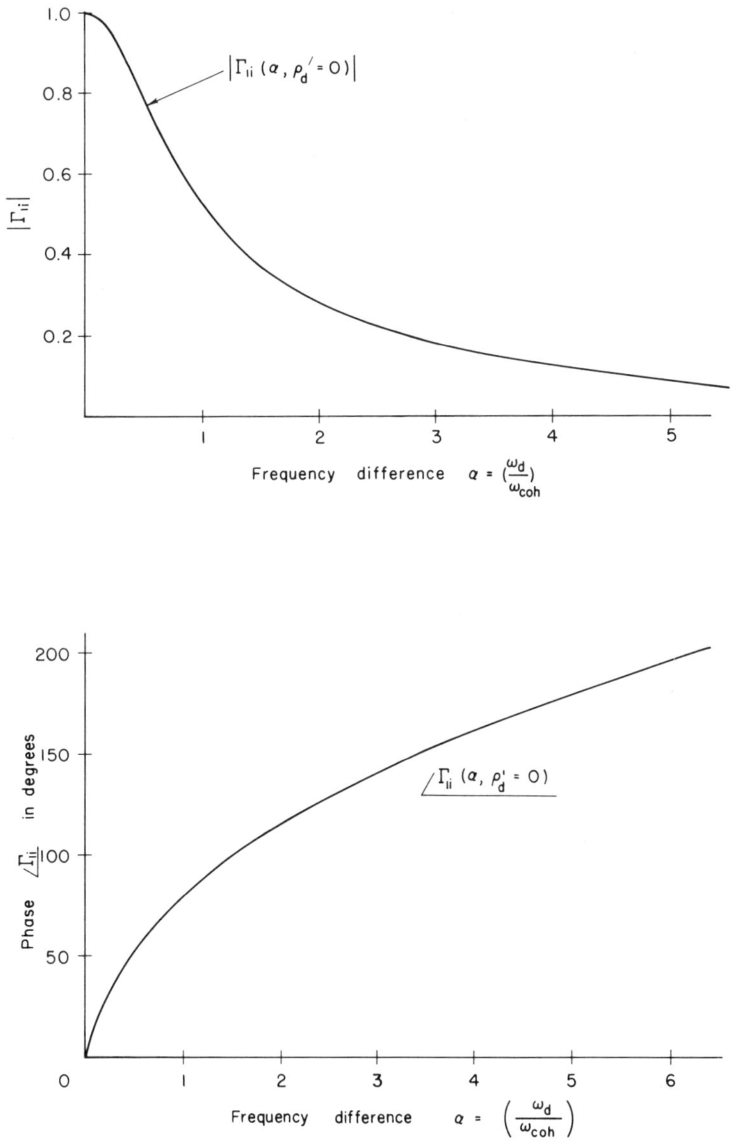

FIG. 15-9 Universal form of the two-frequency mutual coherence function as a function of ω_d/ω_{coh} in the strong fluctuation case. The magnitude and the phase are shown in (a) and (b) respectively.

Therefore we have a series of double poles $\alpha = \alpha_n$:

$$\alpha_n = \frac{[(2n+1)(\pi/2)]^2}{4i}, \qquad n = 0, 1, 2, \ldots, \tag{15-134}$$

since the poles at $n = -1, -2, \ldots,$ coincide with the poles at $n = 0, 1, \ldots$.

These poles are all in the lower half of the α plane along the imaginary axis. Therefore for $t - z/c < 0$, we close the contour in the upper half of the plane and get zero as expected:

$$G(t) = 0 \qquad \text{for} \quad t - z/c < 0. \tag{15-135}$$

For $t - z/c > 0$, we close the contour in the lower half of the plane and obtain a series of residues at the poles. We get

$$G(t) = -2\pi i \sum_{n=0}^{\infty} \text{residues}. \tag{15-136}$$

Calculation of the residues can be performed easily, and we obtain

$$G(t) = \left[\frac{\pi\omega_{\text{coh}}}{4}\right] \sum_{n=0}^{\infty} (2n+1)(-1)^n \exp\left\{-\left[(2n+1)\frac{\pi}{4}\right]^2 T\right\} \tag{15-137}$$

where $T = \omega_{\text{coh}}(t - z/c)$. This impulse response $G(t)$ is shown in Fig. 15-10.

Note that Figs. 15-9 and 15-10 are universal curves applicable to a variety of physical situations. ω_{coh} in (15-122) is the coherence bandwidth.

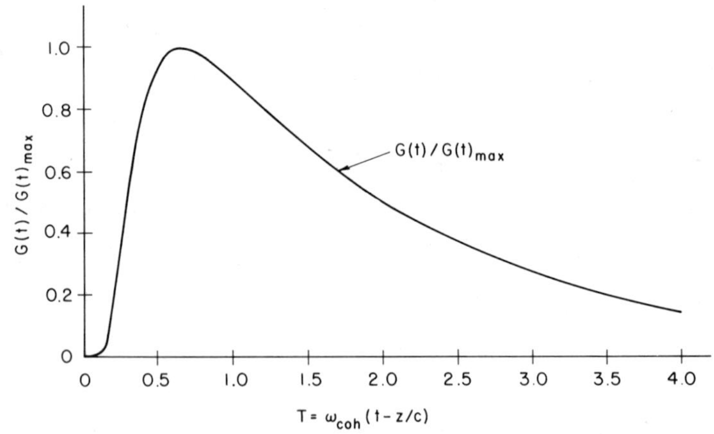

FIG. 15-10 Universal form of the pulse shape of the impulse response.

This should be compared with (15-113) and (15-112). Summarizing them, we write the coherence bandwidth:

$$\begin{aligned}\omega_{\text{coh}} &\approx \frac{10\alpha_{\text{p}} c}{z} && \text{when} \quad \rho_{\text{n}}\sigma_{\text{s}} z < 1 \\ &\approx \left[\frac{3}{\rho_{\text{n}}\sigma_{\text{s}} z}\right]^{1/2} \frac{2\alpha_{\text{p}} c}{z} && \text{when} \quad \rho_{\text{n}}\sigma_{\text{s}} z \sim 1 \\ &= \frac{8\alpha_{\text{p}} c}{\rho_{\text{n}}\sigma_{\text{s}} z^2} && \text{when} \quad \rho_{\text{n}}\sigma_{\text{s}} z \gg 1.\end{aligned} \tag{15-138}$$

The time delay T_{d} is given in (15-115) for the weak fluctuation case. From Fig. 15-9, it is clear that

$$T_{\text{d}} = \rho_{\text{n}}\sigma_{\text{s}} z^2/4\alpha_{\text{p}} c \tag{15-139}$$

is still applicable to the strong fluctuation case and the peak appears approximately at $\frac{1}{4}T_{\text{d}}$.

The analysis presented in this and the preceding section is applicable to millimeter and optical pulse propagation in rain and fog. For example, consider a short optical pulse at $\lambda = 0.6943\ \mu\text{m}$ (ruby laser) with duration on the order of nanoseconds propagating in a typical fog ($\rho_{\text{n}} = 10^8\ \text{m}^{-3}$, median diameter $D = 9\ \mu\text{m}$; (see Section 3-2-2)). Using the approximation $\alpha_{\text{p}} = 2.66(D/\lambda)^2$, and $\sigma_{\text{s}} \approx \sigma_{\text{t}} \approx 2\pi(D/2)^2$, we get from (15-138) the coherence bandwidth ω_{coh}:

$$\omega_{\text{coh}} = 8\alpha_{\text{p}} c/\rho_{\text{n}}\sigma_{\text{s}} z^2 = (8.43 \times 10^{13})/z^2$$

when $\rho_{\text{n}}\sigma_{\text{s}} z \gg 1$. For example, at $z = 5$ km, we have $\omega_{\text{coh}} = 3.37 \times 10^6$ and the spread of the impulse response is 0.3 μsec. This means that the input pulse of nanosecond duration is broadened to 0.3 μsec. Experimental evidence supports this conclusion (Bucher and Lerner, 1973; Ishimaru and Hong, 1975).

The mathematical formulation and technique discussed in this section is also applicable to pulse propagation in a turbulent medium (see Chapter 20) and has been used in the study of the broadening of pulses from pulsars.

PART IV □ WAVES IN RANDOM CONTINUUM AND TURBULENCE

CHAPTER 16 □ SCATTERING OF WAVES FROM RANDOM CONTINUUM AND TURBULENT MEDIA

Random media may be classified into random scatterers, random continua, and rough interfaces. Chapters 2–15 were devoted to the problem of wave propagation and scattering in "random scatterers." In Chapters 16–20, we discuss the problem of wave propagation and scattering in a "random continuum."

In this book, the random continuum is defined as the medium whose dielectric constant $\varepsilon(\mathbf{r}, t)$ is a continuous random function of position and time. Examples of the random continuum are atmospheric and ocean turbulence and biological media.

It is convenient to divide the problem into "scattering" of a wave and "line-of-sight" wave propagation. This chapter is devoted to the scattering problem and the subsequent chapters deal with the line-of-sight problem.

This chapter presents a general formulation of the scattered power from a random medium in the single scattering approximation (Du Castel, 1966; Wheelon and Muchmore, 1955; Wheelon, 1959; Staras, 1952, 1955; Silverman, 1957a, 1958; Pekeris, 1947; Muchmore and Wheelon, 1955, 1963; Booker and Gordon, 1950a,b; Booker, 1956, 1958; Beard, 1961, 1962, 1967; Beard *et al.*, 1969; Villars and Weisskopf, 1955). This applies to many practical problems of tropospheric beyond-the-horizon communication, scattering from a turbulent wake and the plumes of aircraft and rocket engines, and also the detection of clear air turbulence (Pao and Goldburg, 1969). Time variation of the medium causes temporal variation of the scattered field. This is discussed in this chapter as well as the scattering of a pulse from a random medium.

16-1 SINGLE SCATTERING APPROXIMATION AND RECEIVED POWER

Let us consider a random medium illuminated by a transmitter. A part of the transmitted wave is scattered by the randomness of the medium and this scattered wave is detected by a receiver (Fig. 16-1). We wish to find the

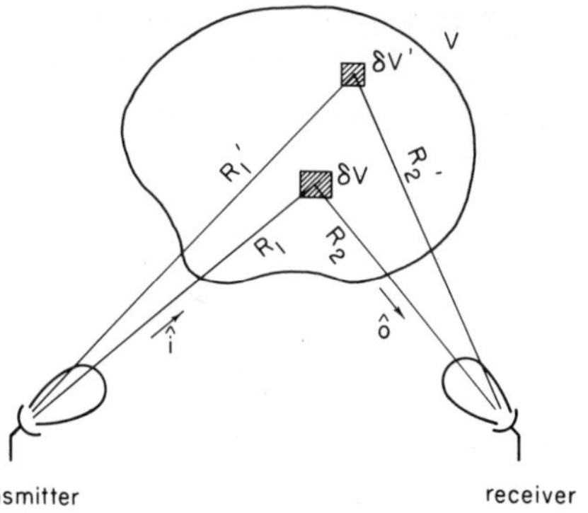

FIG. 16-1 Geometry showing transmitter, receiver, and random medium.

relationship between the received power and the characteristics of the random medium.

Consider a volume δV in the random medium. We assume that the randomness of the medium is so slight that the wave incident on δV is almost equal to the incident wave in the absence of the random medium. If we can describe the amount of scattered power due to the random medium in δV in terms of the equivalent scattering cross section per unit volume $\sigma(\hat{\mathbf{0}}, \hat{\mathbf{i}})$, then the received power P_r is given by the radar equation (see Section 4-1)

$$\frac{P_r}{P_t} = \frac{\lambda^2 G_t(\hat{\mathbf{i}}) G_r(\hat{\mathbf{0}})}{(4\pi)^2 R_1{}^2 R_2{}^2} \sigma(\hat{\mathbf{0}}, \hat{\mathbf{i}})\, \delta V \tag{16-1}$$

where P_t is the transmitted power, and G_t and G_r are the gain functions of the transmitter and receiver in the directions of $\hat{\mathbf{i}}$ and $-\hat{\mathbf{0}}$, respectively.

The elementary volume δV must be small enough so that the incident wave from the transmitter has constant magnitude and can be considered a plane wave throughout δV. At the same time, we take δV to be large enough so that the dimension $D(\delta V \sim D^3)$ is much greater than the correlation distance of the medium. Under this assumption,† the scattered fields from different elementary volumes δV and $\delta V'$ (see Fig. 16-1) are uncorrelated and therefore the scattered powers rather than the scattered fields from different volume elements can be added. We then obtain the total received power:

$$\frac{P_r}{P_t} = \int \frac{\lambda^2 G_t(\hat{\mathbf{i}}) G_r(\hat{\mathbf{0}})}{(4\pi)^2 R_1{}^2 R_2{}^2} \sigma(\hat{\mathbf{0}}, \hat{\mathbf{i}})\, dV. \tag{16-2}$$

† This requires that D is smaller than the Fresnel size $(\lambda R_1)^{1/2}$.

It is clear then that the core of the problem is to find the scattering cross section per unit volume $\sigma(\hat{\mathbf{0}}, \hat{\mathbf{i}})$ of the random medium. In the following section, we will consider this problem further.

16-2 SCATTERING CROSS SECTION PER UNIT VOLUME OF THE STATIONARY RANDOM MEDIUM

Let us consider the random medium in the elementary volume δV. The random medium is characterized by the permittivity $\varepsilon(\mathbf{r}, t)$ which is a random function of time and position.† We write it as the sum of the average $\langle\varepsilon\rangle$ and the fluctuation:

$$\varepsilon(\mathbf{r}, t) = \langle\varepsilon(\mathbf{r}, t)\rangle(1 + \varepsilon_1(\mathbf{r}, t)), \tag{16-3}$$

where ε_1 is the fluctuation with zero average

$$\langle\varepsilon_1\rangle = 0 \tag{16-4}$$

and the angle bracket $\langle\ \rangle$ represents the ensemble average. In terms of the index of refraction, we write

$$n(\mathbf{r}, t) = \sqrt{\varepsilon(\mathbf{r}, t)/\varepsilon_0} = \langle n(\mathbf{r}, t)\rangle(1 + n_1(\mathbf{r}, t)) \tag{16-5}$$

where ε_0 is the free space permittivity and n_1 the fluctuation. For small fluctuation, we have approximately

$$\varepsilon_1(\mathbf{r}, t) \approx 2n_1(\mathbf{r}, t). \tag{16-6}$$

In this section, we assume that the dielectric constant of the medium does not change during the observation time, and therefore the dielectric constant is invariant in time. We also assume that the average dielectric constant is constant and equal to the free space permittivity:

$$\varepsilon(\mathbf{r}) = \varepsilon_0(1 + \varepsilon_1(\mathbf{r})), \qquad n(\mathbf{r}) = 1 + n_1(\mathbf{r}). \tag{16-7}$$

Let us consider a plane monochromatic electromagnetic wave with a unit amplitude incident upon the volume δV of the random medium (Fig. 16-2).

Following a procedure identical to that given in Section 2-4, Eq. (2-20), we obtain the scattered field $\mathbf{E}_s$ at a distance R from the reference point Q_0 in the direction $\hat{\mathbf{0}}$.

$$\mathbf{E}_s = \mathbf{f}(\hat{\mathbf{0}}, \hat{\mathbf{i}})[\exp(ikR)]/R$$

$$\mathbf{f}(\hat{\mathbf{0}}, \hat{\mathbf{i}}) = \frac{k^2}{4\pi}\int_{\delta V}\{-\hat{\mathbf{0}} \times [\hat{\mathbf{0}} \times \mathbf{E}(\mathbf{r}')]\}\varepsilon_1(\mathbf{r}')\exp(-ik\mathbf{r}' \cdot \hat{\mathbf{0}})\,dv', \tag{16-8}$$

† See Section 16-6 for a more precise definition of a time-varying random dielectric constant.

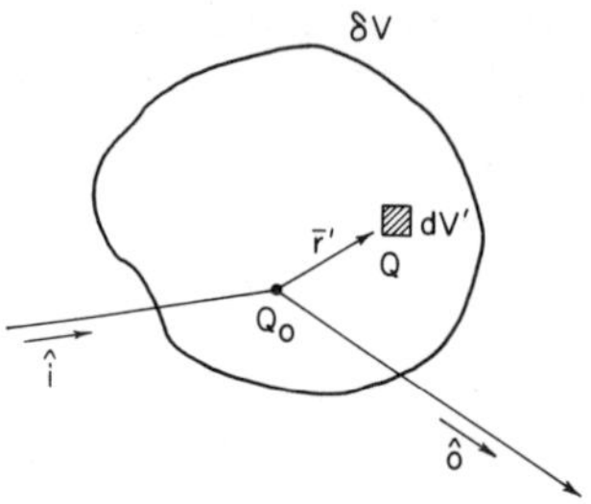

FIG. 16-2

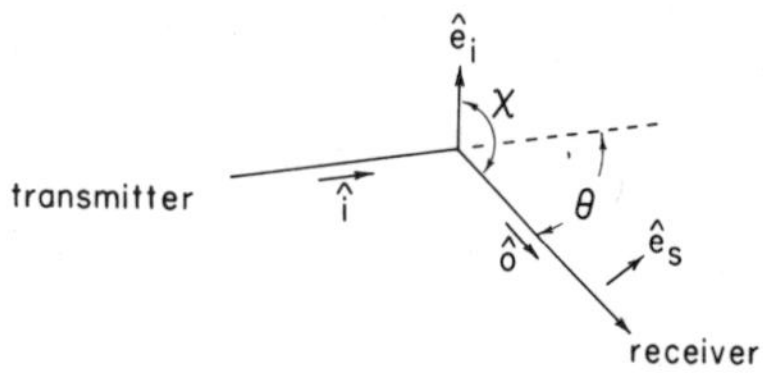

FIG. 16-3

FIG. 16-2 Scattering volume δV illuminated by a plane wave incident in the direction $\hat{\mathbf{i}}$. The scattered field is observed in the direction $\hat{\mathbf{0}}$.

FIG. 16-3 Geometry defining quantities in (16-9).

where $\mathbf{f}(\hat{\mathbf{0}}, \hat{\mathbf{i}})$ is the scattering amplitude and $\mathbf{E}(\mathbf{r}')$ the total field at $\mathbf{r}'$ when the incident field has unit amplitude.

Recognizing that the fluctuation ε_1 is small, we use the Born approximation that $\mathbf{E}(\mathbf{r}')$ is equal to the incident wave $\hat{\mathbf{e}}_i \exp(ik\hat{\mathbf{i}} \cdot \mathbf{r}')$ where $\hat{\mathbf{e}}_i$ is a unit vector representing the polarization. We then obtain

$$\mathbf{f}(\hat{\mathbf{0}}, \hat{\mathbf{i}}) = \hat{\mathbf{e}}_s \sin \chi \frac{k^2}{4\pi} \int_{\delta V} \varepsilon_1(\mathbf{r}') \exp(i\mathbf{k}_s \cdot \mathbf{r}')\, dV', \tag{16-9a}$$

where $\hat{\mathbf{e}}_s \sin \chi = -\hat{\mathbf{0}} \times (\hat{\mathbf{0}} \times \hat{\mathbf{e}}_i)$ and χ is the angle between the polarization direction $\hat{\mathbf{e}}_i$ of the incident wave and the direction of observation $\hat{\mathbf{0}}$ (see Fig. 16-3). The vector $\mathbf{k}_s$ is given by

$$\mathbf{k}_s = k(\hat{\mathbf{i}} - \hat{\mathbf{0}}), \qquad k_s = 2k \sin(\theta/2). \tag{16-9b}$$

The differential cross section per unit volume of the medium according to (2-4) is therefore given by

$$\sigma(\hat{\mathbf{0}}, \hat{\mathbf{i}}) = |\mathbf{f}(\hat{\mathbf{0}}, \hat{\mathbf{i}})|^2/\delta V. \tag{16-10a}$$

However, we note that $\mathbf{f}$ is a random function because $\varepsilon_1(\mathbf{r}')$ is random. Therefore, (16-10a) must be replaced by the ensemble average:

$$\sigma(\hat{\mathbf{0}}, \hat{\mathbf{i}}) = (1/\delta V)\langle f(\hat{\mathbf{0}}, \hat{\mathbf{i}}) f^*(\hat{\mathbf{0}}, \hat{\mathbf{i}})\rangle. \tag{16-10b}$$

Substituting (16-9) into (16-10b), we obtain

$$\sigma(\hat{\mathbf{0}}, \hat{\mathbf{i}}) = \frac{k^4 \sin^2 \chi}{(4\pi)^2\, \delta V} \int_{\delta V} \int_{\delta V} \langle \varepsilon_1(\mathbf{r}_1')\varepsilon_1(\mathbf{r}_2')\rangle \exp[i\mathbf{k}_s \cdot (\mathbf{r}_1' - \mathbf{r}_2')]\, dV_1'\, dV_2', \tag{16-11}$$

where dV_1' and dV_2' are the volume elements at $\mathbf{r}_1'$ and $\mathbf{r}_2'$, respectively.

If the medium is assumed to be statistically homogeneous and isotropic (Davenport and Root, 1958; Feller, 1957; Gnedenko, 1962; Stratonovich, 1963), then the covariance $\langle \varepsilon_1(\mathbf{r}_1')\varepsilon_1(\mathbf{r}_2')\rangle$ is a function of the magnitude of the difference $r_d = |\mathbf{r}_d| = |\mathbf{r}_1' - \mathbf{r}_2'|$:

$$\langle \varepsilon_1(\mathbf{r}_1')\varepsilon_1(\mathbf{r}_2')\rangle = B_\varepsilon(r_d) = 4B_n(r_d), \qquad \langle n_1(\mathbf{r}_1')n_1(\mathbf{r}_2')\rangle = B_n(r_d). \tag{16-12}$$

Let us consider the integral in (16-11) using (16-12):

$$\int_{\delta V}\int_{\delta V} B_n(r_d)\exp(i\mathbf{k}_s\cdot\mathbf{r}_d)\,dV_1'\,dV_2'. \tag{16-13a}$$

We change the variables from $\mathbf{r}_1'$ and $\mathbf{r}_2'$ to the difference $\mathbf{r}_d = \mathbf{r}_1' - \mathbf{r}_2'$ and the average $\mathbf{r}_c = \frac{1}{2}(\mathbf{r}_1' + \mathbf{r}_2')$. Noting that the size of the volume δV is much greater than the correlation distance of the medium and therefore that $B_\varepsilon(r_d)$ is negligibly small for $r_d \gg$ correlation distance, we can approximate (16-13a) by

$$\int_{\delta V} dV_c \int_\infty dV_d\, B_n(r_d)\exp(i\mathbf{k}_s\cdot\mathbf{r}_d) \tag{16-13b}$$

where dV_c and dV_d are the elementary volumes for $\mathbf{r}_c$ and $\mathbf{r}_d$, respectively, and the integral for $\mathbf{r}_d$ is now extended to all space.

Now we recognize that according to the Wiener–Khinchin theorem, the Fourier transform of the correlation function $B_n(\mathbf{r}_d)$ is the spectral density $\Phi_n(\mathbf{K})$:

$$\Phi_n(\mathbf{K}) = \frac{1}{(2\pi)^3}\int_\infty B_n(\mathbf{r}_d)\exp(i\mathbf{K}\cdot\mathbf{r}_d)\,dV_d. \tag{16-14a}$$

Therefore, (16-13b) is proportional to the Fourier transform of B_n evaluated at $\mathbf{K} = \mathbf{k}_s$. We also note that since B_n is a function of the magnitude of r_d only, Φ_n is a function of the magnitude of $\mathbf{K}$ only:

$$\Phi_n(K) = \frac{1}{(2\pi)^3}\int_\infty B_n(r_d)\exp(i\mathbf{K}\cdot\mathbf{r}_d)\,dV_d. \tag{16-14b}$$

Using (16-13b) and (16-14b), (16-11) becomes

$$\sigma(\hat{\mathbf{0}}, \hat{\mathbf{i}}) = 2\pi k^4 \sin^2\chi\,\Phi_n(k_s), \tag{16-15}$$

where $k_s = 2k\sin(\theta/2)$.

Equation (16-15) is the fundamental expression for the scattering cross section per unit volume of the random medium. Let us examine this expression.

First, we note that the dependence $\sin^2\chi$ is due to the radiation pattern of a dipole. The incident wave produces an equivalent dipole source in the random medium, which gives the $\sin\chi$ pattern (see Fig. 16-4).

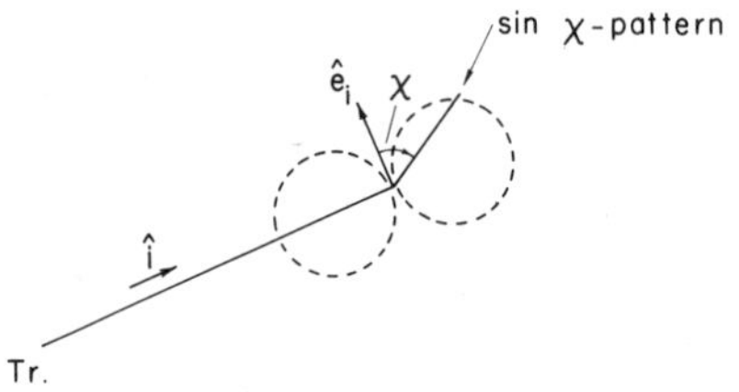

FIG. 16-4

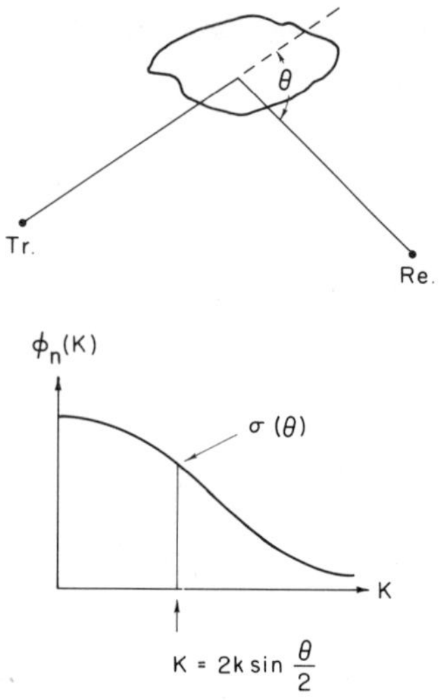

FIG. 16-5

FIG. 16-4 Dipole pattern of the differential cross section.

FIG. 16-5 The differential cross section is proportional to the spectral density evaluated at $2k\ \sin(\theta/2)$.

We note that the cross section σ is proportional to the spectral density Φ_n of the index of refraction evaluated at k_s. Thus, if the spectral density is known, then the cross section in a given direction θ is given by this particular portion of the spectrum at $k_s = 2k\ \sin(\theta/2)$ (see Fig. 16-5).

16-3 BOOKER–GORDON FORMULA

The scattering cross section $\sigma(\hat{\mathbf{0}}, \hat{\mathbf{i}})$ in (16-15) is given in terms of the spectral density function $\Phi_n(k_s)$ of the index of refraction fluctuation. There are several models for Φ_n to represent practical situations. We will discuss three important cases: (a) the Booker–Gordon formula, (b) the Gaussian model, and (c) the Kolmogorov spectrum.

Booker and Gordon (1950a,b) presented a formula based on an assumption that the correlation function is given by an exponential function:

$$B_n(r_d) = \langle n_1{}^2 \rangle \exp(-r_d/l). \tag{16-16}$$

Here the random medium is characterized by two quantities: the "variance" $\langle n_1{}^2 \rangle$ and l. The distance l is called the correlation distance, because at this distance, the correlation function falls off to e^{-1} of the value at $r_d = 0$ and thus this distance indicates an approximate distance within which the medium has some correlation. The correlation distance l is also called the scale of turbulence because this gives an average size of the turbulent blobs or eddies. We note that this choice of the correlation function is based on the ease of mathematical operation and not on consideration of the physical characteristics of the medium. The Kolmogorov spectrum discussed later is based on actual turbulence characteristics.

Let us calculate the spectral density $\Phi_n(K)$ using (16-14b). To evaluate this integral, it is convenient to use the spherical coordinate system and choose the z axis along the direction of $\mathbf{K} = \mathbf{k}_s = k_s \mathbf{i}_s$ (see Fig. 16-6). Then

$$\Phi_n(k_s) = \frac{1}{(2\pi)^3} \int_0^\infty \int_0^{2\pi} \int_0^\pi B_n(r_d) \exp(-ik_s r_d \cos\alpha) r_d^2 \sin\alpha \, d\alpha \, d\phi \, dr_d. \tag{16-17}$$

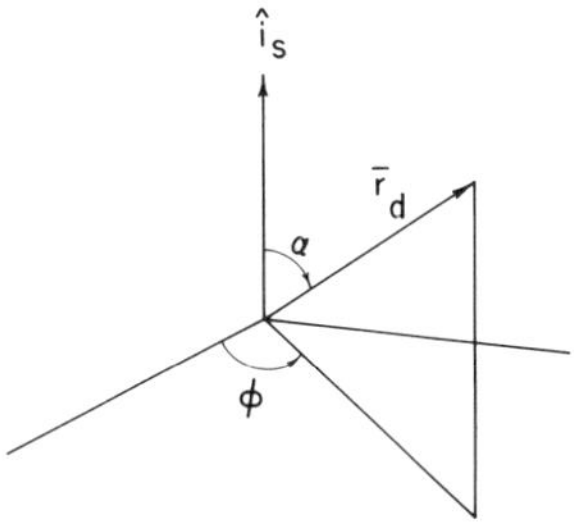

FIG. 16-6 Calculation of (16-14b) and (16-17).

Performing integration with respect to α and ϕ,† we get

$$\Phi_n(k_s) = \frac{1}{(2\pi)^3} \int_0^\infty B_n(r_d) 4\pi \frac{\sin k_s r_d}{k_s} r_d \, dr_d. \tag{16-18}$$

Substituting Eq. (16-16) into Eq. (16-18), we obtain‡

$$\Phi_n(k_s) = \frac{\langle n_1^2 \rangle l^3}{[1 + (k_s l)^2]^2} \left(\frac{1}{\pi^2}\right). \tag{16-19}$$

Thus the cross section σ is given by

$$\sigma(\theta) = \frac{2}{\pi} \frac{k^4 l^3 \sin^2 \chi \langle n_1^2 \rangle}{(1 + 4k^2 l^2 \sin^2(\theta/2))^2} \tag{16-20}$$

† Note that

$$\int \exp(Z\cos\alpha) \sin\alpha \, d\alpha = -\frac{\exp(Z\cos\alpha)}{Z}.$$

‡ Note that

$$\int_0^\infty e^{-Z\rho} \rho \, d\rho = -\frac{e^{-Z\rho}}{Z} \rho \Big|_0^\infty + \int_0^\infty \frac{e^{-Z\rho}}{Z} d\rho = \frac{1}{Z^2}.$$

This is the basic Booker–Gordon formula which is characterized by the two parameters of the medium: the variance $\langle n_1{}^2 \rangle$ and the correlation distance l. We note that if the correlation distance is small compared with the wavelength such that $k_s l \ll 1$, then

$$\sigma(\theta) \approx (2/\pi)k^4 l^3 \sin^2 \chi \langle n_1{}^2 \rangle, \tag{16-21}$$

and thus the scattering is isotropic and the cross section is inversely proportional to λ^4. This is characteristic of Rayleigh scattering and also a basis for Rayleigh's explanation of the blueness of the sky (Born and Wolf, 1964, pp. 99, 652).

If the correlation distance is much greater than the wavelength, as is usually the case with tropospheric propagation, then

$$\sigma(\theta) \approx \frac{\sin^2 \chi \langle n_1{}^2 \rangle}{8\pi l \sin^4(\theta/2)}, \tag{16-22}$$

which shows that σ is independent of frequency and the scattering is sharply directed in the forward direction.

For tropospheric propagation, the variance of the index of refraction fluctuation is of the order of magnitude of $[\langle n_1{}^2 \rangle]^{1/2} = 10^{-6}$, and the correlation distance l is approximately $l \approx 50m$ (20–130 m).

16-4 GAUSSIAN MODEL AND KOLMOGOROV SPECTRUM

In many practical situations, a random medium may be better approximated by the Gaussian correlation function

$$B_n(r_d) = \langle n_1{}^2 \rangle \exp(-r_d{}^2/l^2). \tag{16-23}$$

Using (16-18), we can easily obtain

$$\Phi_n(k_s) = \frac{\langle n_1{}^2 \rangle l^3}{8\pi\sqrt{\pi}} \exp\left[-\frac{(k_s l)^2}{4}\right] \tag{16-24}$$

$$\sigma(\theta) = \frac{\langle n_1{}^2 \rangle k^4 l^3}{4\sqrt{\pi}} \sin^2 \chi \exp\left[-\frac{(k_s l)^2}{4}\right]. \tag{16-25}$$

In atmospheric and ocean turbulence, the preceding two models, exponential and Gaussian, cannot fully explain detailed characteristics of the actual scattering phenomena. Kolmogorov obtained the spectrum of the fully developed turbulence based on the physical consideration of turbulence. This is discussed in more detail in Appendix C. Here, we give a short summary of the Kolmogorov spectrum.

According to the Kolmogorov theory, the turbulence eddies may be characterized by two sizes: the outer scale of turbulence L_0 and the inner scale (also called microscale) of turbulence l_0. We therefore divide the turbulence characteristics according to the size of the eddy or blob into three regions:

(a) *Input range* (eddy size $> L_0$) The energy is introduced into the turbulence in this range of eddy sizes due to wind shear and temperature gradient. In general, the turbulence is anisotropic in this range (see Fig. 16-7). The spectrum in this range depends on how the turbulence is created for the particular case, and thus there is no general formula describing the turbulence characteristics in this range.

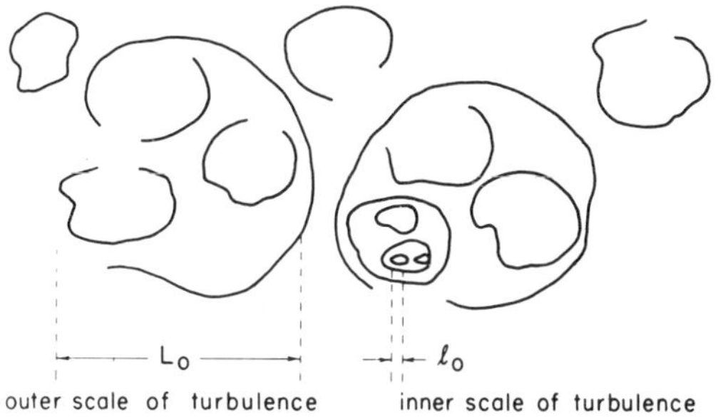

FIG. 16-7 Turbulent eddies showing outer and inner scales.

(b) *Inertial subrange* ($L_0 >$ eddy size $> l_0$) In this range, the kinetic energy of the eddies dominates over the dissipation due to viscosity, and the turbulence is essentially isotropic. The spectrum is proportional to $K^{-11/3}$, where $K = 2\pi/(\text{eddy size})$.

(c) *Dissipation range* ($l_0 >$ eddy size) In this range, the dissipation of energy due to viscosity dominates over the kinetic energy, and therefore, the spectrum is extremely small.

Considering these three ranges, we write the Kolmogorov spectrum as follows. In the input range ($0 \leq K < 2\pi/L_0$), the spectrum is unknown. In the inertial range, we write

$$\Phi_n(K) = 0.033 C_n{}^2 K^{-11/3} \qquad \text{for} \quad 2\pi/L_0 < K < 2\pi/l_0 \tag{16-26}$$

and in the dissipation range

$$\Phi_n(K) = 0 \qquad \text{for} \quad 2\pi/l_0 < K,$$

where C_n is called the structure constant (see Fig. 16-8).

For mathematical convenience, these three regions are often combined to give

$$\Phi_n(K) = 0.033 C_n{}^2 (K^2 + 1/L_0{}^2)^{-11/6} \exp(-K^2/K_m{}^2), \tag{16-27}$$

where $K_m = 5.92/l_0$. This is sometimes called the von Karman spectrum.

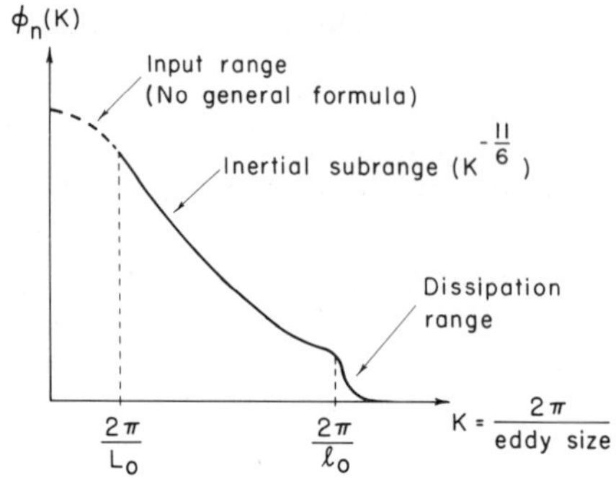

FIG. 16-8 Kolmogorov spectrum of the spectral density of the refractive index fluctuation.

We note here that even though (16-27) describes the entire spectrum, its value in the input range must be considered only approximate because it is in general anisotropic and depends on how the energy is introduced into the turbulence. We also note that Kolmogorov's theory is based on a locally homogeneous medium rather than the homogeneous medium described here. This subtle difference is explained in Appendix B, Section B-3. It will suffice to point out that as long as k_s is in the inertial subrange and dissipation range, there is no need to be concerned with this difference.

The value of the structure constant in the atmosphere is on the order of $10^{-7}\ \mathrm{m}^{-1/3}$ for strong turbulence and $10^{-9}\ \mathrm{m}^{-1/3}$ for weak turbulence.

16-5 ANISOTROPIC RANDOM MEDIUM

If the random medium is homogeneous and anisotropic, then the spectral density Φ_n in the scattering cross section (16-15) must be evaluated at $k(\hat{\mathbf{i}} - \hat{\mathbf{0}}) = \mathbf{k}_s$ rather than $|\mathbf{k}_s|$. Thus we have

$$\sigma(\hat{\mathbf{0}}, \hat{\mathbf{i}}) = 2\pi k^4 \sin^2 \chi \Phi_n(\mathbf{k}_s). \tag{16-28}$$

Let us consider an anisotropic random medium which has different correlation distances in different directions. Let us express this medium by the correlation function:

$$B_n(\mathbf{r}_d) = \langle {n_1}^2 \rangle \exp[-(x^2/{l_1}^2) - (y^2/{l_2}^2) - (z^2/{l_2}^2)] \tag{16-29}$$

where $\mathbf{r}_d = x\hat{\mathbf{x}} + y\hat{\mathbf{y}} + z\hat{\mathbf{z}}$ and l_1, l_2, and l_3 are the correlation distances in the x, y, and z directions (Fig. 16-9). Substituting (16-29) into (16-14a) and evaluating it at $\mathbf{k}_s$, we get

$$\begin{aligned} \Phi_n(\mathbf{k}_s) &= (\langle {n_1}^2 \rangle l_1 l_2 l_3 / 8\pi\sqrt{\pi}) \exp[-\tfrac{1}{4}(k_{s1}^2 {l_1}^2 + k_{s2}^2 {l_2}^2 + k_{s3}^2 {l_3}^2] \\ \sigma(\hat{\mathbf{0}}, \hat{\mathbf{i}}) &= 2\pi k^4 \sin^2 \chi \Phi_n(\mathbf{k}_s) \end{aligned} \tag{16-30}$$

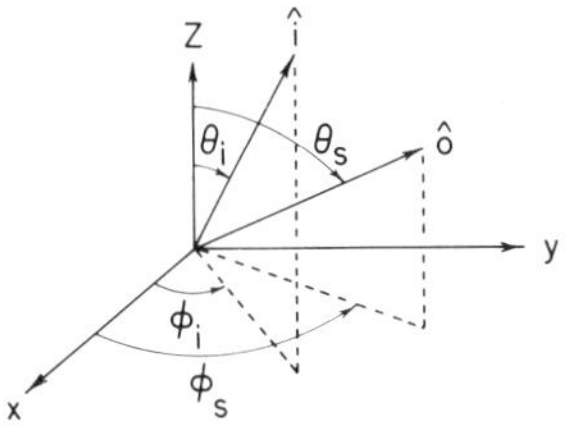

FIG. 16-9 Relationship between three directions (x, y, z) with different correlation distances and the directions of incident wave $\hat{\mathbf{i}}$ and observation $\hat{\mathbf{o}}$.

where

$$\mathbf{k}_s = k_{s1}\hat{\mathbf{x}} + k_{s2}\hat{\mathbf{y}} + k_{s3}\hat{\mathbf{z}}$$

$$k_{s1} = k(\sin\theta_i \cos\phi_i - \sin\theta_s \cos\phi_s)$$

$$k_{s2} = k(\sin\theta_i \sin\phi_i - \sin\theta_s \sin\phi_s)$$

$$k_{s3} = k(\cos\theta_i - \cos\theta_s).$$

In the ionosphere and troposphere, the turbulence is often anisotropic, and the correlation distance in the horizontal plane is much greater than the vertical correlation distance. Therefore, formula (16-30) can be used to describe the effects of anisotropy. The Kolmogorov spectrum for anisotropic turbulence is difficult to obtain since the Kolmogorov spectrum is applicable to a well-developed turbulence with an isotropic inertial subrange. Nevertheless, it may be convenient to express the Kolmogorov spectrum for anisotropic turbulence. The following approximate formula may be used to express the effects of anisotropy:

$$\Phi_n(\mathbf{k}_s) = 0.033C_n^2[k_{s1}^2 l_1^2 + k_{s2}^2 l_2^2 + k_{s3}^2 l_3^2 + 1]^{-11/6}(l_1 l_2 l_3)^{11/3}. \quad (16\text{-}31)$$

l_1, l_2, and l_3 may be considered as the outer scales of turbulence in the x, y, and z directions.

16-6 TEMPORAL FLUCTUATION OF SCATTERED FIELDS DUE TO A TIME-VARYING RANDOM MEDIUM

Up to this point, we assumed that the random medium is time invariant and is characterized by the fluctuation of the dielectric constant $\varepsilon_1(\mathbf{r})$. In this section, we extend our discussion to include the variation of the dielectric constant in time. Equation (16-7) then should be replaced by

$$\varepsilon(\mathbf{r}, t) = \varepsilon_0(1 + \varepsilon_1(\mathbf{r}, t)), \qquad n(r, t) = 1 + n_1(\mathbf{r}, t). \quad (16\text{-}32)$$

We assume here† that the dielectric constant is a slowly varying function of time and can be assumed constant over the period of the carrier frequency.

† In fact, if the medium moves with considerable speed, the dielectric constant cannot be defined simply as (16-32) and the Maxwell equations in the moving medium must be considered.

If we assume that the medium is stationary in time and homogeneous in space, we can write the correlation function in the form (see Appendix A, Section A-7)

$$\begin{aligned} B_n(\mathbf{r}, \tau) &= \langle n_1(\mathbf{r}_1 + \mathbf{r}, t + \tau) n_1(\mathbf{r}_1, t) \rangle \\ &= \iint d\mathbf{K}\, d\omega U(\mathbf{K}, \omega) \exp(-i\mathbf{K} \cdot \mathbf{r} + i\omega\tau) \end{aligned} \tag{16-33}$$

where $U(\mathbf{K}, \omega)$ is the four-dimensional spectral density. Equation (16-33) can also be written as

$$B_n(\mathbf{r}, \tau) = \int d\mathbf{K} \Phi_n(\mathbf{K}, \tau) e^{-i\mathbf{K} \cdot \mathbf{r}} = \int d\omega W_n(\mathbf{r}, \omega) e^{i\omega\tau} \tag{16-34}$$

where Φ_n may be called the time-varying spatial spectral density and W_n may be called the spatially varying temporal spectral density.

Let us assume that the random medium is moving with a velocity $\mathbf{V}$ without changing its spatial variation, and that the velocity $\mathbf{V}$ is almost constant during the correlation time of the scattered field. Then the fluctuation n_1 should have the following functional dependence on $\mathbf{r}$ and t:

$$n_1(\mathbf{r}, t + \tau) = n_1(\mathbf{r} - \mathbf{V}\tau, t), \tag{16-35}$$

which is valid for τ smaller than the correlation time of observation. This is usually called the locally frozen condition.

Under the assumption (16-35), we get

$$\langle n_1(\mathbf{r}_1, t_1) n_1(\mathbf{r}_2, t_2) \rangle = \langle n_1(\mathbf{r}_1 - \mathbf{V}t_1) n_1(\mathbf{r}_2 - \mathbf{V}t_2) \rangle = B_n(\mathbf{r} - \mathbf{V}\tau), \tag{16-36}$$

where $\mathbf{r} = \mathbf{r}_1 - \mathbf{r}_2$ and $\tau = t_1 - t_2$.

We now express (16-36) as

$$B_n(\mathbf{r} - \mathbf{V}\tau) = \int d\mathbf{K} \Phi_n(\mathbf{K}) \exp[-i\mathbf{K} \cdot (\mathbf{r} - \mathbf{V}\tau)], \tag{16-37}$$

where $\Phi_n(\mathbf{K})$ is the spectral density of the medium as viewed from the coordinate system moving with the medium.

Comparing (16-37) with (16-34), we get

$$\Phi_n(\mathbf{K}, \tau) = \Phi_n(\mathbf{K}) e^{i\mathbf{K} \cdot \mathbf{V}\tau}. \tag{16-38}$$

Equations (16-37) and (16-38) are valid when the velocity is constant. If the velocity consists of the average velocity $\mathbf{V}_0$ and the fluctuation $\mathbf{V}_f$, then (16-36) and (16-37) must be averaged over the velocity fluctuation. Then we have

$$\begin{aligned} \langle n_1(\mathbf{r}_1, t_1) n_1(\mathbf{r}_2, t_2) \rangle &= \int d\mathbf{K} \Phi_n(\mathbf{K}, \tau) \exp(-i\mathbf{K} \cdot \mathbf{r}) \\ \Phi_n(\mathbf{K}, \tau) &= \Phi_n(\mathbf{K}) \exp(-\mathbf{K} \cdot \mathbf{V}_0 \tau) \chi(\mathbf{K}\tau), \end{aligned} \tag{16-39}$$

where χ is the characteristic function [see (4-50)] given by

$$\chi(\mathbf{K}\tau) = \langle \exp(i\mathbf{K} \cdot \mathbf{V}_f \tau) \rangle. \qquad (16\text{-}40)$$

If the velocity fluctuation is assumed to be Gaussian with the probability density function

$$p(\mathbf{V}_f) = (1/2\pi\sigma_v^2)^{3/2} \exp(-|V_f|^2/2\sigma_v^2), \qquad (16\text{-}41a)$$

and the variance σ_v^2, the characteristic function in (16-40) becomes

$$\chi(\mathbf{K}\tau) = \exp[-\tfrac{1}{2}(K^2\sigma_v^2\tau^2)]. \qquad (16\text{-}41b)$$

Making use of this and following the procedure described in Section 4-5, we get the radar equation for a time-varying random medium. The temporal correlation function of the scattered field E_s is then given by

$$\frac{1}{2\eta_0}\langle E_s(t+\tau)E_s^*(t)\rangle = P_t \int \frac{\lambda^2 G_t(\hat{\mathbf{i}})G_r(\hat{\mathbf{0}})}{(4\pi)^2 R_1^2 R_2^2}\,\sigma(\hat{\mathbf{0}}, \hat{\mathbf{i}}, \tau)\,dV, \qquad (16\text{-}42)$$

where $\eta_0 = (\mu_0/\varepsilon_0)^{1/2}$ is the free space characteristic impedance and

$$\sigma(\hat{\mathbf{0}}, \hat{\mathbf{i}}, \tau) = 2\pi k^4 \sin^2 \chi \Phi_n(\mathbf{k}_s) \exp(i\mathbf{k}_s \cdot \mathbf{V}_0 \tau)\chi(\mathbf{k}_s \tau). \qquad (16\text{-}43)$$

The temporal frequency spectrum $W_s(\omega)$ is then given by

$$\begin{aligned} W_s(\omega) &= 2\int_{-\infty}^{\infty} e^{i\omega\tau}\langle E_s(t+\tau)E_s^*(t)\rangle\,d\tau \\ &= 2\eta_0 P_t \int \frac{\lambda^2 G_t(\hat{\mathbf{i}})G_r(\hat{\mathbf{0}})}{(4\pi)^2 R_1^2 R_2^2}\,W_\sigma(\hat{\mathbf{0}}, \hat{\mathbf{i}}, \omega)\,dV, \end{aligned} \qquad (16\text{-}44)$$

where

$$W_\sigma(\hat{\mathbf{0}}, \hat{\mathbf{i}}, \omega) = 2\int_{-\infty}^{\infty} e^{i\omega\tau}\sigma(\hat{\mathbf{0}}, \hat{\mathbf{i}}, \tau)\,d\tau. \qquad (16\text{-}45)$$

In this, the definition of the spectrum is made in conformity with the convention that if the correlation function $B(\tau)$ is real, the spectrum is given by

$$W(\omega) = 4\int_0^{\infty} \cos\omega\tau B(\tau)\,d\tau. \qquad (16\text{-}46)$$

As an example, let us take the Gaussian velocity fluctuation given by (16-41a) and (16-41b). We then have

$$W_\sigma(\hat{\mathbf{0}}, \hat{\mathbf{i}}, \omega) = 2\pi k^4 \sin^2 \chi \Phi_n(k_s)\frac{2\sqrt{2\pi}}{k_s^2\sigma_v^2}\exp\left[-\frac{(\omega + \mathbf{k}_s \cdot \mathbf{V}_0)^2}{2k_s^2\sigma_v^2}\right]. \qquad (16\text{-}47)$$

It shows the Doppler shift $\mathbf{k}_s \cdot \mathbf{V}_0$ due to the constant velocity $\mathbf{V}_0$ and the frequency spread due to the velocity fluctuation (see Fig. 16-10). We note

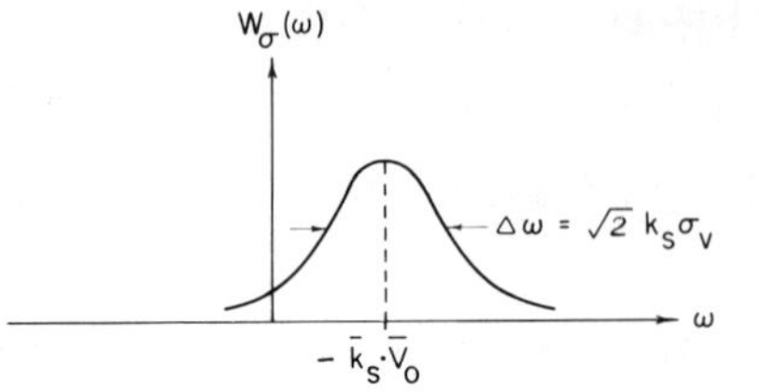

FIG. 16-10 Temporal frequency spectrum showing Doppler shift $-\mathbf{k}_s \cdot \mathbf{V}_o$ and the frequency spread $\Delta\omega$.

that (16-47) represents the spectrum of the fluctuation only, and therefore $\omega = 0$ in Fig. 16-10 corresponds to the carrier frequency f_0. Note also that the Doppler shift is given by

$$f_d = -f_0(\hat{\mathbf{i}} - \hat{\mathbf{0}}) \cdot \mathbf{V}_0 \tag{16-48}$$

and the frequency spread Δf is given by

$$\Delta f = [\sqrt{2}2 \sin(\theta/2)](\sigma_v/c). \tag{16-49}$$

16-7 STRONG FLUCTUATIONS

Up to this point, we have assumed that the fluctuation of the random medium is weak and that the single scattering approximation can be used. In some practical situations, this assumption is no longer valid and we need to take into account multiple scattering effects.

We will not go into a detailed discussion of the strong fluctuation problem in this section since this is discussed in Chapter 20. Here we indicate the first order modification of the single scattering approximation.

In the single scattering approximation, the field incident on the scattering random medium is assumed to be equal to the incident wave in free space. The first order multiple scattering approximation takes into account the attenuation of the incident wave due to the randomness of the medium.

In place of (16-2), the first order solution is given by

$$\frac{P_r}{P_t} = \int \frac{\lambda^2 G_t(\hat{\mathbf{i}}) G_r(\hat{\mathbf{0}})}{(4\pi)^2 R_1{}^2 R_2{}^2} \sigma(\hat{\mathbf{0}}, \hat{\mathbf{i}}) \exp(-\tau_1 - \tau_2)\, dV, \tag{16-50}$$

where τ_1 and τ_2 are given by

$$\tau_1 = \int_0^{R_1} \alpha(R')\, dR' \qquad \text{and} \qquad \tau_2 = \int_0^{R_2} \alpha(R'')\, dR''. \tag{16-51}$$

τ_1 and τ_2 represent the total attenuation along the path from the transmitter to dV and from dV to the receiver, respectively. The attenuation constant

$\alpha(R')$ consists of α_ε due to the dielectric loss in the medium and α_r due to the randomness of the medium

$$\alpha = \alpha_\varepsilon + \alpha_r \tag{16-52}$$

$$\alpha_\varepsilon = k\varepsilon_i \varepsilon_r^{-1/2} \quad \text{where} \quad \varepsilon = \varepsilon_r + i\varepsilon_i . \tag{16-53}$$

The attenuation constant α_r is obtained by calculating the total scattered power by integrating $\sigma(\hat{\mathbf{0}}, \hat{\mathbf{i}})$ in (16-15) over 4π:

$$\alpha_r = \int_{4\pi} \sigma(\hat{\mathbf{0}}, \hat{\mathbf{i}})\, d\Omega. \tag{16-54}$$

For an electromagnetic wave we use (16-15), and noting that

$$d\Omega = \sin\theta\, d\theta\, d\phi = (k_s\, dk_s\, d\phi)/k^2, \qquad k_s = 2k\sin(\theta/2),$$

we get

$$\alpha_r = 4\pi^2 k^2 \int_0^{2k} \left[1 - \frac{k_s}{k} + \frac{1}{4}\left(\frac{k_s}{k}\right)^3\right] \Phi_n(k_s) k_s\, dk_s, \tag{16-55}$$

where we have used

$$\int_0^{2\pi} \sin^2\chi\, d\phi = 2\pi\left(1 - \frac{\sin^2\theta}{2}\right).$$

For a scalar wave, the dipole pattern $\sin^2\chi$ in the cross section $\sigma(\hat{\mathbf{0}}, \hat{\mathbf{i}})$ is absent, and therefore we have

$$\alpha_r = 4\pi^2 k^2 \int_0^{2k} \Phi_n(k_s) k_s\, dk_s . \tag{16-56}$$

The attenuation constant α_r given in (16-55) and (16-56) is caused by the randomness of the medium, and it will be shown in Chapter 20 (see Keller, 1964; Keller and Karal, 1966) that this attenuation constant is indeed consistent with that obtained by a more exact analysis of the strong fluctuation problem.

16-8 SCATTERING OF A PULSE BY A RANDOM MEDIUM

Chapter 5 was devoted to the problem of scattering of a pulse by randomly distributed scatterers. The mathematical formulations of pulse scattering from a random continuum are identical to those given in Chapter 5 with appropriate substitutions.

The correlation function of the output pulse as given in (5-31) and (5-32) is equally applicable to a random continuum case if $\rho\sigma_{bi}(\hat{\mathbf{0}}, \hat{\mathbf{i}})$, γ_1, and γ_2 in (5-31) are replaced by

$$\rho\sigma_{bi}(\hat{\mathbf{0}}, \hat{\mathbf{i}}) \to 4\pi\sigma(\hat{\mathbf{0}}, \hat{\mathbf{i}}) \quad \text{in (16-15)}$$

$$\gamma_1 \text{ and } \gamma_2 \to \tau_1 \text{ and } \tau_2 \quad \text{in (16-51).}$$

Since all the mathematical formulations in Chapter 5 are valid for a random continuum with these substitutions, and the details of pulse scattering are already discussed in Chapter 5, we do not repeat them here.

16-9 ACOUSTIC SCATTERING CROSS SECTION PER UNIT VOLUME

In contrast to the scattering cross section $\sigma(\hat{\mathbf{0}}, \hat{\mathbf{i}})$ given in (16-15) for electromagnetic waves, the acoustic scattering cross section per unit volume of a random medium (Tatarski, 1971, p. 160; Derr and Little, 1970; for more rigorous theory, see Wenzel and Keller, 1971) is given by

$$\sigma(\hat{\mathbf{0}}, \hat{\mathbf{i}}) = 2\pi k^4 \cos^2 \theta \Phi_n(k_s) \tag{16-57}$$

where Φ_n is the spectral density of the equivalent acoustic refractive index fluctuation and $k_s = 2k \sin(\theta/2)$. The spectral density $\Phi_n(k_s)$ is given by

$$\Phi_n(k_s) = (1/c_0^2)E(k_s)\cos^2(\theta/2) + (1/4T_0^2)\Phi_T(k_s) \tag{16-58}$$

where c_0 and T_0 are the mean values of the sound velocity (m/sec) and the temperature (°K), respectively. E and Φ_T are the spectral densities of the velocity field and the temperature field, respectively, and are given by

$$E(k_s) = 0.061 C_v^2 k_s^{-11/3}, \qquad \Phi_T(k_s) = 0.033 C_T^2 k_s^{-11/3} \tag{16-59}$$

where C_v and C_T are the structure constants for velocity and temperature, respectively. Typical ranges of the values of these constants are

$$C_T = 10^{-3}\text{–}10^{-1}\ \text{deg/m}^{1/3}, \qquad C_v = 10^{-2}\text{–}1\ \text{m}^{2/3}/\text{sec},$$

$$c_0 = 340\ \text{m/sec}, \qquad T_0 = 288°\text{K}.$$

We note from (16-57) that the scattering at $\theta = 90°$ is zero. Also from (16-58), the first term is zero at $\theta = 180°$ (backscattering), and therefore the velocity fluctuation $E(k_s)$ has no effect on the backscattering. For radio and optical scattering, a typical value of C_n in the atmosphere is 10^{-7}–10^{-9} m$^{-1/3}$. For acoustic scattering, a typical value of C_n may be 10^{-3}–10^{-6} m$^{-1/3}$. Therefore the acoustic scattering cross section is several orders of magnitude larger than that of optical and radio waves. Because of this large difference, acoustic scattering is more sensitive to sense the atmospheric turbulence. However, because of this large scattering, the acoustic wave attenuates greatly in a short distance and cannot penetrate far into the atmosphere.

16-10 NARROW BEAM EQUATION

In Section 4-3, we gave a narrow beam equation for randomly distributed particles. This is directly applicable to the random continuum. From (4-16), we have (Fig. 4-3a)

$$\frac{P_r}{P_t} = \frac{\lambda^2 G_t(\hat{\mathbf{i}}) G_r(\hat{\mathbf{0}})}{(4\pi)^3 R_1{}^2 R_2{}^2} \sigma(\hat{\mathbf{0}}, \hat{\mathbf{i}}) \exp(-\gamma_1 - \gamma_2) V_c \tag{16-60}$$

where the gain functions G_t and G_r can be expressed in terms of the aperture areas (4-20) or the beamwidths (4-22). The cross section $\sigma(\hat{\mathbf{0}}, \hat{\mathbf{i}})$ is given in (16-15) or (16-57). γ_1 and γ_2 are τ_1 and τ_2 in (16-51) and the common volume V_c is given in (4-17).

The backscattering from the random medium located from $R = R_1$ and R_2 is given by (4-19):

$$\frac{P_r}{P_t} = (2.855 \times 10^{-4}) \lambda^2 G_t(\hat{\mathbf{i}})^2 \theta_1 \phi_1 \int_{R_1}^{R_2} \frac{\sigma(-\hat{\mathbf{i}}, \hat{\mathbf{i}})}{R^2} \exp(-2\gamma)\, dR. \tag{16-61}$$

For a square pulse of duration T_0, $R_2 - R_1$ should be the radar cell length $R_2 - R_1 = cT_0/2$ where c is the velocity of wave propagation.

CHAPTER 17 □ LINE-OF-SIGHT PROPAGATION OF A PLANE WAVE THROUGH A RANDOM MEDIUM—WEAK FLUCTUATION CASE

In the preceding chapter we discussed the scattering of a wave from a random medium. In this and subsequent chapters, we deal with the problem of line-of-sight wave propagation through a random medium.†

In contrast with scattering problems, the observed wave in line-of-sight propagation is always a mixture of the incident and the scattered wave, and therefore we need to deal with the coherent field (average field) and the incoherent field (fluctuating field). Examples of line-of-sight propagation are microwave and optical propagation in atmospheric turbulence, and acoustic wave propagation through a biological medium.

When an electromagnetic wave is incident upon a turbulent medium, the amplitude and phase of the wave experience fluctuations due to the fluctuation of the index of refraction of the medium.

This chapter deals specifically with the characteristics of a plane wave as it propagates through such a turbulent medium. Because of its relative mathematical simplicity, the plane wave problem provides us with an opportunity to clarify some fundamental relationships between the wave and the fluctuating medium without undue complications. In later chapters, we extend the results to other wave types such as spherical and beam waves.

In this chapter, we are interested mainly in the correlation functions and the variances of the amplitude and phase fluctuations of a wave. We are specifically interested in their dependence on frequency, distance, and the characteristics of the turbulence. The later chapters deal with other aspects of the wave fluctuations such as structure functions, variations in time, and temporal frequency spectra.

Within a given distance, if the intensity of the turbulence is weaker than a certain level, some mathematical simplifications are possible, and this situation is usually referred to as the "weak fluctuation" approximation. This chapter is devoted to the weak fluctuation case. In many microwave applica-

† The most important books are Tatarski (1961, 1971), as well as Chernov (1960). For review papers, see Strohbehn (1968), Lawrence and Strobhen (1970), and Brookner (1970). Also see Chapter 6 for line-of-sight propagation in randomly distributed scatterers.

tions for the earth's atmosphere, this approximation is valid for distances as long as several tens of kilometers. However, for optical frequencies, the distance is usually limited to a few kilometers. Beyond these ranges, we must deal with the "strong fluctuations," and these are discussed in Chapter 20.

We make extensive use of spectral representations of random functions. This has definite advantages over other direct approaches, and it is also most commonly used in the description of turbulence.

Many practical problems require consideration of the variation of the turbulence intensity along the path of wave propagation. This aspect is also discussed in the last section.

Scintillation of radio waves propagating through the ionosphere has been observed at frequencies ranging from 10 MHz to 6 GHz. See Crane (1977) and Rino (1976) for reviews of the current status, of the study of ionospheric scintillation.

17-1 MAXWELL'S EQUATIONS FOR A FLUCTUATING MEDIUM

In a turbulent medium, the relative dielectric constant ε_r or the index of refraction n varies from point to point and at different instants of time. In general, however, these variations cannot be predicted and, furthermore, even if they were known, it would be practically impossible to describe their values at all times and at all locations. It is therefore necessary to describe the medium statistically and to seek statistical information about the behavior of the wave in it.

The dielectric constant ε_r should, therefore, be described by a random function of position and time:

$$\varepsilon_r = \varepsilon_r(\mathbf{r}, t) = n^2(\mathbf{r}, t). \tag{17-1}$$

For the moment, we shall assume that ε_r is a function of $\mathbf{r}$ only, independent of time:

$$\varepsilon_r = \varepsilon_r(\mathbf{r}) = n^2(\mathbf{r}). \tag{17-2}$$

This means that the wave maintains a single frequency and that no frequency shift or spectrum is considered here. The frequency spectra are examined separately in Chapter 19.

Let us write Maxwell's equations for the medium given in (17-2):

$$\nabla \times \mathbf{E}(\mathbf{r}) = i\omega\mu_0\mathbf{H}(\mathbf{r}), \qquad \nabla \times \mathbf{H}(\mathbf{r}) = -i\omega\varepsilon_0\,\varepsilon_r(\mathbf{r})\mathbf{E}(\mathbf{r}) \tag{17-3}$$

where the time dependence of $\exp(-i\omega t)$ is assumed. Combining these two equations, we obtain

$$\nabla \times \nabla \times \mathbf{E}(\mathbf{r}) - \omega^2\mu_0\,\varepsilon_0\,\varepsilon_r(\mathbf{r})\mathbf{E}(\mathbf{r}) = 0.$$

Noting that $\nabla \times \nabla \times \mathbf{E} = -\nabla^2\mathbf{E} + \nabla(\nabla \cdot \mathbf{E})$ and $\nabla \cdot [\varepsilon_r(\mathbf{r})\mathbf{E}(\mathbf{r})] = 0$, we get

$$\nabla^2\mathbf{E}(\mathbf{r}) + \omega^2\mu_0\varepsilon_0\varepsilon_r(\mathbf{r})\mathbf{E} - \nabla\left(\frac{\nabla\varepsilon_r}{\varepsilon_r} \cdot \mathbf{E}\right) = 0. \tag{17-4}$$

In terms of the refractive index n, we write

$$\nabla^2\mathbf{E}(\mathbf{r}) + k_0{}^2n^2\mathbf{E}(\mathbf{r}) - 2\nabla\left(\frac{\nabla n}{n} \cdot \mathbf{E}\right) = 0. \tag{17-5}$$

The refractive index n can be written in terms of the average $\langle n \rangle$ and the fluctuation n_1. Then the y component of the second term of (17-5) is on the order of $k_0{}^2E_y + k_0{}^2 2n_1 E_y$ while the last term is on the order of $(2n_1/l_0{}^2)E_y$. Here l_0 is the correlation distance of the refractive index fluctuation n_1 and the x axis is chosen to be in the direction of propagation. It is then clear that the last term of (17-5) can be neglected as long as the wavelength λ is much smaller than the scale size l_0 of the random medium. We note that by neglecting the last term of (17-5) the depolarization effect is also neglected (for the polarization effect, see Strohbehn and Clifford, 1967; Strohbehn, 1971).

We shall now consider a wave propagation in the direction of the x axis. The y component of the electric field $U(\mathbf{r}) = E_y(\mathbf{r})$ satisfies

$$(\nabla^2 + k_0{}^2n^2)U(\mathbf{r}) = 0. \tag{17-6}$$

The index of refraction n fluctuates about the average value $\langle n \rangle$, and thus using the wave number for the average $k^2 = k_0{}^2\langle n \rangle^2$, we can write

$$[\nabla^2 + k^2(1 + n_1)^2]U(\mathbf{r}) = 0, \tag{17-7}$$

where n_1 represents the fluctuation of the index of refraction.

For weak fluctuation, it is necessary to obtain an approximate solution of (17-7) for small n_1. This can be done in two ways: One is to expand U in a series:

$$U = U_0 + U_1 + U_2 + \cdots, \tag{17-8}$$

and the other is to expand the exponent of U in a series:

$$U = \exp(\psi_0 + \psi_1 + \psi_2 + \cdots). \tag{17-9}$$

The expansion (17-8) is the Born approximation and the other (17-9) is called the Rytov solution. Their first iteration solutions are discussed in the following section.

17-2 BORN AND RYTOV METHODS

17-2-1 Born Approximation

We write (17-7) as

$$(\nabla^2 + k^2)U = -k^2\,\delta n U, \qquad \delta n = (1 + n_1)^2 - 1 = 2n_1 + n_1{}^2. \quad (17\text{-}10)$$

This can be converted into the following integral equation for U:

$$U(\mathbf{r}) = U_0(\mathbf{r}) + k^2 \int_V G(\mathbf{r} - \mathbf{r}')\,\delta n(\mathbf{r}')U(\mathbf{r}')\,dV', \qquad (17\text{-}11)$$

where $U_0(\mathbf{r})$ is the field in the absence of the fluctuation ($\delta n = 0$), and

$$G(\mathbf{r} - \mathbf{r}') = \frac{\exp(ik|\mathbf{r} - \mathbf{r}'|)}{4\pi|\mathbf{r} - \mathbf{r}'|}. \qquad (17\text{-}12)$$

If we substitute U_0 in the integral, we obtain the first iteration solution U_1. This is commonly known as the Born approximation. Repeating this iteration, we can develop a series expression for U. The first iteration U_1 has already been used for scattering problems.

17-2-2 Rytov Transformation

It is possible to write $U(\mathbf{r})$ as

$$U(\mathbf{r}) = e^{\psi(\mathbf{r})} \qquad (17\text{-}13)$$

and to develop a series solution for $\psi(\mathbf{r})$. This technique, known as the Rytov method, is widely used in line-of-sight propagation problems, partly because it simplifies the procedure of obtaining both amplitude and phase fluctuations and partly because its exponential representation is thought to represent a propagation wave better than the algebraic series representation of the Born method. There has been some theoretical and experimental evidence (Keller, 1969; Fried, 1967d; Brown, 1966, 1967) which shows that, for line-of-sight propagation, the first term of the Rytov representation is superior to the first Born approximation.

Starting with (17-13), we note that

$$\nabla^2 U = \nabla \cdot \nabla U = \nabla \cdot (U\nabla\psi) = U[\nabla\psi \cdot \nabla\psi + \nabla^2\psi],$$

and therefore (17-10) becomes

$$\nabla^2\psi + \nabla\psi \cdot \nabla\psi + k^2(1 + \delta n) = 0. \qquad (17\text{-}14)$$

This is a nonlinear first order differential equation for $\nabla\psi$ and is known as the Riccati equation.

In the absence of the fluctuation ($\delta n = 0$), we have

$$\nabla^2\psi_0 + \nabla\psi_0 \cdot \nabla\psi_0 + k^2 = 0. \tag{17-15}$$

Taking the difference between (17-14) and (17-15) and writing

$$\psi = \psi_0 + \psi_1, \tag{17-16}$$

we obtain

$$\nabla^2\psi_1 + 2\nabla\psi_0 \cdot \nabla\psi_1 = -(\nabla\psi_1 \cdot \nabla\psi_1 + k^2\,\delta n). \tag{17-17}$$

Noting the identity

$$\nabla^2(U_0\psi_1) = (\nabla^2 U_0)\psi_1 + 2U_0\nabla\psi_0 \cdot \nabla\psi_1 + U_0\nabla^2\psi_1,$$

the left-hand side of (17-17) becomes

$$\nabla^2\psi_1 + 2\nabla\psi_0 \cdot \nabla\psi_1 = (1/U_0)[\nabla^2(U_0\psi_1) + k^2 U_0\psi_1],$$

and we obtain the following equation for $U_0\psi_1$:

$$(\nabla^2 + k^2)(U_0\psi_1) = [\nabla\psi_1 \cdot \nabla\psi_1 + k^2\,\delta n]U_0.$$

This is an inhomogeneous wave equation, and we can convert it to the integral equation for ψ_1:

$$\psi_1(\mathbf{r}) = \frac{1}{U_0(\mathbf{r})}\int_{V'} G(\mathbf{r}-\mathbf{r}')[\nabla\psi_1 \cdot \nabla\psi_1 + k^2\,\delta n]U_0(\mathbf{r}')\,dV'. \tag{17-18}$$

By iteration, we can obtain a series solution of (17-18). As a first iteration, we let $\psi_1 = 0$ inside the integral and obtain

$$\psi_{10}(\mathbf{r}) = \frac{1}{U_0(\mathbf{r})}\int_{V'} G(\mathbf{r}-\mathbf{r}')\,\delta n(\mathbf{r}')U_0(\mathbf{r}')\,dV'. \tag{17-19}$$

This is the first Rytov solution and is extensively used in weak fluctuation theory.

The first Rytov solution can be written as

$$U(\mathbf{r}) = \exp(\psi_0 + \psi_{10}) = U_0(\mathbf{r})\exp(\psi_{10}(\mathbf{r})). \tag{17-20}$$

We note that if the exponential in (17-20) is expanded in a series, its first two terms, given by $U_0(\mathbf{r})(1 + \psi_{10}(\mathbf{r}))$, are identical to the first Born approximation. However, (17-20) includes more terms than the first Born approximation, and thus it is generally regarded as superior to the Born approximation.

The iterated series representation for ψ_1 can be developed from the following, which is obtained from (17-18) and (17-19):

$$\psi_1(\mathbf{r}) = \psi_{10}(\mathbf{r}) + \frac{1}{U_0(\mathbf{r})}\int_{V'} G(\mathbf{r}-\mathbf{r}')\nabla\psi_1(\mathbf{r}') \cdot \nabla\psi_1(\mathbf{r}')U_0(\mathbf{r}')\,dV'. \tag{17-21}$$

Substituting ψ_{10} in the integral, we get the next iteration, and this process may be repeated to obtain a complete series expression for ψ_1.

17-3 LOG-AMPLITUDE AND PHASE FLUCTUATIONS

Let us start with Rytov's first iteration solution for a weakly turbulent medium. In this case, it is convenient to use

$$\delta n = 2n_1 + n_1^2 \simeq 2n_1 \tag{17-22}$$

and write

$$U(\mathbf{r}) = U_0(\mathbf{r}) \exp(\psi_1(\mathbf{r})) \tag{17-23a}$$

$$\psi_1(\mathbf{r}) = \int_{V'} h(\mathbf{r}, \mathbf{r}')n_1(\mathbf{r}')\, dV', \tag{17-23b}$$

where

$$h(\mathbf{r}, \mathbf{r}') = 2k^2 G(\mathbf{r} - \mathbf{r}')U_0(\mathbf{r}')/U_0(\mathbf{r}). \tag{17-23c}$$

Equation (17-23a) gives the relationship between the field $\psi_1(\mathbf{r})$ and the index of refraction fluctuation $n_1(\mathbf{r})$.

Let us now find expressions for the amplitude A and the phase S of the field $U(\mathbf{r})$. Writing

$$U(\mathbf{r}) = A(\mathbf{r})e^{iS(\mathbf{r})}, \qquad U_0(\mathbf{r}) = A_0(\mathbf{r}) \exp(iS_0(\mathbf{r})), \tag{17-24}$$

we get

$$\psi_1(\mathbf{r}) = \chi + iS_1 = \ln(A/A_0) + i(S - S_0). \tag{17-25}$$

The real part of ψ_1 denoted by χ represents the fluctuation of the logarithm of the amplitude and is called the log-amplitude fluctuation. The imaginary part denoted by S_1 is called the phase fluctuation. In this chapter, we study the statistical characteristics of χ and S_1 for plane wave propagation. It should be noted here that χ is different from the amplitude fluctuation $(A - \langle A \rangle)/\langle A \rangle$, where $\langle A \rangle$ is the average amplitude. χ is also different from the fluctuation of the log-amplitude $\ln A$: $\ln A - \langle \ln A \rangle$. However, for $|\chi| \ll 1$, χ is approximately equal to $(A - A_0)/A_0$.

17-4 PLANE WAVE FORMULATION

We now consider a plane wave incident upon a turbulent medium at $x = 0$ (see Fig. 17-1). The observation point is a distance L away from $x = 0$. For a plane wave, the incident wave in the absence of the random medium is given by

$$U_0(\mathbf{r}) = e^{ikx}. \tag{17-26}$$

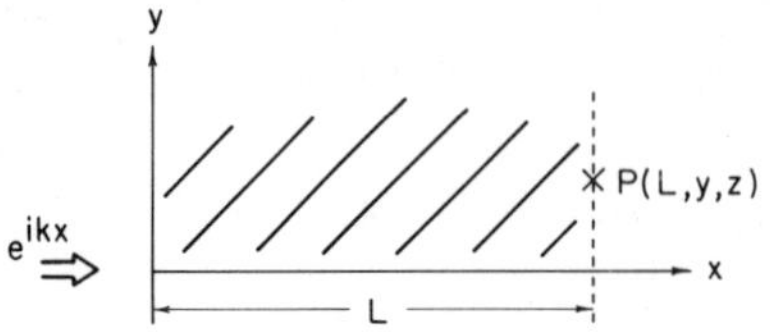

FIG. 17-1 Plane wave incident on random medium and the observation point $P(L, y, z)$.

We substitute (17-26) into (17-23b) and (17-23c) and obtain the expression for ψ_1. The subsequent mathematical operations can be considerably simplified if we can make a few reasonable assumptions for our problem. First, we assume that the backscattering is negligible and, therefore, the integration in (17-23b) will be limited to the region between $x = 0$ and $x = L$. Also, we note that the major contribution to $\psi_1(\mathbf{r})$ comes from the region $y' \sim y$ and $z' \sim z$ and thus $|y' - y|$ and $|z' - z|$ are assumed to be much smaller than the distance $|x - x'|$. This is especially true for $\lambda < l_0$, because then the scattering from an inhomogeneity of size l_0 is confined within the forward angle of the order of λ/l_0. However, the condition $\lambda < l_0$ may be too stringent, and it can be relaxed for spherical and beam wave cases.

Under these assumptions, we write

$$\psi_1(L, y, z) = \int_0^L dx' \int_{-\infty}^{\infty} dy' \int_{-\infty}^{\infty} dz' h(L - x', y - y', z - z') n_1(x', y', z'), \tag{17-27a}$$

where

$$h(L - x', y - y', z - z') = \frac{k^2}{2\pi} \frac{1}{(L - x')} \exp\left[i \frac{k}{2} \frac{(y' - y)^2 + (z' - z)^2}{L - x'} \right]. \tag{17-27b}$$

This is obtained by approximating $G(\mathbf{r} - \mathbf{r}')$ in (17-23c) by

$$\frac{\exp(ik|\mathbf{r} - \mathbf{r}'|)}{4\pi|\mathbf{r} - \mathbf{r}'|}$$

$$\simeq \left\{ \exp\left[ik\left((x - x') + \frac{(y - y')^2 + (z - z')^2}{2(x - x')} \right) \right] \right\} \Big/ 4\pi(x - x'). \tag{17-27c}$$

Equations (17-27a) and (17-27b) constitute the basic starting point in our analysis of amplitude and phase fluctuations for a plane wave.

17-5 DIRECT METHOD AND SPECTRAL METHOD

From (17-27a), we can proceed in one of two ways: the direct approach or the spectral approach. The direct calculations can be made in the following manner: We can write the real and imaginary parts of (17-27),

$$\chi(\mathbf{r}) = \int h_r(\mathbf{r} - \mathbf{r}')n_1(\mathbf{r}')\,dV', \qquad S_1(\mathbf{r}) = \int h_i(\mathbf{r} - \mathbf{r}')n_1(\mathbf{r}')\,dV', \quad \text{(17-28a)}$$

and directly form the correlation functions:

$$B_\chi(\mathbf{r}_1, \mathbf{r}_2) = \langle \chi(\mathbf{r}_1)\chi(\mathbf{r}_2)\rangle, \qquad B_S(r_1, r_2) = \langle S_1(r_1)S_1(r_2)\rangle. \quad \text{(17-28b)}$$

This can be expressed as

$$B_\chi(\mathbf{r}_1, \mathbf{r}_2) = \int dV_1' \int dV_2' h_r(\mathbf{r}_1 - \mathbf{r}_1')h_r(\mathbf{r}_2 - \mathbf{r}_2')B_n(\mathbf{r}_1', \mathbf{r}_2'), \quad \text{(17-29)}$$

where $B_n(\mathbf{r}_1', \mathbf{r}_2') = \langle n_1(\mathbf{r}_1')n_1(\mathbf{r}_2')\rangle$ is the correlation function of the fluctuation of the index of refraction.

By calculating the integral in (17-29), it is possible to obtain the amplitude correlation function B_χ from B_n.

However, this direct approach, is generally inferior to the spectral approach described in the following section. The reason is that the calculation of the integrals in (17-29) is usually very difficult and can be performed only for a special form of the correlation function B_n. Even in such a case, complicated mathematical operations are needed to obtain useful solutions.

The spectral approach, on the other hand, often simplifies mathematical operations and also provides different and useful points of view. This situation is similar to two approaches used to solve transient problems: a direct time domain method and the use of the frequency spectrum.

17-6 SPECTRAL REPRESENTATION OF THE AMPLITUDE AND PHASE FLUCTUATIONS

Let us consider our fundamental equation (17-27) and obtain its spectral representation. We assume here that n_1 is a homogeneous and isotropic random function and is real. We may be tempted to express n_1 in a three-dimensional spectral representation. However, this is not convenient because the wave ψ_1 is expected to be homogeneous and isotropic only in the two-dimensional yz plane and its two-dimensional correlation characteristics depend very much on the two-dimensional correlation characteristics of n_1 in the yz plane, but very little on the correlation distance of n_1 in the x direction. For this reason, we need to express both n_1 and ψ_1 in two-dimensional spectral representations in the yz plane.

Therefore, we write†

$$n_1(x, \boldsymbol{\rho}) = \int e^{i\boldsymbol{\kappa}\cdot\boldsymbol{\rho}}\, dv(x, \boldsymbol{\kappa}), \tag{17-30}$$

where $\boldsymbol{\kappa} = K_y\hat{\mathbf{y}} + K_z\hat{\mathbf{z}}$ and $\boldsymbol{\rho} = y\hat{\mathbf{y}} + z\hat{\mathbf{z}}$. Substituting this into (17-27) and performing integration with respect to $\boldsymbol{\rho}' = y'\hat{\mathbf{y}} + z'\hat{\mathbf{z}}$, we obtain

$$\psi_1(L, \boldsymbol{\rho}) = \int_0^L dx' \int e^{i\boldsymbol{\kappa}\cdot\boldsymbol{\rho}} H(L - x', \boldsymbol{\kappa})\, dv(x', \boldsymbol{\kappa}), \tag{17-31}$$

where H is a Fourier transform of h given by

$$H(L - x', \boldsymbol{\kappa}) = \int d\boldsymbol{\rho} e^{-i\boldsymbol{\kappa}\cdot\boldsymbol{\rho}} h(L - x', \boldsymbol{\rho}) = ik \exp\left[-i\frac{(L - x')}{2k}\kappa^2\right], \tag{17-32}$$

where $d\boldsymbol{\rho} = dy\, dz$ and $\kappa^2 = |\boldsymbol{\kappa}|^2 = K_y{}^2 + K_z{}^2$. Equation (17-31) indicates that the spectrum of the index of refraction dv at x' is modified by H to produce the random amplitude $H\, dv$ for the fluctuation ψ_1. The total fluctuation ψ_1 is the sum of all the contributions from $x' = 0$ to $x' = L$. The function H, therefore, represents the effect of wave propagation from x' to L. This situation is similar to the network shown in Fig. 17-2, where the spectrum of the index of refraction dv plays the role of an input, H is a network, and the output $H\, dv$ is the random amplitude for ψ_1.

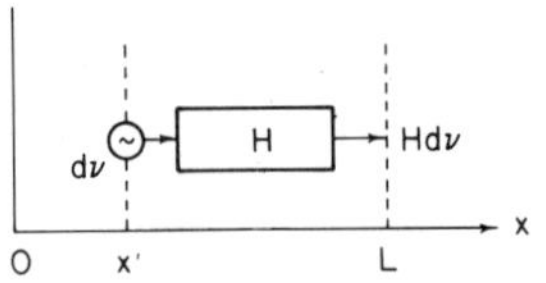

FIG. 17-2 Random amplitude of of the output ψ_1 is given by the product of H and the random amplitude dv of the refractive index fluctuation.

Now, we wish to find the log-amplitude and phase fluctuations, which are the real and imaginary parts of $\psi_1(L, y, z)$. It is not advisable to take the real and imaginary parts of (17-31) directly because in general dv is complex and its real and imaginary parts are not explicitly known. Instead, we use the relationships

$$\chi(L, \boldsymbol{\rho}) = \tfrac{1}{2}[\psi_1(L, \boldsymbol{\rho}) + \psi_1{}^*(L, \boldsymbol{\rho})],$$

$$S_1(L, \boldsymbol{\rho}) = (1/2i)[\psi_1(L, \boldsymbol{\rho}) - \psi_1{}^*(L, \boldsymbol{\rho})] \tag{17-33}$$

† For the spectral representation of a random function, see Appendix A.

and express $\psi_1{}^*$ in the same spectral form as ψ_1. $\psi_1{}^*$ is given by

$$\psi_1{}^*(L, \boldsymbol{\rho}) = \int_0^L dx' \iint_{-\infty}^{\infty} e^{-i\boldsymbol{\kappa}\cdot\boldsymbol{\rho}} H^*(L - x', \boldsymbol{\kappa})\, dv^*(x', \boldsymbol{\kappa}). \tag{17-34}$$

To transform this into the form of (17-31), we let $\boldsymbol{\kappa} \to -\boldsymbol{\kappa}$ and note that since n_1 is real [see Eq. (A-34) in Appendix A],

$$dv(x', \boldsymbol{\kappa}) = dv^*(x', -\boldsymbol{\kappa}), \tag{17-35}$$

and that H is an even function of $\boldsymbol{\kappa}$:

$$H(L - x', -\boldsymbol{\kappa}) = H(L - x', \boldsymbol{\kappa}). \tag{17-36}$$

Therefore, we get

$$\psi_1{}^*(L, \boldsymbol{\rho}) = \int_0^L dx' \iint_{-\infty}^{\infty} e^{i\boldsymbol{\kappa}\cdot\boldsymbol{\rho}} H^*(L - x', -\boldsymbol{\kappa})\, dv(x', \boldsymbol{\kappa}). \tag{17-37}$$

Now we combine (17-31) and (17-37) to obtain χ and S_1:

$$\chi(L, \boldsymbol{\rho}) = \int_0^L dx' \iint_{-\infty}^{\infty} e^{i\boldsymbol{\kappa}\cdot\boldsymbol{\rho}} H_r(L - x', \boldsymbol{\kappa})\, dv(x', \boldsymbol{\kappa}) \tag{17-38}$$

$$S_1(L, \boldsymbol{\rho}) = \int_0^L dx' \iint_{-\infty}^{\infty} e^{i\boldsymbol{\kappa}\cdot\boldsymbol{\rho}} H_i(L - x', \boldsymbol{\kappa})\, dv(x', \boldsymbol{\kappa}), \tag{17-39}$$

where H_r and H_i are the real and imaginary parts of H:

$$H_r(L - x', \boldsymbol{\kappa}) = k \sin\left[\frac{(L - x')}{2k} \kappa^2\right] \tag{17-40}$$

$$H_i(L - x', \boldsymbol{\kappa}) = k \cos\left[\frac{(L - x')}{2k} \kappa^2\right]. \tag{17-41}$$

Equations (17-38) and (17-39) together with (17-40) and (17-41) constitute the spectral representation of the amplitude and phase fluctuations.

17-7 AMPLITUDE AND PHASE CORRELATION FUNCTIONS

We now consider the amplitude and phase correlation functions at a plane $x = L$. From (17-38), we write

$$B_\chi(L; \boldsymbol{\rho}_1, \boldsymbol{\rho}_2) = \langle \chi(L, \boldsymbol{\rho}_1)\chi(L, \boldsymbol{\rho}_2)\rangle = \langle \chi(L, \boldsymbol{\rho}_1)\chi^*(L, \boldsymbol{\rho}_2)\rangle. \tag{17-42}$$

Making use of the relationship [see Eq. (A-21), in Appendix A]

$$\langle dv(x', \boldsymbol{\kappa})\, dv^*(x'', \boldsymbol{\kappa}')\rangle = F_n(|x' - x''|, \boldsymbol{\kappa})\, \delta(\boldsymbol{\kappa} - \boldsymbol{\kappa}')\, d\boldsymbol{\kappa}\, d\boldsymbol{\kappa}', \qquad (17\text{-}43)$$

where $\delta(\boldsymbol{\kappa} - \boldsymbol{\kappa}') = \delta(K_y - K_y')\,\delta(K_z - K_z')$, $d\boldsymbol{\kappa} = dK_y\, dK_z$, and $d\boldsymbol{\kappa}' = dK_y'\, dK_z'$, and using $\boldsymbol{\rho} = \boldsymbol{\rho}_1 - \boldsymbol{\rho}_2$, we obtain

$$B_\chi = \int_0^L dx' \int_0^L dx'' \iint_{-\infty}^{\infty} e^{i\boldsymbol{\kappa}\cdot\boldsymbol{\rho}} H_r(L - x', \boldsymbol{\kappa}) H_r^*(L - x'', \boldsymbol{\kappa}) F_n(|x' - x''|, \boldsymbol{\kappa})\, d\boldsymbol{\kappa}. \qquad (17\text{-}44)$$

Since the integrand contains the spectrum F_n as a function of the difference coordinate $x' - x''$, it is convenient to transform the integration with respect to x' and x'' into the integration involving the difference coordinate $x_d = x' - x''$, and the center-of-mass coordinate $\eta = \frac{1}{2}(x' + x'')$:

$$\int_0^L dx' \int_0^L dx'' = \int_0^L d\eta \int_{\xi_1(\eta)}^{\xi_2(\eta)} dx_d \qquad (17\text{-}45)$$

where $\xi_1(\eta)$ and $\xi_2(\eta)$ are shown in Fig. 17-3. Now, the integrand in (17-44) contains the spectrum of the index of refraction $F_n(|x_d|, \boldsymbol{\kappa})$. Therefore, it is clear that F_n has some appreciable value only when $|x_d|$ is within the

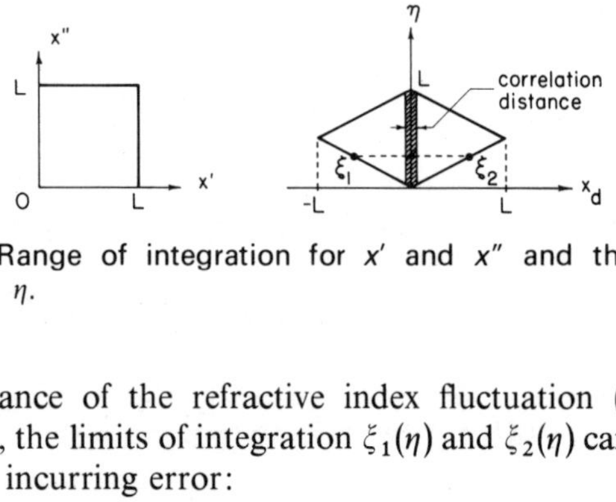

FIG. 17-3 Range of integration for x' and x'' and the corresponding range for x_d and η.

correlation distance of the refractive index fluctuation (shaded area in Fig. 17-3). Thus, the limits of integration $\xi_1(\eta)$ and $\xi_2(\eta)$ can be extended to infinity without incurring error:

$$\int_0^L dx' \int_0^L dx'' f \approx \int_0^L d\eta \int_{-\infty}^{\infty} dx_d\, f, \qquad (17\text{-}46)$$

if $|f| \approx 0$ for $x_d > l$ (correlation distance) and $l \ll L$.

Also, we note that H_r is a slowly varying function of x' and can be approximated by the value at the center-of-mass coordinate η:

$$H_r(L - x', \boldsymbol{\kappa}) \approx H_r(L - \eta, \boldsymbol{\kappa}). \qquad (17\text{-}47)$$

Also, we note [see Eq. (A-27) in Appendix A] that

$$\int_{-\infty}^{\infty} dx_d\, F_n(|x_d|, \boldsymbol{\kappa}) = 2\pi\Phi_n(\kappa). \tag{17-48}$$

Using (17-46)–(17-48), we can write (17-44) as

$$B_\chi = 2\pi \int_0^L d\eta \iint_{-\infty}^{\infty} e^{i\boldsymbol{\kappa}\cdot\boldsymbol{\rho}} H_r(L-\eta, \boldsymbol{\kappa})^2 \Phi_n(\kappa)\, d\boldsymbol{\kappa}. \tag{17-49}$$

Since B_χ is a function of the transverse distance ρ, it is convenient to convert (17-49) into an integral in the cylindrical coordinate system. To do this, we let $\kappa_y = \kappa \cos\phi$ and $\kappa_z = \kappa \sin\phi$, and $y_d = \rho\cos\phi'$ and $z_d = \rho \sin\phi'$. Then we use J_0, the Bessel function of order zero.

$$\iint_{-\infty}^{\infty} d\boldsymbol{\kappa}\, e^{i\boldsymbol{\kappa}\cdot\boldsymbol{\rho}} = \int_0^{\infty} \kappa\, d\kappa \int_0^{2\pi} d\phi \exp[i\kappa\rho\cos(\phi-\phi')] = 2\pi \int_0^{\infty} \kappa\, d\kappa J_0(\kappa\rho). \tag{17-50}$$

We also used the fact that H_r and Φ_n are independent of ϕ.

Using (17-50), (17-49) becomes

$$B_\chi(L, \rho) = (2\pi)^2 \int_0^L d\eta \int_0^{\infty} \kappa\, d\kappa J_0(\kappa\rho) H_i(L-\eta, \kappa)^2 \Phi_n(\kappa). \tag{17-51}$$

An identical procedure can be followed for the phase correlation function $B_S(L, \rho)$ with H_r replaced by H_i:

$$\begin{aligned} B_S(L, \rho) &= \langle S_1(L, \boldsymbol{\rho}_1) S_1(L, \boldsymbol{\rho}_2)\rangle \\ &= (2\pi)^2 \int_0^L d\eta \int_0^{\infty} \kappa\, d\kappa J_0(\kappa\rho) H_i(L-\eta, \kappa)^2 \Phi_n(\kappa). \end{aligned} \tag{17-52}$$

Similarly, the cross correlation between the log-amplitude and the phase fluctuations is given by

$$\begin{aligned} B_{\chi S}(L, \rho) &= \langle \chi(L, \boldsymbol{\rho}_1) S_1(L, \boldsymbol{\rho}_2)\rangle \\ &= (2\pi)^2 \int_0^L d\eta \int_0^{\infty} \kappa\, d\kappa J_0(\kappa\rho) H_r(L-\eta, \kappa) H_i(L-\eta, \kappa) \Phi_n(\kappa). \end{aligned} \tag{17-53}$$

These three equations constitute the basic expressions for the correlation functions in the weak fluctuation approximation.

Let us consider the convergence of the integrals in (17-51)–(17-53). We note that as $\kappa \to 0$, $H_r \to \kappa^4$ and $H_i \to$ const. Therefore, if the spectral density

Φ_n behaves as κ^p for small κ, then B_χ is convergent as long as $p > -6$ whereas the convergences of B_S and $B_{\chi S}$ require $p > -2$ and $p > -4$, respectively. The Kolmogorov spectrum for turbulence has $p = -11/3$ in the inertial subrange [see Eq. (B-33) in Appendix B] and it is not homogeneous, but locally homogeneous. Even if the turbulence is locally homogeneous, we note that B_χ is still finite and, therefore, the amplitude correlation function exists. On the other hand, if Φ_n is represented by $\kappa^{-11/3}$, then the phase correlation function does not exist.

17-8 AMPLITUDE AND PHASE STRUCTURE FUNCTIONS

In the preceding section, we obtained the correlation function when the refractive index fluctuation is homogeneous and isotropic. We now consider the amplitude and phase structure functions (see also Appendix B) defined by

$$D_\chi(L, \boldsymbol{\rho}) = \langle [\chi(L, \boldsymbol{\rho}_1) - \chi(L, \boldsymbol{\rho}_2)]^2 \rangle,$$

$$D_S(L, \boldsymbol{\rho}) = \langle [S_1(L, \boldsymbol{\rho}_1) - S_1(L, \boldsymbol{\rho}_2)]^2 \rangle \tag{17-54}$$

where $\boldsymbol{\rho} = \boldsymbol{\rho}_1 - \boldsymbol{\rho}_2$.

For a homogeneous and isotropic random medium, we follow the procedure shown in the preceding section and obtain

$$D_\chi(L, \rho) = 8\pi^2 \int_0^L d\eta \int_0^\infty \kappa\, d\kappa [1 - J_0(\kappa\rho)] H_r^2(L - \eta, \kappa) \Phi_n(\kappa)$$

$$D_S(L, \rho) = 8\pi^2 \int_0^L d\eta \int_0^\infty \kappa\, d\kappa [1 - J_0(\kappa\rho)] H_i^2(L - \eta, \kappa) \Phi_n(\kappa). \tag{17-55}$$

Let us examine the convergence of these integrals. For small κ, $H_r^2 \to \kappa^4$ and $H_i^2 \to$ const and, therefore, for the spectrum $\Phi_n(\kappa) \sim \kappa^p$ for small κ, B_χ exists if $p > -8$ while B_S exists if $p > -4$. This means that even if the medium is locally homogeneous and the form of the spectrum Φ_n is such that the correlation function does not exist (e.g., $p = -11/3$), the structure function is finite. Because of this, the phase structure function D_S is often used in turbulence study.

17-9 SPECTRAL AND SPATIAL FILTER FUNCTIONS

The three basic equations (17-51)–(17-53) are of the general form

$$B(L, \rho) = \int_0^L d\eta \int_0^\infty \kappa\, d\kappa F(\eta, \kappa, \rho) \Phi_n(\kappa). \tag{17-56}$$

The spectral density of the index of refraction fluctuation $\Phi_n(\kappa)$ in this expression is evaluated at the location η, and it may vary along the path of wave propagation. To include this effect, we should write more generally

$$B(L, \rho) = \int_0^L d\eta \int_0^\infty \kappa \, d\kappa F(\eta, \kappa, \rho)\Phi_n(\eta, \kappa). \tag{17-57}$$

Here we ignored the variation of Φ_n in the transverse direction. This is permissible (see Section 19-8) as long as the lateral variation is negligible over the distance $\sqrt{\lambda L}$. The function $F(\eta, \kappa, \rho)$ plays the role of filtering a certain portion of the spectrum Φ_n to produce the correlation function B, and thus is called a filter function (Ishimaru, 1969a,b). It is often convenient to consider the following two special cases.

17-9-1 Spectral Filter Function

If Φ_n does not depend on the location η, then we can write

$$\begin{aligned} B(L, \rho) &= \int_0^\infty \kappa \, d\kappa \left[\int_0^L d\eta F(\eta, \kappa, \rho) \right] \Phi_n(\kappa) \\ &= 2\pi^2 k^2 L \int_0^\infty \kappa \, d\kappa J_0(\kappa\rho) f(\kappa) \Phi_n(\kappa). \end{aligned} \tag{17-58a}$$

Similarly, the structure functions are given by

$$D(L, \rho) = 4\pi^2 k^2 L \int_0^\infty \kappa \, d\kappa [1 - J_0(\kappa\rho)] f(\kappa) \Phi_n(\kappa). \tag{17-58b}$$

The function $f(\kappa)$ may be called the spectral filter function since it filters a portion of the spectrum Φ_n. The filter function $f(\kappa)$ can be easily obtained. For the log-amplitude fluctuation,

$$f_\chi(\kappa) = 1 - \frac{\sin(\kappa^2 L/k)}{\kappa^2 L/k}. \tag{17-59a}$$

For the phase fluctuation,

$$f_S(\kappa) = 1 + \frac{\sin(\kappa^2 L/k)}{\kappa^2 L/k}. \tag{17-59b}$$

For the cross correlation between the amplitude and the phase fluctuation,

$$f_{\chi S}(\kappa) = \frac{\sin^2(\kappa^2 L/2k)}{\kappa^2 L/2k}. \tag{17-59c}$$

17-9-2 Spatial Filter Function

In some situations, the spectrum Φ_n may be expressed as the product of a function of η and a function of κ:

$$\Phi_n(\eta, \kappa) = \Phi_{n1}(\eta)\Phi_{n2}(\kappa). \tag{17-60}$$

The function $\Phi_{n1}(\eta)$ represents the variation of the strength of the refractive index fluctuation along the path and is proportional to the variance of the refractive index fluctuation at η. In this case, we write

$$\begin{aligned} B(L, \rho) &= \int_0^L d\eta \left[\int_0^\infty \kappa \, d\kappa F(\eta, \kappa, \rho)\Phi_{n2}(\kappa) \right] \Phi_{n1}(\eta) \\ &= \int_0^L d\eta G(\eta, \rho)\Phi_{n1}(\eta). \end{aligned} \tag{17-61}$$

The function $G(\eta, \rho)$ filters the magnitude of $\Phi_{n1}(\eta)$ in a different portion of the path and emphasizes or deemphasizes the effect of the random medium along the path. For this reason, $G(\eta, \rho)$ may be called the spatial filter function.

17-10 HOMOGENEOUS RANDOM MEDIA AND SPECTRAL FILTER FUNCTION

If the random medium can be considered homogeneous, then the spectrum Φ_n is a function of $\boldsymbol{\kappa}$ only, and we can make use of the formulations given in (17-58) with (17-59a)–(17-59c). Let us examine the filter functions. The general shapes of the filter functions $f_\chi(\kappa)$, $f_S(\kappa)$, and $f_{\chi S}(\kappa)$ are shown in Fig. 17-4. Note that $f_\chi(\kappa)$ emphasizes the portion of $\Phi_n(\kappa)$ for which $\kappa \gtrsim \sqrt{2\pi}/\sqrt{\lambda L}$. The spatial wave number κ represents 2π/(correlation distance), and thus the physical meaning of $f_\chi(\kappa)$ is that the amplitude fluctuations are affected mostly by blobs of size $\sqrt{\lambda L}$ or smaller. The situation is different for the phase fluctuation, because $f_S(\kappa)$ tends to emphasize the region in which $\kappa \lesssim \sqrt{2\pi}/\sqrt{\lambda L}$. This means that the phase fluctuations are

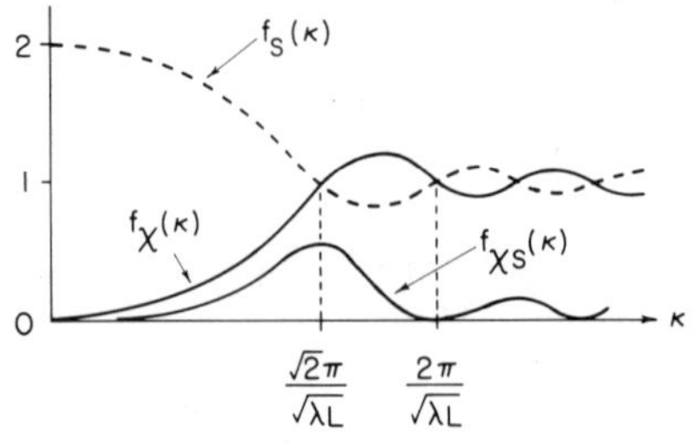

FIG. 17-4 General shapes of filter functions.

affected by all sizes of blobs and in particular by the blobs of size $\sqrt{\lambda L}$ or greater.

From the preceding it is clear that the fluctuation characteristics are greatly different depending on whether the correlation distance l is much smaller or greater than the Fresnel zone size $\sqrt{\lambda L}$. In the following sections, we examine the general characteristics of these two different regions.

17-11 GEOMETRIC OPTICAL REGION $L \ll l^2/\lambda$

Let us start with the case $L \ll l^2/\lambda$. The spectral density $\Phi_n(\kappa)$ and the filter functions $f_\chi(\kappa)$ and $f_S(\kappa)$ should appear as shown in Fig. 17-5, and the total spectrum is the product of f_χ and Φ_n for the amplitude correlation and f_S and Φ_n for the phase correlation. Note that $\Phi_n(\kappa)$ extends up to $\kappa \approx 2\pi/l$ and is negligible for $\kappa \gg 2\pi/l$.

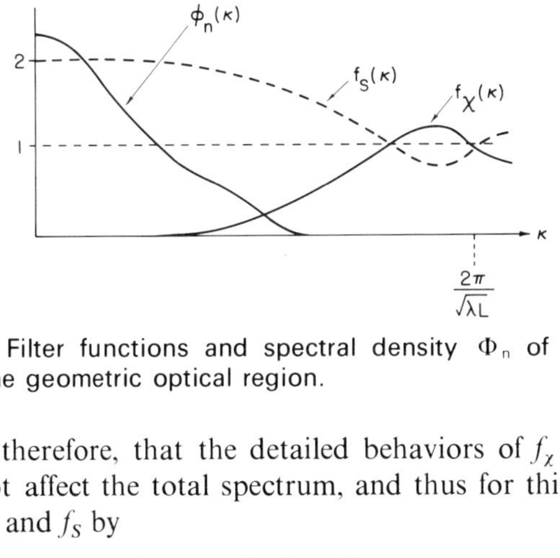

FIG. 17-5 Filter functions and spectral density Φ_n of refractive index fluctuation in the geometric optical region.

It is clear, therefore, that the detailed behaviors of f_χ and f_S beyond $\kappa \simeq 2\pi/l$ do not affect the total spectrum, and thus for this range we can approximate f_χ and f_S by

$$f_\chi(\kappa) \simeq 1 - \frac{(\kappa^2 L/k) - \frac{1}{6}(\kappa^2 L/k)^3 + \cdots}{(\kappa^2 L/k)} = \tfrac{1}{6}(\kappa^2 L/k)^2, \tag{17-62}$$

$$f_S(\kappa) \simeq 2. \tag{17-63}$$

The correlation functions are then given by

$$B_\chi(L, \rho) = \frac{\pi^2 L^3}{3} \int_0^\infty \kappa^5 J_0(\kappa\rho)\Phi_n(\kappa)\, d\kappa \tag{17-64}$$

$$B_S(L, \rho) = 4\pi^2 k^2 L \int_0^\infty \kappa J_0(\kappa\rho)\Phi_n(\kappa)\, d\kappa \tag{17-65}$$

for $L \ll l^2/\lambda$.

Let us examine (17-64). Several characteristics are immediately deduced from this: (a) $B_\chi(L, \rho)$ is independent of frequency (k); (b) it is proportional to L^3; (c) because of κ^5, the spectrum Φ_n in the range of small $\kappa(\ll 2\pi/l)$ has no significant effect on B_χ; and (d) since the peak of $f_\chi \Phi_n$ occurs at $\kappa > 2\pi/l$, the correlation distance of B_χ should be smaller than the correlation distance of n_1.

We deduce from (17-65) the following characteristics of the phase correlation function: (a) It is proportional to the square of the frequency (k^2); (b) it is proportional to the distance L; (c) it depends on the complete spectrum $\Phi_n(\kappa)$; and (d) its correlation distance is substantially the same as that of the index of refraction.

If the correlation function of the refractive index fluctuations $B_n(r_d)$ is known, then it is possible to derive the following alternative expression for (17-65). We note [see Eq. (A-28) in Appendix A] that

$$\Phi_n(\kappa) = \frac{1}{2\pi^2\kappa}\int_0^\infty B_n(r_d) r_d \sin(\kappa r_d)\, dr_d.$$

Substituting this into (17-65) and noting the formula (Magnus and Oberhettinger, 1954, p. 37)

$$\int_0^\infty J_0(\kappa\rho)\sin(\kappa r_d)\, d\kappa = \begin{cases} 0 & \text{for} \quad \rho > r_d \\ \dfrac{1}{\sqrt{r_d^2 - \rho^2}} & \text{for} \quad \rho < r_d, \end{cases} \tag{17-66}$$

we get

$$B_S(L, \rho) = 2k^2 L \int_0^\infty B_n(r_d) \frac{r_d\, dr_d}{\sqrt{r_d^2 - \rho^2}} = 2k^2 L \int_0^\infty B_n(\sqrt{\rho^2 + x^2})\, dx. \tag{17-67}$$

The mean square phase fluctuation is given by

$$B_S(L, 0) = \langle S_1^2 \rangle = 2k^2 L \langle n_1^2 \rangle L_n, \tag{17-68}$$

where

$$L_n = \int_0^\infty \frac{B_n(x)}{B_n(0)} dx = \frac{1}{\langle n_1^2 \rangle}\int_0^\infty B_n(x)\, dx$$

is called the integral scale and can be considered as a definition of the "correlation distance."

The two formulas (17-64) and (17-65) can also be obtained from geometric optical theory. We now sketch this derivation briefly.

When a wave is incident on an object of size l, the diffracted wave is mostly confined within a conical sector of angle λ/l, and thus at a distance L

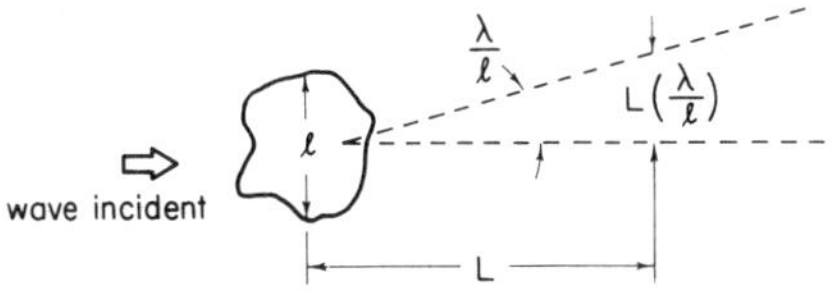

FIG. 17-6 Wave scattered from a blob of size l is mostly confined within a conical sector of angle λ/l.

the diffracted wave has a spread of $(\lambda/l)L$ (see Fig. 17-6). As long as this spread is smaller than the size of the shadow l $(L\lambda/l \ll l)$, the diffraction effect is small, and the geometric optical solution should adequately describe the field.

Therefore, even though our solutions (17-64) and (17-65) are obtained from a wave equation, it is expected that in this region $L \ll l^2/\lambda$, the geometric optical approach will give the same solution. This is indeed the case.

We do not intend to describe the derivation in detail. We only indicate some well-known relationships which have been used in the past. We write

$$U(\mathbf{r}) = A(\mathbf{r})e^{iS(\mathbf{r})}, \tag{17-69}$$

where $A(\mathbf{r})$ is the amplitude and $S(\mathbf{r})$ is the phase. According to geometric optics, $S(\mathbf{r})$ obeys the equation (Born and Wolf, 1964)

$$|\nabla S|^2 = k^2 n^2, \tag{17-70}$$

which is called the eikonal equation. The solution is given by

$$S(\mathbf{r}) - S(\mathbf{r}_0) = k \int_{\mathbf{r}_0}^{\mathbf{r}} n\, ds, \tag{17-71}$$

where the integration is taken along the ray path. The amplitude $A(\mathbf{r})$ is given by

$$A(\mathbf{r}) = A(\mathbf{r}_0) \exp\left(-\frac{1}{2} \int_{\mathbf{r}_0}^{\mathbf{r}} \frac{\nabla^2 S}{kn}\, ds \right). \tag{17-72}$$

Equations (17-71) and (17-72) do not provide complete information unless the ray path is known. However, for the weak random medium under consideration, we can approximate the ray path by a straight line along the x axis. Thus, the integral $\int_{\mathbf{r}_0}^{\mathbf{r}} ds$ may be approximated by $\int_{x_0}^{x} dx'$. Under this approximation, we get the phase fluctuation

$$S_1(L, y, z) = S(L, y, z) - kx = k \int_0^L n_1(x', y', z')\, dx' \tag{17-73}$$

and the log-amplitude fluctuation

$$\chi(L, y, z) = -\frac{1}{2k}\int_0^L \nabla^2 S\, dx' \qquad (17\text{-}74)$$

where $n \simeq 1$ is assumed. Noting that

$$\int_0^L \frac{\partial^2}{\partial x'^2} S\, dx' = \int_0^L k \frac{\partial n_1}{\partial x'} dx' = kn_1(L) - kn_1(0),$$

which may be set equal to zero, we get

$$\chi(L, y, z) = -\frac{1}{2}\int_0^L dx \int_0^x d\xi \left(\frac{\partial^2}{\partial y^2} + \frac{\partial^2}{\partial z^2}\right) n_1(\xi, y, z). \qquad (17\text{-}75)$$

Equations (17-73) and (17-75) are often used as a starting point to derive the correlation functions.

We will not pursue this any further, but we should mention that if we apply the spectral technique to (17-73) and (17-75), we can get solutions identical to (17-64) and (17-65).

17-12 THE REGION IN WHICH $L \gg l^2/\lambda$

For this case, the spectral density $\Phi_n(\kappa)$ and the filter functions $f_S(\kappa)$ and $f_\chi(\kappa)$ should appear as shown in Fig. 17-7.

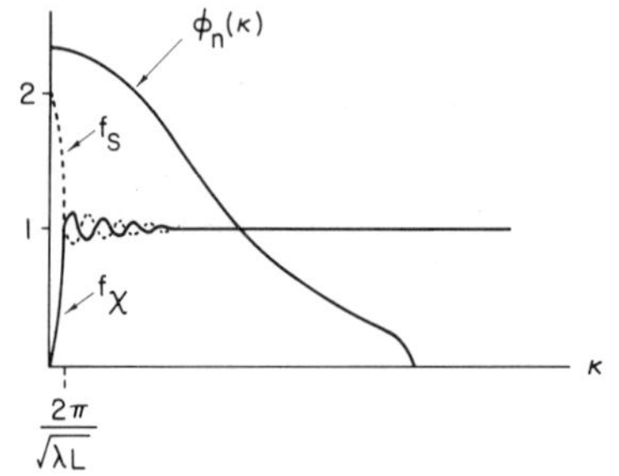

FIG. 17-7 Filter functions and spectral density ϕ_n in the region $L \ll l^2/\lambda$.

It is obvious from the figure that

$$f_\chi(\kappa) \simeq f_S(\kappa) \simeq 1 \qquad (17\text{-}76)$$

and, therefore, the correlation functions become

$$B_\chi(L, \rho) = B_S(L, \rho) = 2\pi^2 k^2 L \int_0^\infty \kappa\, d\kappa J_0(\kappa\rho)\Phi_n(\kappa). \qquad (17\text{-}77)$$

We note that this is exactly half the phase correlation function for the geometric optical case. If the correlation function $B_n(\mathbf{r})$ is known, we can follow the same procedure as in (17-67) and obtain

$$B_\chi(L, \rho) = B_S(L, \rho) = k^2 L \int_0^\infty B_n(\sqrt{\rho^2 + x^2})\, dx. \qquad (17\text{-}78)$$

Equation (17-78) gives the same expression for B_χ and B_S, but we must keep in mind that this is an approximation, and there is an important difference between B_χ and B_S. This difference arises from the fact that even though $f_\chi(\kappa) \simeq f_S(\kappa)$ for $\kappa \neq 0$, $f_\chi(0) = 0$ while $f_S(0) = 2$. Because of this, the amplitude correlation function must change its sign while the phase correlation function does not do so. The proof of this can be seen from the general expression (17-56). Since this is a form of the Fourier–Bessel transform, we can write the inverse transform:

$$2\pi^2 k^2 L f_\chi(\kappa)\phi_n(\kappa) = \int_0^\infty \rho\, d\rho J_0(\kappa\rho) B_\chi(L, \rho).$$

We note that since $f_\chi(0) = 0$,

$$\int_0^\infty \rho\, d\rho B_\chi(L, \rho) = 0. \qquad (17\text{-}79)$$

Equation (17-79) means that the correlation function integrated over the transverse plane must be zero and, therefore, the amplitude correlation function $B_\chi(L, \rho)$ must be negative in some part of ρ.

The relationship (17-79) does not obviously apply to the phase correlation function $B_S(L, \rho)$, because $f_S(0)$ is not zero.

17-13 GENERAL CHARACTERISTICS OF THE FLUCTUATIONS IN A HOMOGENEOUS RANDOM MEDIUM

In the preceding two sections, we described the general formulas for the correlation functions B_χ and B_S in two different regions. When $L \ll l^2/\lambda$, we have (17-64) and (17-65); however, when $L \gg l^2/\lambda$, we have (17-77).

These equations exhibit a certain characteristic common to all wave fluctuation phenomena irrespective of the form of the correlation function. If the correlation distance is much smaller than the Fresnel zone size $\sqrt{\lambda L}$, then in general, the amplitude and phase variances are equal and are proportional to the square of the frequency k^2 and to the distance L. For example,

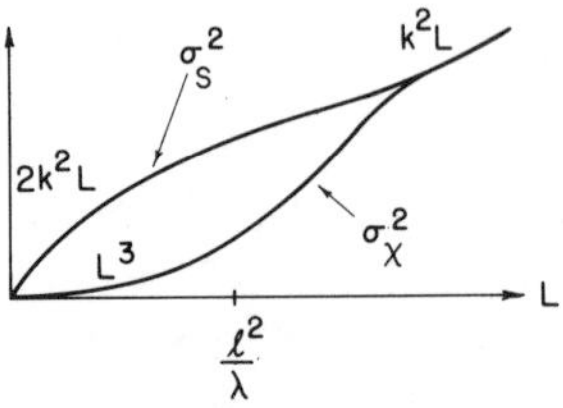

FIG. 17-8

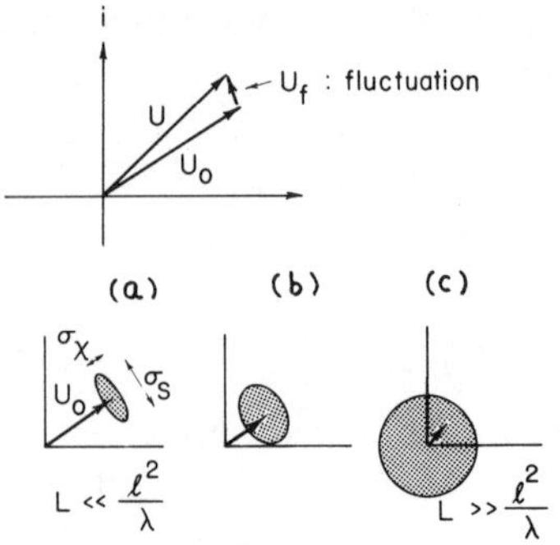

FIG. 17-9

FIG. 17-8 General characteristics of the log-amplitude variance σ_χ^2 and the phase variance σ_S^2.

FIG. 17-9 (a) Geometric optical region ($\sigma_\chi \ll \sigma_S$), (b) region where $L \gg l^2/\lambda$ ($\sigma_\chi \approx \sigma_S$), and (c) strong fluctuation region.

the amplitude variance $\sigma_\chi^2 = B_\chi(L, 0)$ and the phase variance $\sigma_S^2 = B_S(L, 0)$ are

$$\sigma_\chi^2 = \sigma_S^2 \sim k^2 L. \tag{17-80}$$

On the other hand, if the correlation distance l is much greater than the Fresnel zone size $\sqrt{\lambda L}$, then σ_χ^2 is independent of frequency and is proportional to L^3, and σ_S^2 is equal to twice the value given in (17-80):

$$\sigma_\chi^2 \sim L^3, \qquad \sigma_S^2 \sim 2k^2 L. \tag{17-81}$$

This is illustrated in Fig. 17-8. Physically this means that for a short distance, the amplitude does not vary much whereas the phase is directly affected by the distance of wave propagation. As the distance is increased, the amplitude fluctuation increases and eventually there is no distinction between the amplitude and phase fluctuation. This is shown in Fig. 17-9 expressing the field $U = U_0 + U_f$ in a complex plane showing the amplitude and phase fluctuation. It also shows the strong fluctuation region where the average (coherent) field is negligible and the field is almost totally incoherent. This case is discussed in Chapter 20.

17-14 HOMOGENEOUS RANDOM MEDIUM WITH GAUSSIAN CORRELATION FUNCTION

In many practical problems, the form of the correlation function is not known. However, its general shape is often approximated by a Gaussian function. We represent the correlation function by

$$\langle n_1(\mathbf{r}_1) n_1(\mathbf{r}_2)\rangle = \langle n_1^2\rangle \exp(-|\mathbf{r}_1 - \mathbf{r}_2|^2/l^2). \tag{17-82}$$

The spectral density for this is given by [see Eq. (A-33) in Appendix A]

$$\Phi_n(\kappa) = (\langle n_1{}^2\rangle l^3/8\pi\sqrt{\pi}) \exp(-\kappa^2 l^2/4). \tag{17-83}$$

Substituting these in (17-64), (17-65), and (17-77), we obtain for $L \ll l^2/\lambda$,

$$B_\chi(\rho) = \langle n_1{}^2\rangle(8\sqrt{\pi}/3)(L/l)^3\,{}_1F_1(3, 1; -\rho^2/l^2), \tag{17-84a}$$

$$B_S(\rho) = \langle n_1{}^2\rangle k^2 Ll\sqrt{\pi}\exp(-\rho^2/l^2). \tag{17-84b}$$

and for $L \gg l^2/\lambda$,

$$B_\chi(\rho) = B_S(\rho) = \langle n_1{}^2\rangle \tfrac{1}{2}k^2 Ll\sqrt{\pi}\exp(-\rho^2/l^2). \tag{17-85}$$

In deriving (17-84a), we have used the formula [Magnus and Oberhettinger (1954, p. 88), see Eq. (D-1) in Appendix D]

$$\int_0^\infty \kappa^\mu J_{c-1}(\kappa\rho)\exp\left(-\frac{\kappa^2}{\kappa_m{}^2}\right)d\kappa$$

$$= \frac{\Gamma[(c+\mu)/2]\kappa_m^{(\mu+c)}}{4^{c/2}\rho^{1-c}\Gamma(c)}\,{}_1F_1\left(\frac{\mu+c}{2}, c; -\frac{\kappa_m{}^2\rho^2}{4}\right), \tag{17-86}$$

which is valid for $\mu + c > 0$.

17-15 HOMOGENEOUS AND LOCALLY HOMOGENEOUS TURBULENCE

The wave fluctuation in homogeneous and locally homogeneous turbulence has been extensively investigated (Tatarski, 1961, 1971), and therefore in this section we present a summary of the results with a brief explanation of the derivation.

As was discussed in Section 16-4, the general shape of the spectrum $\Phi_n(\kappa)$ is characterized by the smallest size l_0 of blobs called the inner scale and the largest size L_0 called the outer scale of turbulence. See Fig. 17-10. The region of spectrum $\Phi_n(\kappa)$ for which $\kappa < 2\pi/L_0$ is called the input range. The shape of the spectrum in this region depends on how the particular turbulence is

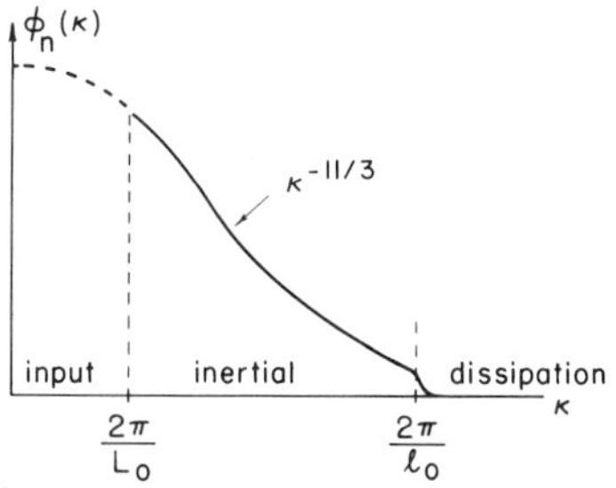

FIG. 17-10 Kolmogorov spectrum.

being created, and therefore no universal spectrum can be assigned and it is usually anisotropic. The region between $2\pi/L_0$ and $2\pi/l_0$ is called the inertial subrange. In this region, the kinetic energy of the eddies dominates over the energy dissipation and the spectrum should have the general shape

$$\Phi_n(\kappa) \sim \kappa^{-11/3}. \tag{17-87}$$

The region for which $\kappa > 2\pi/l_0$ is called the dissipation range, where the energy dissipation dominates the kinetic energy, and consequently there is very little energy in this range. Mathematically, we may write

$$\Phi_n(\kappa) = \begin{cases} 0.033C_n^{\,2}\kappa^{-11/3} & \text{for} \quad 2\pi/L_0 < \kappa < 2\pi/l_0 \\ 0 & \text{for} \quad 2\pi/l_0 < \kappa. \end{cases} \tag{17-88}$$

Or, we may combine these two:

$$\Phi_n(\kappa) = 0.033C_n^{\,2}\kappa^{-11/3}\exp(-\kappa^2/\kappa_m^{\,2}) \qquad \text{for} \quad 2\pi/L_0 < \kappa \tag{17-89}$$

with $\kappa_m = 5.92/l_0$. For mathematical convenience,

$$\Phi_n(\kappa) = 0.033C_n^{\,2}(\kappa_L^{\,2} + \kappa^2)^{-11/6}\exp(-\kappa^2/\kappa_m^{\,2}), \tag{17-90}$$

with $\kappa_L = 1/L_0$, is often used to describe the entire range of κ, even though as we already indicated, the input range cannot be simply described in an isotropic form as in the foregoing.

Let us consider the total spectrum which is the product of $f_\chi(\kappa)$ [or $f_S(\kappa)$] and $\Phi_n(\kappa)$. It is clear that the product shows quite different characteristics, depending on whether $2\pi/\sqrt{\lambda L}$ is greater than $2\pi/l_0$, or between $2\pi/L_0$ and

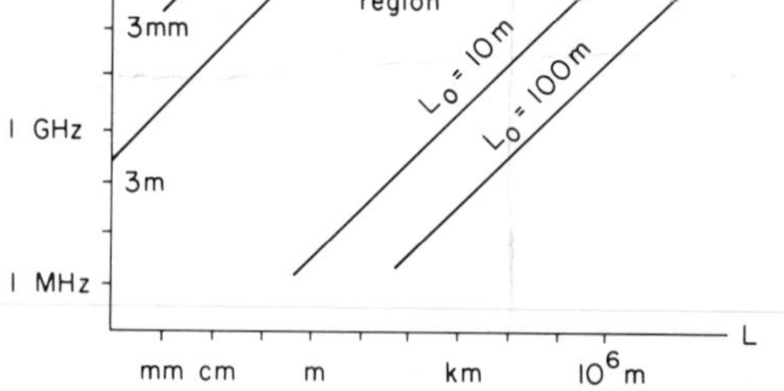

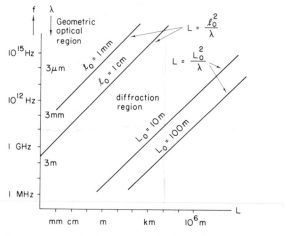

FIG. 17-11 Geometric optical region and diffraction region in the earth's atmosphere.

$2\pi/l_0$, or smaller than $2\pi/L_0$. These three cases represent the following ranges of the distance L:

$$L < l_0{}^2/\lambda \qquad \text{(geometric optical region)} \qquad (17\text{-}91a)$$

$$l_0{}^2/\lambda < L < L_0{}^2/\lambda \qquad \text{(diffraction region)} \qquad (17\text{-}91b)$$

$$L_0{}^2/\lambda < L. \qquad (17\text{-}91c)$$

In the earth's atmosphere, the inner scale l_0 is on the order of millimeters and the outer scale L_0 is on the order of 10–100 m. The boundaries between different ranges are of course not definite, but Fig. 17-11 gives some idea of the order of magnitude of the distance at different frequencies.

We substitute (17-88), (17-89), or (17-90) into the general formula (17-58) and obtain expressions for B_χ, B_S, and $B_{\chi S}$. In the following we give a summary of the results.

17-15-1 When $L \ll l_0{}^2/\lambda$

This is the geometric optical region discussed in Section 17-10, and we can use Eqs. (17-64) and (17-65).

Because of κ^5 in the integrand of (17-64), $\Phi_n(\kappa)$ for small κ does not contribute to the amplitude fluctuation B_χ. Therefore, we use (17-89) for Φ_n and the integral formula (17-86). We obtain

$$\begin{aligned} B_\chi(L, \rho) &= 0.033C_n{}^2(\pi^2L^3/6)\Gamma(\tfrac{7}{6})\kappa_m^{7/3}\,{}_1F_1(\tfrac{7}{6}, 1; -\kappa_m{}^2\rho^2/4) \\ &= 0.0504C_n{}^2L^3\kappa_m^{7/3}\,{}_1F_1(\tfrac{7}{6}, 1; -\kappa_m{}^2\rho^2/4). \end{aligned} \qquad (17\text{-}92)$$

For the phase correlation function, we cannot use (17-88) or (17-89) for κ small. However, we can use (17-90) with (17-65) and obtain

$$\begin{aligned} \sigma_S{}^2 = B_S(L, 0) &= 4\pi^2k^2L\int_0^\infty \kappa\, d\kappa\left(\kappa^2 + \frac{1}{L_0{}^2}\right)^{-11/6}\exp\left(-\frac{\kappa^2}{\kappa_m{}^2}\right) \\ &= 0.6514C_n{}^2k^2LL_0^{5/3}\Psi(1, \tfrac{1}{6}; 1/\kappa_m{}^2L_0{}^2), \end{aligned} \qquad (17\text{-}93)$$

where $\psi(a, b; z)$ is a confluent hypergeometric function which is independent of the Kummer function, and we use the integral formula (Abramowitz and Stegun, 1964, p. 505)

$$\int_0^\infty e^{-zt}t^{a-1}(1 + t)^{b-a-1}\, dt = \Gamma(a)\psi(a, b; z), \qquad \operatorname{Re} a > 0, \quad \operatorname{Re} z > 0. \qquad (17\text{-}94)$$

Noting that $\kappa_m{}^2L_0{}^2 \gg 1$, and using the asymptotic form† of ψ, we get

$$\sigma_S{}^2 = 0.7817C_n{}^2k^2LL_0^{5/3}. \qquad (17\text{-}95)$$

† $\psi(a, b; z) \to z^{-a}$ for $|z|$ large.

17-15-2 When $l_0{}^2/\lambda \ll L \ll L_0{}^2/\lambda$

This is the diffraction region, and therefore, in general, both l_0 and L_0 affect the fluctuations of a wave. For the amplitude fluctuation, the filter function $f_\chi \sim \kappa^2$ for κ small, and, therefore, the effect of the outer scale L_0 is negligibly small. If $L \gg l_0{}^2/\lambda$, then we can use

$$\Phi_n(\kappa) = 0.033 C_n{}^2 \kappa^{-11/3} \tag{17-96}$$

for all ranges of κ, and we obtain

$$\begin{aligned}\sigma_\chi{}^2 &= 2\pi^2 k^2 L \int_0^\infty \kappa\, d\kappa f_\chi(\kappa)\Phi_n(\kappa) \\ &= 0.033 C_n{}^2 \pi^2 k^{7/6} L^{11/6}[-\Gamma(-\tfrac{5}{6})]\tfrac{6}{11} \cos(5\pi/12) \\ &= 0.307 C_n{}^2 k^{7/6} L^{11/6},\end{aligned} \tag{17-97}$$

which is given by Tatarski (1961, p. 153) [see also Eq. (D-10) in Appendix D].

In contrast to the geometric optical case given in (17-92), where $\sigma_\chi{}^2$ is independent of k and proportional to L^3, $\sigma_\chi{}^2$ in the diffraction region is proportional to $k^{7/6}$ and $L^{11/6}$. These powers ($\frac{7}{6}$ and $\frac{11}{6}$) are directly related to the shape of the Kolmogorov spectrum $\kappa^{-11/3}$ and thus they are often used to verify the validity of the Kolmogorov representation.

We can take into account the effect of the inner scale by using (17-89). Then we obtain

$$\begin{aligned}\sigma_\chi{}^2 &= 0.033\pi^2 C_n{}^2 k^2 L[-\Gamma(-\tfrac{5}{6})] \\ &\quad \left\{-\kappa_m{}^{-5/3} + \frac{6}{11}\frac{k}{L}\,\mathrm{Im}\left[\exp\left(j\frac{\pi}{12}\right)\left(\frac{L}{k} + i\frac{1}{\kappa_m{}^2}\right)^{11/6}\right]\right\}.\end{aligned} \tag{17-98}$$

The correlation function $B_\chi(L, \rho)$ is given by

$$B_\chi(L, \rho) = \sigma_\chi{}^2 b_\chi(L, \rho), \qquad b_\chi(L, \rho) = \frac{B_1 - B_2}{1 - B_3}, \tag{17-99}$$

where

$$B_1 = {}_1F_1(-\tfrac{5}{6}, 1; -\kappa_m{}^2\rho^2/4)$$

$$B_2 = \frac{6}{11}\left(\frac{L\kappa_m{}^2}{k}\right)^{5/6} \mathrm{Im}\left[\exp\left(i\frac{\pi}{12}\right)\left(1 + i\frac{k}{L\kappa_m{}^2}\right)^{11/6} {}_1F_1\left(-\frac{11}{6}, 1; -i\frac{\rho^2}{4}\frac{k}{L}\left(1 + i\frac{k}{L\kappa_m{}^2}\right)^{-1}\right)\right]$$

$$B_3 = \frac{6}{11}\left(\frac{L\kappa_m{}^2}{k}\right)^{5/6} \mathrm{Im}\left[\exp\left(j\frac{\pi}{12}\right)\left(1 + i\frac{k}{L\kappa_m{}^2}\right)^{11/6}\right].$$

This expression shows that the correlation distance depends on both $\sqrt{\lambda L}$ and l_0. Some numerical calculations show that the correlation distance is on the order of magnitude of $\sqrt{\lambda L}$, and it is slightly dependent on l_0.

The evaluation of the integrals for (17-98) and (17-99) is made by the integral formula†

$$\int_0^\infty \left[1 - \frac{\sin(A\kappa^2)}{A\kappa^2}\right]\kappa^{n+1} \exp(-B\kappa^2)\, d\kappa$$

$$= \frac{\Gamma(n/2)}{2}\left\{\frac{n}{2} B^{-(n/2+1)} - \frac{1}{A} \operatorname{Im}\left[\exp\left(j\frac{n\pi}{4}\right)(A + iB)^{-n/2}\right]\right\} \quad (17\text{-}100)$$

which is valid for $n > -6$.

The filter function $f_S(\kappa)$ for the phase fluctuation becomes close to 2 when κ is in the input range ($< 2\pi/L_0$), and therefore we may expect that the eddies of size L_0 or greater will affect the phase correlation function. In this range, the turbulence is in general anisotropic and the shape of the spectrum depends on how the turbulence is created. For this reason, it is not possible to obtain a general expression for the phase correlation function. However, the phase fluctuation is useful in deciding the characteristics of large-scale turbulence. In many situations, it is more practical to measure a phase difference between two points or two instants of time. Here we must deal with the structure function rather than the correlation function.

17-16 INHOMOGENEOUS RANDOM MEDIUM WITH GAUSSIAN CORRELATION FUNCTION AND THE SPATIAL FILTER FUNCTION

As discussed in Section 17-9-2, the random medium may not be uniform, and we need to take into account the variation of the fluctuation characteristics of the refractive index along the path of propagation. Let us assume that the variance $\sigma_n^2 = \langle n_1^2 \rangle$ is not constant, but varies along the propagation path. Also, we assume that the correlation function is described by a Gaussian function. Then we write

$$\langle n_1(\mathbf{r}_1) n_1(\mathbf{r}_2)\rangle = \left\langle n_1^2\left(\frac{\mathbf{r}_1 + \mathbf{r}_2}{2}\right)\right\rangle \exp\left(-\frac{|\mathbf{r}_1 - \mathbf{r}_2|^2}{l^2}\right), \quad (17\text{-}101)$$

where the variance $\langle n_1^2 \rangle$ varies as a function of the center coordinate and l is the correlation distance.

† To obtain this formula, we use (17-86) with $c = 1$, $p = 0$, and $\mu = n + 1$. Then we evaluate (17-100) term by term. Each term is valid only for $n > -2$, and it diverges at $n = -2, -4, -6, \ldots$. However, the sum of these two terms is regular at $n = -2$ and -4, and, therefore, the sum given in (17-100) is valid for $n > -6$. This reasoning is called the analytic continuation.

The spectral density is easily obtained [see Eq. (A-33) in Appendix A]:

$$\Phi_n(\kappa) = \frac{\langle n_1{}^2(\eta)\rangle l^3}{8\pi\sqrt{\pi}} \exp\left(-\frac{\kappa^2 l^2}{4}\right). \tag{17-102}$$

Substituting this into (17-51), we get the form given in (17-61):

$$B_\chi(L, \rho) = \int_0^L d\eta G_\chi(\eta, \rho)\langle n_1{}^2(\eta)\rangle. \tag{17-103}$$

Similarly, we have B_S and $B_{\chi S}$ with G_S and $G_{\chi S}$ in the integrand, respectively. The spatial filter function $G(\eta, \rho)$ is then given by

$$\begin{aligned} G_\chi(\eta, \rho) &= \frac{\sqrt{\pi}k^2 l^3}{2} \int_0^\infty \kappa\, d\kappa J_0(\kappa\rho) \sin^2\left(\frac{L - x'}{2k}\kappa^2\right) \exp\left(-\frac{\kappa^2 l^2}{4}\right) \\ &= \frac{\sqrt{\pi}k^2 l^3}{8} \left\{\left[\frac{1}{A}\,{}_1F_1\left(1, 1; -\frac{\rho^2}{4A}\right)\right] - \text{Re}\left[\frac{1}{B}\,{}_1F_1\left(1, 1; -\frac{\rho^2}{4B}\right)\right]\right\}, \end{aligned} \tag{17-104}$$

where $A = l^2/4$ and $B = l^2/4 - i(L - \eta)/k$, and ${}_1F_1(a, b; z)$ is a Kummer function.† In deriving this we have made use of (17-86) with $\mu = c = 1$. Similarly, for the spatial filter function for the phase fluctuation, we have

$$G_S(\eta, \rho) = \frac{\sqrt{\pi}k^2 l^3}{8} \left\{\left[\frac{1}{A}\,{}_1F_1\left(1, 1; -\frac{\rho^2}{4A}\right) + \text{Re}\left[\frac{1}{B}\,{}_1F_1\left(1, 1; -\frac{\rho^2}{4B}\right)\right]\right]\right\}, \tag{17-105}$$

and for the cross correlation, we have

$$G_{\chi S}(\eta, \rho) = \frac{\sqrt{\pi}k^2 l^3}{8} \text{Im}\left[\frac{1}{B}\,{}_1F_1\left(1, 1; -\frac{\rho^2}{4B}\right)\right]. \tag{17-106}$$

Let us consider these filter functions for the following two extreme cases, and obtain expressions for the variances of fluctuations at $\boldsymbol{\rho} = \boldsymbol{\rho}_1 - \boldsymbol{\rho}_2 = 0$:

$$\begin{aligned} B_\chi(L, 0) &= \langle\chi^2\rangle = \sigma_\chi{}^2, \qquad B_S(L, 0) = \langle S_1{}^2\rangle = \sigma_S{}^2, \\ B_{\chi S}(L, 0) &= \langle\chi S_1\rangle = \sigma_{\chi S}^2. \end{aligned} \tag{17-107}$$

† ${}_1F_1(1, 1, z)$ can also be expressed as an exponential integral (Abramowitz and Stegun, 1964, p. 510).

(a) *When* $l \ll \sqrt{\lambda L}$ In this case, we have

$$\langle \sigma_\chi^2 \rangle = \langle \sigma_S^2 \rangle = \frac{\sqrt{\pi} k^2 l}{2} \int_0^L d\eta \langle n_1^2(\eta) \rangle$$

$$\langle \sigma_{\chi S}^2 \rangle = \frac{\sqrt{\pi} k^2 l^3}{8} \int_0^L d\eta \frac{((L-\eta)/k)}{(l/2)^4 + ((L-\eta)/k)^2} \langle n_1^2(\eta) \rangle. \quad (17\text{-}108)$$

(b) *When* $l \gg \sqrt{\lambda L}$ In this case, we have

$$\langle \sigma_\chi^2 \rangle = \frac{8\sqrt{\pi}}{l^3} \int_0^L (L-\eta)^2 \langle n_1^2(\eta) \rangle, \qquad \langle \sigma_S^2 \rangle = \sqrt{\pi} k^2 l \int_0^L d\eta \langle n_1^2(\eta) \rangle$$

$$\langle \sigma_{\chi S}^2 \rangle = \frac{2\sqrt{\pi} k}{l} \int_0^L d\eta (L-\eta) \langle n_1^2(\eta) \rangle. \quad (17\text{-}109)$$

17-17 VARIATIONS OF THE INTENSITY OF TURBULENCE ALONG THE PROPAGATION PATH

In Section 17-15, the turbulence is assumed to be homogeneous and isotropic. However, we realize that turbulence can be homogeneous and isotropic only within the distance on the order of the outer scale of turbulence L_0. At different locations which are separated beyond L_0, we can expect that the turbulence intensity can be quite different as in the case of vertical wave propagation through the atmosphere. In the case of propagation at a fixed height above the ground, the intensity of the turbulence can be approximately uniform.

We can characterize this situation by writing the spectrum of the index of refraction as

$$\Phi_n(\kappa, \mathbf{r}) = C_n^2(\mathbf{r}) \Phi_n^{(0)}(\kappa)$$

$$\Phi_n^{(0)}(\kappa) = 0.033 \kappa^{-11/3} \quad \text{in the inertial subrange.} \quad (17\text{-}110)$$

The structure constant C_n is here considered to be a function of position.

For plane wave propagation, $C_n(\mathbf{r})$ can be assumed to be a function of distance x only. This is the case discussed in Section 17-9-2. Using (17-110) and performing the integration in (17-61), we get from (17-51)

$$\sigma_\chi^2 = 0.563 k^{7/6} \int_0^L d\eta C_n^2(\eta)(L-\eta)^{5/6}, \quad (17\text{-}111)$$

which is valid for $\lambda L \ll l_0^2$.

If the condition $\lambda L \ll l_0{}^2$ is not satisfied, we need to take into account the effect of the inner scale by using (17-89). We then get

$$B_\chi(L, \rho) = 0.033\pi^2 k^{7/6} L^{11/6}[-\Gamma(-\tfrac{5}{6})]\frac{1}{L}\int_0^L d\eta C_n{}^2(\eta) G_\chi(\eta),$$

where

$$G_\chi(\eta) = \mathrm{Re}\{D^{5/6}\,{}_1F_1(-\tfrac{5}{6}, 1; -\rho^2/4D) - A^{5/6}\,{}_1F_1(-\tfrac{5}{6}, 1; -\rho^2/4A)\}$$

$$D = i(1 - \eta/L) + (k/L)(1/\kappa_m{}^2), \qquad A = (k/L)(1/\kappa_m{}^2). \tag{17-112}$$

Note that when $\rho = 0$, ${}_1F_1(a, b; 0) = 1$ and, therefore, we have (Ishimaru, 1969a,b)

$$G_\chi(\eta) = \mathrm{Re}\{D^{5/6} - A^{5/6}\}. \tag{17-113}$$

17-18 RANGE OF VALIDITY OF THE WEAK FLUCTUATION THEORY

The range of validity of the Rytov first iteration solution used in this chapter is that the log-amplitude variance is small compared with unity and no higher than about 0.2–0.5:

$$\sigma_\chi{}^2 < 0.2\text{–}0.5. \tag{17-114}$$

Although this requirement is reasonable for amplitude fluctuation, it is generally agreed that the weak fluctuation theory for the phase fluctuation is valid beyond the range (17-114). In fact the Rytov solution for the phase fluctuation is considered valid in the strong fluctuation region where the Rytov solution for the log-amplitude fluctuation is no longer valid (Barabanenkov *et al.*, 1971).

17-19 RELATED PROBLEMS

There have been numerous investigations related to wave fluctuations in a line-of-sight propagation. The effects of random media on the characteristics of communication channels are discussed by many workers (Brookner, 1970; Gallager, 1964; Crane, 1971; Kennedy and Karp, 1969; Kennedy and Hoverstein, 1968; Hoffman, 1960). There have been considerable studies made on microwave fluctuation through the turbulence (Lee and Waterman, 1966; Clifford and Strohbehn, 1970; Wheelon and Muchmore, 1955; Muchmore *et al.*, 1955, 1963; Norton *et al.*, 1955) and planetary atmospheres (Woo and Ishimaru, 1973, 1974; Woo *et al.*, 1974). Extensive studies (Fried, 1966, 1967a–c, 1968; Fried and Cloud, 1966; Ackley, 1971; Fitsmaurice *et al.*, 1969; Gurvich *et al.*, 1968; Izyumov, 1968; Mitchell, 1968; Pokasov and Khmelevtsov, 1968; Yura, 1969; Borisov *et al.*, 1969; Bouricius and Clifford, 1970; Buser, 1971; Davis, 1966; Ryzhov *et al.*, 1965; Sergeyenko, 1970) on optical propagation through the atmosphere have been conducted. Other studies include aperture averaging and resolution limits (Titterton, 1973; Kaydanovskiy and Smirnova, 1968; Homstad *et al.*, 1974; Heidbreder and Mitchell, 1966; Fried, 1964, 1966; Coulman, 1966), probability density of fluctuation (Wang and Strohbehn, 1974), scintillation of waves from stellar radio sources (Kuriksha, 1968; Roddier and Roddier, 1973), turbulence effects on synthetic aperture radars (Porcello, 1970), and imaging in water (Yura, 1973a).

CHAPTER 18 □ LINE-OF-SIGHT PROPAGATION OF SPHERICAL AND BEAM WAVES THROUGH A RANDOM MEDIUM—WEAK FLUCTUATION CASE

In Chapter 17 we discussed the fluctuations of a plane wave in a random medium. In many practical situations, however, it is necessary to consider the relative positions of transmitter and receiver, and whether the receiver is in the near field or far field of the transmitter.

Consider a transmitting aperture of diameter D. If most of the random medium between the transmitter and the receiver is located in the far zone of the transmitter, then the radiation can be well approximated by a spherical wave. On the other hand, if the random medium is in the near field of the transmitter, we need to take into account the beam nature of the wave. This is often the case in optical propagation. In this chapter, we consider the problem of spherical and beam waves in a random medium.

18-1 RYTOV SOLUTION FOR THE SPHERICAL WAVE

As shown in Section 17-3, Rytov's first iteration solution is given by

$$U(\mathbf{r}) = U_0(\mathbf{r}) \exp[\psi_1(\mathbf{r})], \qquad \psi_1(\mathbf{r}) = \chi(\mathbf{r}) + iS(\mathbf{r}) \tag{18-1a}$$

$$\psi_1(\mathbf{r}) = \int_{v'} h(\mathbf{r}, \mathbf{r}')n_1(\mathbf{r}')\, dv' \tag{18-1b}$$

where

$$h(\mathbf{r}, \mathbf{r}') = 2k^2 G(\mathbf{r} - \mathbf{r}')U_0(\mathbf{r}')/U_0(\mathbf{r}). \tag{18-1c}$$

$U_0(\mathbf{r})$ is the incident wave in the absence of the turbulent medium and $G(\mathbf{r} - \mathbf{r}')$ is the free space Green's function.

Let us consider a spherical wave radiated from the origin $U_0(\mathbf{r}) = (1/4\pi r) \exp(ikr)$. Since the scattering takes place in the neighborhood of the

x axis (see Section 17-4), we approximate $U_0(\mathbf{r})$ and $G(\mathbf{r}-\mathbf{r}')$ by the following:

$$U_0(\mathbf{r}) \approx \frac{1}{4\pi x} \exp\left[ik\left(x + \frac{\rho^2}{2x}\right)\right]$$

$$G(\mathbf{r}-\mathbf{r}') \approx \frac{1}{4\pi|x-x'|} \exp\left\{ik\left[(x-x') + \frac{|\boldsymbol{\rho}-\boldsymbol{\rho}'|^2}{2(x-x')}\right]\right\}, \tag{18-2}$$

where $\boldsymbol{\rho} = y\hat{\mathbf{y}} + z\hat{\mathbf{z}}$ and $\boldsymbol{\rho}' = y'\hat{\mathbf{y}} + z'\hat{\mathbf{z}}$.

Substituting (18-2) into (18-1c), we obtain

$$h(\mathbf{r}, \mathbf{r}') = \frac{k^2}{2\pi} \frac{1}{\gamma(x-x')} \exp\left[i\frac{k}{2}\frac{|\boldsymbol{\rho}'-\gamma\boldsymbol{\rho}|^2}{\gamma(x-x')}\right] \tag{18-3}$$

where $\gamma = x'/x$.

We note that if $\gamma = 1$, then (18-3) reduces to an equation for a plane wave. We note also that the factor $\boldsymbol{\rho}' - \gamma\boldsymbol{\rho}$ indicates the expansion of the wave from x' to x (see Fig. 18-1).

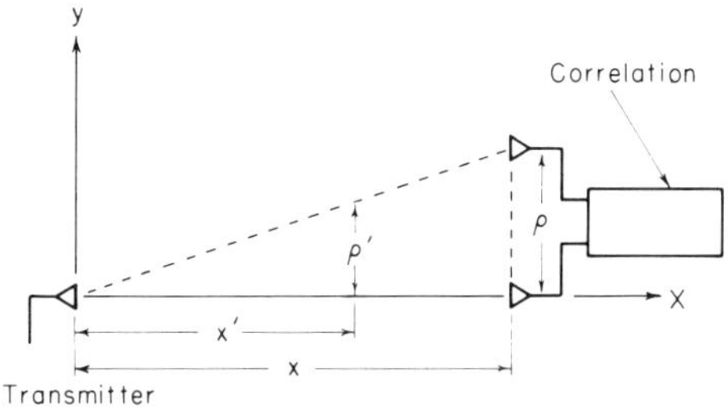

FIG. 18-1 Spherical wave showing the correspondence between ρ' and $(x'/x)\,\rho$.

Since (18-3) can be obtained from the plane wave case (17-27b) by changing $\boldsymbol{\rho}$ to $\gamma\boldsymbol{\rho}$ and $x - x'$ to $\gamma(x - x')$, we can conclude that all the results for the plane wave case can be converted to the spherical wave case by the replacement

$$\boldsymbol{\rho} \to \gamma\boldsymbol{\rho} \qquad \text{and} \qquad x - x' \to \gamma(x - x'). \tag{18-4}$$

We therefore obtain the following correlation functions from (17-51)–(17-53):

$$B_\chi(L, \rho) = \langle\chi(L, \boldsymbol{\rho}_1)\chi(L, \boldsymbol{\rho}_2)\rangle = (2\pi)^2 \int_0^L d\eta \int_0^\infty \kappa\, d\kappa\, J_0(\kappa\gamma\rho)|H_r|^2\Phi_n(\kappa) \tag{18-5}$$

where

$$|H_r|^2 = k^2 \sin^2\left[\frac{\gamma(L-\eta)}{2k}\kappa^2\right], \qquad \gamma = \frac{\eta}{L}, \qquad \text{and} \qquad \boldsymbol{\rho} = |\boldsymbol{\rho}_1 - \boldsymbol{\rho}_2|;$$

$$B_S(L, \rho) = \langle S_1(L, \boldsymbol{\rho}_1)S_1(L, \boldsymbol{\rho}_2)\rangle$$

$$= (2\pi)^2 \int_0^L d\eta \int_0^\infty \kappa\, d\kappa\, J_0(\kappa\gamma\rho)|H_i|^2\Phi_n(\kappa) \tag{18-6}$$

where

$$|H_i|^2 = k^2 \cos^2\left[\frac{\gamma(L-\eta)}{2k}\kappa^2\right];$$

and

$$B_{\chi S}(L, \rho) = \langle x(L, \boldsymbol{\rho}_1)S_1(L, \boldsymbol{\rho}_2)\rangle$$

$$= (2\pi)^2 \int_0^L d\eta \int_0^\infty \kappa\, d\kappa\, J_0(\kappa\gamma\rho)H_r H_i \Phi_n(\kappa) \tag{18-7}$$

where

$$H_r H_i = \frac{k^2}{2}\sin\left[\frac{\gamma(L-\eta)}{k}\kappa^2\right].$$

The structure functions are given by

$$D_\chi(L, \rho) = \langle|\chi(L, \boldsymbol{\rho}_1) - \chi(L, \boldsymbol{\rho}_2)|^2\rangle$$

$$= 8\pi^2 \int_0^L d\eta \int_0^\infty \kappa\, d\kappa\,[1 - J_0(\kappa\gamma\rho)]|H_r|^2\Phi_n(\kappa), \tag{18-8}$$

$$D_S(L, \rho) = \langle|S_1(L, \boldsymbol{\rho}_1) - S_1(L, \boldsymbol{\rho}_2)|^2\rangle$$

$$= 8\pi^2 \int_0^L d\eta \int_0^\infty \kappa\, d\kappa\,[1 - J_0(\kappa\gamma\rho)]|H_i|^2\Phi_n(\kappa). \tag{18-9}$$

18-2 VARIANCE FOR THE KOLMOGOROV SPECTRUM

Let us consider the variance of the log-amplitude fluctuation when the spectrum $\Phi_n(\kappa)$ of the refractive index fluctuation is given by the Kolmogorov spectrum. We can conveniently consider the three cases: (a) $l_0{}^2/\lambda \ll L_s \ll L_0{}^2/\lambda$; (b)$L_s \sim L_0{}^2/\lambda$, (c) $L_s \sim l_0{}^2/\lambda$ where L_s is $(\eta/L)(L-\eta)$ and η is the distance from the transmitter to the medium.†

† Since the major contribution to the received wave fluctuation comes from the region near the midpoint ($\eta = L/2$), L_s is on the order of L (or $L/4$).

(a) *When the distance L_s is in the range $l_0^2/\lambda \ll L_s \ll L_0^2/\lambda$*, most contributions to the amplitude fluctuation come from the inertial subrange ($1/L_0 \ll \kappa \ll 1/l_0$) of the spectrum $\Phi_n(\kappa)$. Therefore, we use

$$\Phi_n(\kappa) = 0.033C_n^2(\eta)\kappa^{-11/3}. \tag{18-10}$$

Substituting (18-10) into (18-5) and letting $\rho = 0$, we get the variance:

$$\sigma_\chi^2 = 0.563k^{7/6}\int_0^L d\eta C_n^2(\eta)\left[\frac{\eta(L-\eta)}{L}\right]^{5/6}. \tag{18-11a}$$

If the structure constant C_n^2 is constant throughout the path, then we have

$$\sigma_\chi^2 = 0.124C_n^2k^{7/6}L^{11/6}. \tag{18-11b}$$

(b) *When $L_s \sim L_0^2/\lambda$, (18-10) is no longer valid* Here, the effect of the outer scale must be taken into account. One convenient way is to use the modified von Karman spectrum given by

$$\Phi_n(\kappa) = 0.033C_n^2(\eta)(\kappa^2 + 1/L_0^2)^{-11/6}. \tag{18-12}$$

Substituting (18-12) into (18-5), we obtain

$$\sigma_\chi^2 = 0.326k^2\int_0^L d\eta C_n^2(\eta)L_0^{5/3}\left[\frac{6}{5} - \mathrm{Re}\,\psi\left(1, \frac{1}{6}; i\frac{\gamma(L-\eta)}{kL_0^2}\right)\right]. \tag{18-13}$$

(c) *When $L_s \sim l_0^2/\lambda$, the effect of the inner scale cannot be ignored* We therefore use the spectrum

$$\Phi_n(\kappa) = 0.033C_n^2(\eta)\kappa^{-11/3}\exp(-\kappa^2/k_m^2) \tag{18-14}$$

where $\kappa_m = 5.91/l_0$, and obtain

$$\sigma_\chi^2 = 2.1755k^{7/6}L^{5/6}\int_0^L d\eta C_n^2(\eta)G_\chi(\eta) \tag{18-15}$$

where

$$G_\chi(\eta) = \mathrm{Re}\{D^{5/6} - A^{5/6}\}, \qquad D = i\gamma\frac{(L-\eta)}{L} + \left(\frac{k}{L}\right)\frac{1}{\kappa_m^2}, \qquad A = \left(\frac{k}{L}\right)\frac{1}{\kappa_m^2}.$$

Let us next consider the variance σ_S^2 of the phase fluctuation. For this case, we cannot use (18-10) since it leads to a divergent integral. This is because the phase fluctuations are greatly influenced by the input range of the spectrum $\Phi_n(\kappa)$ and thus the outer scale plays an important role. We therefore need to use (18-12). We then obtain

$$\sigma_S^2 = 0.326k^2\int_0^L d\eta C_n^2(\eta)L_0^{5/3}\left[\frac{6}{5} + \mathrm{Re}\,\psi\left(1, \frac{1}{6}; i\frac{\gamma(L-\eta)}{kL_0^2}\right)\right]. \tag{18-16}$$

18-3 CORRELATION AND STRUCTURE FUNCTIONS FOR THE KOLMOGOROV SPECTRUM

The amplitude correlation function is obtained for the spectrum (18-10):

$$B_\chi(L, \rho) = 2.1755k^{7/6}L^{5/6} \int_0^L d\eta C_n^2(\eta) G_\chi(\eta, \rho) \tag{18-17}$$

where

$$G_\chi(\eta, \rho) = \mathrm{Re}\left[D^{5/6}\,{}_1F_1\left(-\frac{5}{6}, 1; i\frac{k\gamma\rho^2}{4(L-\eta)}\right)\right].$$

Note that for $\rho \gg \sqrt{\lambda L}$, G_χ approaches $G_\chi(\eta, \rho) = 1.063(k\gamma^2/4L)^{5/6}\rho^{5/3}$.

Let us next consider the structure functions D_χ and D_S. Using the spectrum (18-10), we obtain (Ishimaru, 1969a,b)

$$\left.\begin{matrix} D_\chi \\ D_S \end{matrix}\right\} = 2.1755k^{7/6}L^{5/7} \int_0^L d\eta C_n^2(\eta) G_d(\eta, \rho) \tag{18-18}$$

$$G_d(\eta, \rho) = 0.6697\left(\frac{k}{L}\right)^{5/6}(\gamma\rho)^{5/3} \pm 2D^{5/6}\left[1 - {}_1F_1\left(-\frac{5}{6}, 1; -\frac{(\gamma\rho)^2}{4D}\left(\frac{k}{L}\right)\right)\right]$$

where $D = i\gamma(1 - \eta/L)$ and the upper and lower signs are for D_χ and D_S, respectively. Note that for $\rho \gg \sqrt{\lambda L}$, G_d for the phase structure function reduces to $G_d(\eta, \rho) \to 1.339(k/L)^{5/6}(\gamma\rho)^{5/3}$.

For $C_n^2(\eta) = \text{const}$, the phase structure function when $\rho \gg \sqrt{\lambda L}$ is given by

$$D_S(L, \rho) = 1.0924k^2LC_n^2\rho^{5/3}. \tag{18-19}$$

18-4 BEAM WAVE

Let us first consider a beam wave in free space. At the aperture ($x = 0$), we assume that the amplitude distribution is Gaussian with a beam size W_0 and the phase distribution is parabolic with a radius of curvature R_0. Such a phase distribution should produce a beam focused at $x = R_0$ (Fig. 18-2).

The field at $x = 0$ is therefore given by

$$U_0(0, \rho) = \exp[-(1/W_0^2 + ik/2R_0)\rho^2]. \tag{18-20a}$$

For convenience, we write this as

$$U_0(0, \rho) = \exp[-\tfrac{1}{2}(k\alpha)\rho^2] \tag{18-20b}$$

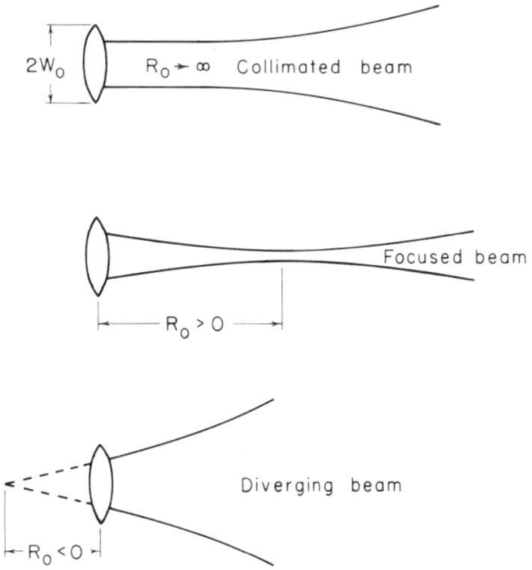

FIG. 18-2 Beam wave.

where $\alpha = \alpha_r + i\alpha_i = (\lambda/\pi W_0^2) + i(1/R_0)$. At an arbitrary point $(x, \boldsymbol{\rho})$, the beam wave is given by (Ishimaru, 1969a,b)

$$U_0(x, \rho) = \frac{1}{1 + i\alpha x} \exp\left[ikx - \frac{k\alpha}{2} \frac{\rho^2}{(1 + i\alpha x)}\right]. \tag{18-21}$$

This is valid within a distance

$$x \ll \pi^3 W_0^4/\lambda^3. \tag{18-22}$$

This is always satisfied in most practical cases. The power at (x, ρ) is therefore given by

$$|U_0(x, \rho)|^2 = (W_0^2/W^2) \exp(-2\rho^2/W^2) \tag{18-23a}$$

where W is the beam size at x and is given by

$$W^2 = W_0^2[(1 - \alpha_i x)^2 + (\alpha_r x)^2]. \tag{18-23b}$$

The total transmitted power should be independent of the distance and is given by

$$P_t = 2\pi \int_0^\infty \rho \, d\rho \, |U_0(x, \rho)|^2 = \frac{\pi}{2} W_0^2. \tag{18-23c}$$

We note that for a collimated beam,

$$R_0 \to \infty \qquad \text{and} \qquad \alpha_i = 0. \tag{18-24}$$

For divergent and focused beams, we have, respectively,

$$R_0 < 0 \qquad \text{and} \qquad R_0 > 0. \tag{18-25}$$

If the beam is focused at the observation point, $x = R_0$ and $\alpha_i x = 1$.

Now, we substitute (18-21) into (18-1c) and obtain

$$h(\mathbf{r}, \mathbf{r}') = \frac{k^2}{2\pi} \frac{1}{\gamma(x - x')} \exp\left[i \frac{k}{2} \frac{|\boldsymbol{\rho}' - \gamma\boldsymbol{\rho}|^2}{\gamma(x - x')} \right] \tag{18-26}$$

where $\gamma = (1 + i\alpha x')/(1 + i\alpha x)$. We note here that when $\gamma = 1$, (18-26) reduces to the plane wave case and when $\gamma = x'/x$, it reduces to the spherical wave case. In terms of α, the plane wave case is obtained by letting $\alpha \to 0$ ($W_0 \to \infty$ and $R_0 \to \infty$) and the spherical wave case is obtained by letting $\alpha \to \infty$ ($W_0 \to 0$ and $R_0 \to \infty$). We note that γ in (18-26) is complex and therefore the procedure of obtaining correlation functions and structure functions must be modified accordingly. The details of the derivation are discussed in a number of references (Ishimaru, 1969a,b). Here, we present the general expressions of the correlation functions, the variances, and the structure functions:

$$B_\chi = B_\chi(L, \boldsymbol{\rho}_1, \boldsymbol{\rho}_2) = \langle \chi(L, \boldsymbol{\rho}_1)\chi(L, \boldsymbol{\rho}_2) \rangle$$

$$B_S = B_S(L, \boldsymbol{\rho}_1, \boldsymbol{\rho}_2) = \langle S_1(L, \boldsymbol{\rho}_1)S_1(L, \boldsymbol{\rho}_2) \rangle \tag{18-27a}$$

$$\sigma_\chi^2(L, \boldsymbol{\rho}) = B_\chi(L, \boldsymbol{\rho}, \boldsymbol{\rho}), \qquad \sigma_S^2(L, \boldsymbol{\rho}) = B_S(L, \boldsymbol{\rho}, \boldsymbol{\rho})$$

$$D_\chi(L, \boldsymbol{\rho}_1, \boldsymbol{\rho}_2) = \langle |\chi(L, \boldsymbol{\rho}_1) - \chi(L, \boldsymbol{\rho}_2)|^2 \rangle$$

$$D_S(L, \boldsymbol{\rho}_1, \boldsymbol{\rho}_2) = \langle |S_1(L, \boldsymbol{\rho}_1) - S_1(L, \boldsymbol{\rho}_2)|^2 \rangle \tag{18-27c}$$

$$\left.\begin{matrix} B_\chi(L, \boldsymbol{\rho}_1, \boldsymbol{\rho}_2) \\ B_S(L, \boldsymbol{\rho}_1, \boldsymbol{\rho}_2) \end{matrix}\right\} = \operatorname{Re}\left\{ 4\pi^2 \int_0^L d\eta \int_0^\infty \kappa\, d\kappa \tfrac{1}{2}[J_0(\kappa P)|H|^2 \pm J_0(\kappa Q)H^2]\Phi_n(\eta, \kappa) \right\} \tag{18-28a}$$

$$\left.\begin{matrix} \sigma_\chi^2(L, \boldsymbol{\rho}) \\ \sigma_S^2(L, \boldsymbol{\rho}) \end{matrix}\right\} = 2\pi^2 \int_0^L d\eta \int_0^\infty \kappa\, d\kappa [I_0(2\gamma_i \kappa \rho)|H|^2 \pm \operatorname{Re}(H^2)]\Phi_n(\eta, \kappa) \tag{18-28b}$$

where $I_0(z)$ is the modified Bessel function:

$$\left.\begin{matrix} D_\chi(L, \boldsymbol{\rho}_1, \boldsymbol{\rho}_2) \\ D_S(L, \boldsymbol{\rho}_1, \boldsymbol{\rho}_2) \end{matrix}\right\} = 4\pi^2 \int_0^L d\eta \int_0^\infty \kappa\, d\kappa \operatorname{Re}\{[\tfrac{1}{2}I_0(2\gamma_i\kappa\rho_1) + \tfrac{1}{2}I_0(2\gamma_i\kappa\rho_2) - J_0(\kappa P)]|H|^2 \pm [1 - J_0(\kappa Q)]H^2\}\Phi_n(\eta, \kappa). \tag{18-28c}$$

In these expressions, the upper and lower signs correspond to the upper and lower functions, respectively. The following notations are used:

$$P = [(\gamma y_1 - \gamma^* y_2)^2 + (\gamma z_1 - \gamma^* z_2)^2]^{1/2}$$

$$Q = \gamma[(y_1 - y_2)^2 + (z_1 - z_2)^2]^{1/2}$$

$$|H|^2 = k^2 \exp\left[-\frac{\gamma_i(L-\eta)}{k}\kappa^2\right], \qquad H^2 = -k^2 \exp\left[-i\frac{\gamma(L-\eta)}{k}\kappa^2\right]$$

$$\gamma = (1 + i\alpha\eta)/(1 + i\alpha L) = \gamma_r - i\gamma_i$$

$$\rho = (y^2 + z^2)^{1/2}, \qquad \rho_1 = (y_1{}^2 + z_1{}^2)^{1/2}, \qquad \rho_2 = (y_2{}^2 + z_2{}^2)^{1/2}$$

$$\alpha = \alpha_r + i\alpha_i = \lambda/\pi W_0{}^2 + i1/R_0.$$

18-5 VARIANCE FOR A BEAM WAVE AND THE VALIDITY OF THE RYTOV SOLUTION

The variance of the log-amplitude fluctuation on the beam axis $\rho = 0$ is obtained from (18-28b). We have

$$\sigma_\chi{}^2 = 2.1755 k^{7/6} L^{5/6} \int_0^L d\eta C_n{}^2(\eta) G_\chi(\eta)$$

$$G_\chi(\eta) = \mathrm{Re}[\gamma(1 - \eta/L)]^{5/6} - [\gamma_i(1 - \eta/L)]^{5/6}. \tag{18-29}$$

The variances for collimated and diverging beams are similar to those of plane and spherical waves, respectively. For a focused beam, the variance is considerably smaller at the focal point, indicating the reduction of scintillation. Experimental results indicate, however, that this reduction of scintillation either does not exist or is difficult to measure. A part of this discrepancy may be attributed to the limited range of validity of (18-29). The Rytov solution is valid only when

$$\sigma_\chi{}^2 < 0.2\text{–}0.5. \tag{18-30}$$

This is true for collimated and diverging beams. However, for a focused beam, recent study (Ishimaru, 1977b) shows that the range of validity is

$$1.36 C_n{}^2 k^{7/6} L^{11/6} \left[\frac{\alpha_r L}{(\alpha_r L)^2 + (1 - \alpha_i L)^2}\right]^{5/6} \ll 1, \tag{18-31}$$

in addition to (18-30). When the beam is focused at the observation point ($L = R_0$), (18-31) becomes

$$0.762 C_n{}^2 k^2 L W_0^{5/3} \ll 1. \tag{18-32}$$

This is much more severe than (18-30) and is often difficult to satisfy in many practical situations.

This range of validity of the Rytov solution is applicable to the log-amplitude fluctuation. However, there seems to be evidence that the Rytov solution for the phase fluctuation may be valid well beyond the weak fluctuation range (18-30). In fact, the Rytov solution for the phase fluctuation is considered valid even in the strong fluctuation region (Barabanenkov *et al.*, 1971).

18-6 REMOTE PROBING OF PLANETARY ATMOSPHERES

As an example, we consider the problem of remote sensing of the turbulence characteristics of a planetary atmosphere by radio occultation of a space probe (Woo and Ishimaru, 1973, 1974; Woo *et al.*, 1974; Gurvich, 1969a). Assume that a spacecraft is at a distance L_1 from the point of closest

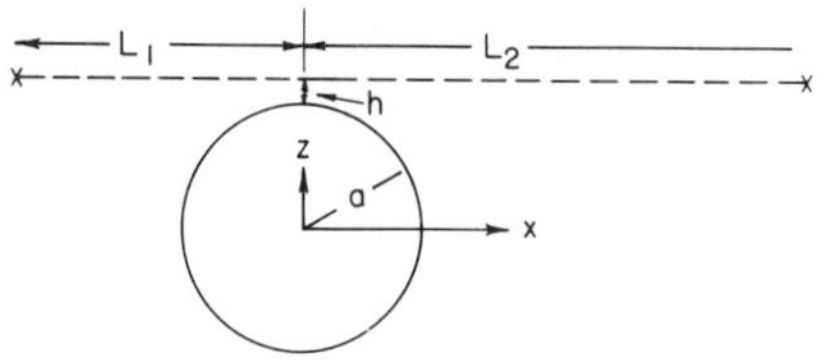

FIG. 18-3 Radio occultation technique of remote sensing the planetary atmosphere.

approach of the ray path to the planet and the earth is at a distance L_2 (see Fig. 18-3). We also assume that the structure constant $C_n{}^2(r)$ is a function of radial distance r from the center of the planet, and is approximated by an exponential profile with the scale height H:

$$C_n{}^2(r) = C_{n0}^2 \exp[-(r-a)/H] \tag{18-33}$$

where C_{n0}^2 is the value of the structure constant at $r = a$.

The ray path is given by $z = a + h$ and $y = 0$ (see Fig. 18-3). Therefore we have

$$r = [x^2 + (a+h)^2]^{1/2}. \tag{18-34}$$

Since the random medium is mostly located near the planet, we can assume that $|x| \ll a + h$ and therefore approximately we have

$$r = (a+h) + x^2/2(a+h). \tag{18-35}$$

Now we assume that $L_s = L_1 L_2/L$ is in the range of $l_0{}^2/\lambda \ll L_s \ll L_0{}^2/\lambda$, and therefore we can use (18-11a). We then have

$$\sigma_\chi{}^2 = 0.563 k^{7/6} \int_0^L d\eta C_n{}^2(\eta) \left[\frac{\eta(L-\eta)}{L}\right]^{5/6} \tag{18-36}$$

where

$$C_n^{\,2}(\eta) = C_{n0}^2 \exp\left[-\frac{h}{H} - \frac{(\eta - L_1)^2}{2(a+h)H}\right].$$

Assuming that $A^2 = 2(a+h)H \ll L_s^{\,2}$, we can use approximation

$$\int d\eta \exp\left[-\frac{(\eta - L_1)^2}{A^2}\right] f(\eta)\, d\eta \simeq (\pi)^{1/2} A f(L_1) \qquad (18\text{-}37)$$

and obtain

$$\sigma_\chi^{\,2} = 0.563 k^{7/6} [C_{n0}^2 \exp(-h/H)][2\pi(a+h)H]^{1/2}[(L_1 L_2)/(L_1 + L_2)]^{5/6}. \qquad (18\text{-}38)$$

This is valid in the range $l_0^{\,2}/\lambda \ll L_1 L_2/(L_1 + L_2) \ll L_0^{\,2}/\lambda$. If the distance $L_1 L_2/(L_1 + L_2)$ becomes greater than $L_0^{\,2}/\lambda$ or HL_0/λ, then the Fresnel size $[\lambda(L_1 L_2)/(L_1 + L_2)]^{1/2}$ becomes comparable to or greater than H and we need to take into account the effect of this inhomogeneity in the z direction. This is discussed in Section 19-8.

18-7 SOME RELATED PROBLEMS

Spherical and beam wave propagation in a random medium have been studied extensively in the past. In this section, we outline developments of these studies together with other related problems.

The beam wave theory was first studied by Kon and Tatarski (1965) for a collimated beam. Subsequently Schmeltzer (1967), Fried and Seidman (1967), Kinoshita *et al.* (1968), and Ishimaru (1969a,b) made extensive studies on a focused beam as well as on a collimated beam. Gebhardt and Collins (1969) considered the mean of the log-amplitude fluctuation using the second iteration term of the Rytov solution.

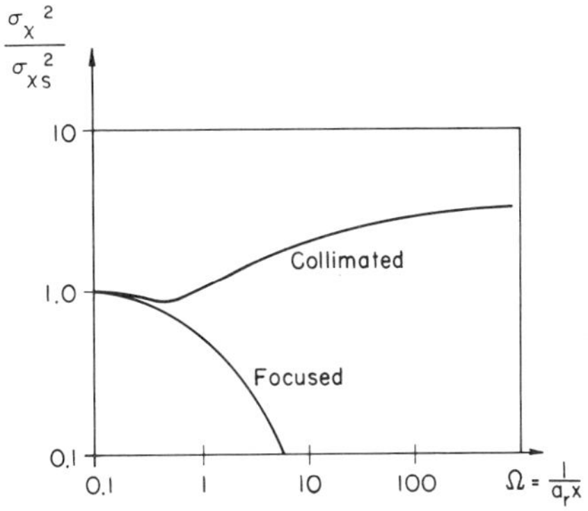

FIG. 18-4 The variance σ_χ^2 of the log-amplitude fluctuation of a beam wave normalized to the variance $\sigma_{\chi S}^2$ of a spherical wave as a function of $\Omega = 1/(\alpha_r x) = \pi W_0^2/\lambda x$.

The variation of the log-amplitude fluctuation given in (18-29) has been calculated by Fried and Seidman (see Fig. 18-4). The variance for a collimated beam approaches that of a spherical wave as the aperture size reduces to zero ($\alpha_r x \to \infty$) and approaches that of a plane wave as the aperture size is increased to infinity ($\alpha_r x \to 0$). This behavior is experimentally verified by Khmelevtsov and Tsvik (1970).

For a focused beam, the variance is shown to be extremely small when the aperture is large ($\alpha_r x \ll 1$) (see Fig. 18-4). Several experimental studies (Kerr *et al.*, 1972, 1973) have been conducted to verify this reduction of scintillation predicted by (18-29). However, the experimental variance for a focused beam appears to be closer to the case of a collimated beam than the value calculated from (18-29). Further study (Ishimaru, 1977b) indicates that although the variance on the beam axis is reduced considerably, the variance off the beam axis becomes large for a focused beam, and the condition (18-30) applied to the variance off the axis leads to the range of validity (18-31). Within this range, the average intensity of a beam wave in a random medium is approximately given by (Ishimaru, 1977b)

$$\langle I(L,\rho)\rangle = \langle |U(L,\rho)|^2\rangle = \frac{W_0{}^2}{W_b{}^2}\exp\left(-\frac{2\rho^2}{W_b{}^2}\right)$$

$$W_b{}^2 = W^2/(1-f), \qquad W^2 = W_0{}^2[(\alpha_r L)^2 + (1-\alpha_i L)^2] \qquad (18\text{-}39)$$

$$f = 1.36 C_n{}^2 k^{7/6} L^{11/6}\left[\frac{\alpha_r L}{(\alpha_r L)^2 + (1-\alpha_i L)^2}\right]^{5/6}.$$

Buck (1967) reported experimental results for a beam focused at an observation point. He showed that at a fixed distance and a given frequency, the beam size W_b decreases at first with the increase of the transmitter aperture size W_0, reaches a minimum, and then increases. He also noted that the beam size W_b increases with the distance L at a rate slightly greater than linear. These results can be explained by examining (18-39) for a focused case ($\alpha_i L = 1$). Other experimental studies on a beam wave were reported by Fried *et al.* (1967) and Furuhama *et al.* (1973). Dowling and Livingston (1973) examined various theoretical and experimental studies on beam wave. Beam wander and the cancellation of beam wander by a fast-tracking transmitter were discussed by Dunphy and Kerr (1974). Other studies on a beam wave include those by Ochs *et al.* (1969), Gurvich (1969b), Komisarov (1967), Kon (1970), and Khmelevtsov (1973). Spherical wave propagation (Kon and Feizulin, 1970; Carlson and Ishimaru, 1969) has been studied and its use in turbulence diagnostics has been reported by Carlson (1969). The finite size of a receiving aperture reduces the fluctuation in the received signal. This effect is called the aperture averaging and numerous studies have been conducted (Fried, 1967b; Titterton, 1973; Tatarski, 1971; Lutomirski and Yura, 1969).

Kon and Tatarski (1965) obtained expressions for the angle of arrival $\Delta\alpha$ in the weak fluctuation case. If a wave is incident on a constant x plane making an angle $\Delta\alpha$ with the x axis, the phase difference ΔS_1 at two points separated by ρ in the constant x plane is given by $\Delta S_1 = (k\rho)\,\Delta\alpha$. Therefore the variance of the fluctuation of the angle of arrival is given by

$$\langle(\Delta\alpha)^2\rangle = \lim_{\rho\to 0} D_S(\rho)/k^2\rho^2 \tag{18-40}$$

where $D_S(\rho) = \langle(\Delta S_1)^2\rangle$ is the phase structure function. The theoretical results given by Kon and Tatarski were verified experimentally by Borisov *et al.* (1969). In this chapter we discussed the weak fluctuation theory of spherical and beam wave fluctuations. Strong fluctuation problems are discussed in Chapter 20.

CHAPTER 19 □ TEMPORAL CORRELATION AND FREQUENCY SPECTRA OF WAVE FLUCTUATIONS IN A RANDOM MEDIUM AND THE EFFECTS OF AN INHOMOGENEOUS RANDOM MEDIUM

In Chapters 17 and 18 we assumed that the refractive index fluctuation is a random function of position only and independent of time. This is of course not the case in many practical problems. For example, the atmospheric turbulence is in constant motion and, therefore, a wave in the turbulence fluctuates with time.

A statistical description of the temporal fluctuations of the wave in a moving random medium may be given in terms of the temporal correlation functions or the temporal frequency spectra. In this chapter, we present the theory of temporal fluctuations and frequency spectra. In addition to the fluctuations at one operating frequency, we discuss correlations between fluctuations at two operating frequencies. We also discuss the correlation of two crossed beams. The effects of inhomogeneities of a random medium are also included.

19-1 TEMPORAL FREQUENCY SPECTRA OF A PLANE WAVE

We will first consider a plane wave incident upon a random medium characterized by the fluctuation $n_1(\mathbf{r}, t)$ of the refractive index. We assume that the medium is moving with a wind velocity $\mathbf{V}$ and that the velocity $\mathbf{V}$ is expressed as the sum of the average wind velocity $\mathbf{U}$ and the velocity fluctuation $\mathbf{V}_f$:

$$\mathbf{V} = \mathbf{U} + \mathbf{V}_f. \tag{19-1}$$

We assume that the wind velocity is a slowly varying function of time. For example, in turbulence, we assume that the eddies of a certain size l do not appreciably change their shape within the time required for the eddies to move the distance l. This means that the medium is considered "frozen" and is transported by the wind without changing its detailed variation. This is called Taylor's frozen-in hypothesis and is considered valid in most atmospheric turbulence. We express this "frozen-in" condition by

$$n_1(\mathbf{r}, t) = n_1(\mathbf{r} - \mathbf{V}t, 0). \tag{19-2}$$

The fluctuations of a plane wave are given by the same expression as the time-independent case except for the introduction of (19-2):

$$U(\mathbf{r}, t) = U_0(\mathbf{r}) \exp[\psi_1(\mathbf{r}, t)], \qquad \psi_1(\mathbf{r}, t) = \int dV' h(\mathbf{r} - \mathbf{r}') n_1(\mathbf{r}', t). \tag{19-3}$$

Using (19-2), and writing $\mathbf{r}' - \mathbf{V}t = \mathbf{r}''$, we obtain

$$\psi_1(\mathbf{r}, t) = \int dV' h(\mathbf{r} - \mathbf{V}t - \mathbf{r}'') n_1(\mathbf{r}'', 0). \tag{19-4}$$

Comparing (19-4) with the previous time-independence case,

$$\psi_1(\mathbf{r}) = \int dV' h(\mathbf{r} - \mathbf{r}') n_1(\mathbf{r}'), \tag{19-5}$$

we note that $\mathbf{r}$ in the integrand in (19-5) is simply replaced by $\mathbf{r} - \mathbf{V}t$. We will make use of this correspondence in the following sections.

19-2 WHEN THE AVERAGE WIND VELOCITY U IS TRANSVERSE AND THE WIND FLUCTUATION $\mathbf{V}_f$ IS NEGLIGIBLE

Let us first examine the simplest case when the fluctuating component $\mathbf{V}_f$ is negligible compared with the average velocity $\mathbf{U}_t$, and $\mathbf{U}_t$ is entirely transverse to the direction of wave propagation. Equation (19-4) becomes in this case

$$\psi_1(\mathbf{r}, t) = \int dV' h(\mathbf{r} - \mathbf{U}_t t - \mathbf{r}'') n_1(\mathbf{r}'', 0), \qquad \mathbf{U}_t = U_2 \hat{\mathbf{y}} + U_3 \hat{\mathbf{z}}. \tag{19-6}$$

The integration with respect to $\mathbf{r}'$ can be replaced by integration with respect to $\mathbf{r}''$ because the ranges of integration for x', y', z' are identical to those for x'', y'', z'':

$$\psi_1(\mathbf{r}, t) = \int dV'' h(\mathbf{r} - \mathbf{U}_t t - \mathbf{r}'') n_1(\mathbf{r}'', 0). \tag{19-7}$$

Comparing (19-7) with the time-independent case (19-5) we note that $\psi_1(\mathbf{r}, t)$ is identical to ψ_1 in (19-5) if $\mathbf{r}$ is replaced by $\mathbf{r} - \mathbf{U}_t t$:

$$\psi_1(\mathbf{r}, t) = \psi_1(\mathbf{r} - \mathbf{U}_t t). \tag{19-8}$$

Therefore, the fluctuation in this case can be obtained by the time-independent case simply by replacing $\mathbf{r}$ by $\mathbf{r} - \mathbf{U}_t t$. The physical significance

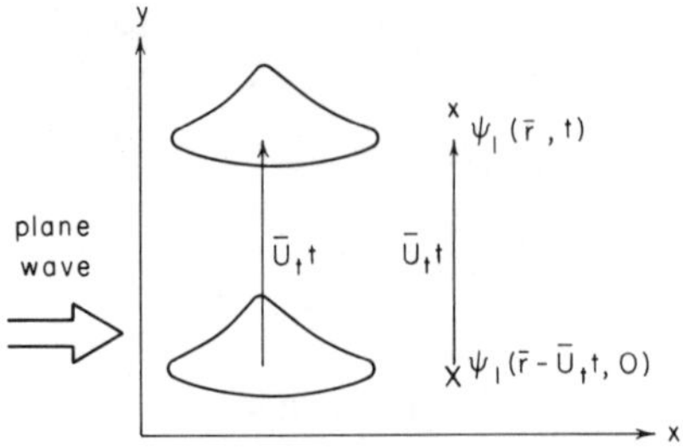

FIG. 19-1 Parallel shift of wave fluctuation in space and time.

of (19-8) is that under the frozen-in hypothesis, the medium moves transversely with the velocity $\mathbf{U}_t$, and the fluctuation at $\mathbf{r}$ and t is identical to the fluctuation at $\mathbf{r} - \mathbf{U}_t t$ and $t = 0$ (see Fig. 19-1).

We must caution here that this parallel shift of fluctuation with the wind velocity is valid only for a plane wave case. For spherical and beam waves, the amount of shift depends on the distance from the source.

We now consider the correlation function B_χ of the log-amplitude fluctuation χ at two different points in a plane $x = L$ and at two different times:

$$B_\chi(\mathbf{r}_1, t_1; \mathbf{r}_2, t_2) = \langle \chi(\mathbf{r}_1, t_1)\chi(\mathbf{r}_2, t_2)\rangle, \tag{19-9}$$

where $\mathbf{r}_1 = L\hat{\mathbf{x}} + y_1\hat{\mathbf{y}} + z_1\hat{\mathbf{z}}$ and $\mathbf{r}_2 = L\hat{\mathbf{x}} + y_2\hat{\mathbf{y}} + z_2\hat{\mathbf{z}}$. Employing (19-8), the correlation function for the time-dependent case is obtained from that of the time-independent case by replacing y_d and z_d by $y_d - U_2\tau$ and $z_d - U_3\tau$, where $y_d = y_1 - y_2$, $z_d = z_1 - z_2$, and $\tau = t_1 - t_2$:

$$\begin{aligned} B_\chi &= (\mathbf{r}_1, t_1; \mathbf{r}_2, t_2) = B_\chi(L, y_d - U_2\tau, z_d - U_3\tau) \\ &= 2\pi^2 k^2 L \int_0^\infty \kappa\, d\kappa J_0(\kappa\rho) f_\chi(\kappa)\Phi_n(\kappa) \end{aligned} \tag{19-10}$$

where $f_\chi(\kappa) = 1 - \sin(\kappa^2 L/k)/(\kappa^2 L/k)$ and

$$\rho = [(y_d - U_2\tau)^2 + (z_d - U_3\tau)^2]^{1/2}.$$

The temporal correlation function $B_\chi(L, \tau)$ at one location $y_d = 0$ and $z_d = 0$ is therefore given by

$$B_\chi(L, \tau) = 2\pi^2 k^2 L \int_0^\infty \kappa\, d\kappa\, J_0(\kappa U_t \tau) f_\chi(\kappa)\Phi_n(\kappa) \tag{19-11}$$

where $(U_2{}^2 + U_3)^{1/2}$ is the transverse average wind velocity. We note here that the temporal correlation function $B_\chi(L, \tau)$ at a given point $(L, 0, 0)$ is identical to the spatial correlation function at two points separated by $\boldsymbol{\rho}$ at the same time $B_\chi(L, \boldsymbol{\rho})$ provided $\mathbf{U}_t\tau$ is replaced by $\boldsymbol{\rho}$. This correspondence between time and space ($\mathbf{U}_t \tau \sim \boldsymbol{\rho}$) is, of course, the direct consequence of the frozen-in hypothesis.

The temporal frequency spectrum $W_\chi(\omega)$ is the Fourier transform of the correlation function (19-11):

$$W_\chi(\omega) = 2\int_{-\infty}^{\infty} B_\chi(L, \tau)e^{-i\omega\tau}\, d\tau = 4\int_0^{\infty} B_\chi(L, \tau)\cos(\omega\tau)\, d\tau. \tag{19-12}$$

We note that the integral of the temporal frequency spectrum $W_\chi(\omega)$ is proportional to the variance:

$$\sigma_\chi{}^2 = B_\chi(0) = \frac{1}{4\pi}\int_{-\infty}^{\infty} W_\chi(\omega)\, d\omega. \tag{19-13}$$

Equation (19-12) can be further simplified by using the following definite integral:

$$\int_0^{\infty} \cos(\omega\tau)J_0(\kappa U_t\tau)\, d\tau = \begin{cases} [(\kappa U_t)^2 - \omega^2]^{1/2} & \text{for} \quad \kappa U_t > \omega \\ 0 & \text{for} \quad \kappa U_t < \omega. \end{cases} \tag{19-14}$$

Then, we obtain

$$\begin{aligned} W_\chi(\omega) &= 8\pi^2 k^2 L\int_{\omega/U_t}^{\infty} \kappa\, d\kappa\, f_\chi(\kappa)\Phi_n(\kappa)[(\kappa U_t)^2 - \omega^2]^{1/2} \\ &= \frac{8\pi^2 k^2 L}{U_t}\int_0^{\infty} f_\chi(\kappa)\Phi_n(\kappa)\, d\kappa', \qquad \kappa = \left(\kappa'^2 + \frac{\omega^2}{U_t}\right)^{1/2}. \end{aligned} \tag{19-15}$$

Similarly, we obtain the phase frequency spectrum by replacing $f_\chi(\kappa)$ in (19-15) by $f_S(\kappa)$.

We note from (19-15) that the frequency spectrum $W_\chi(\omega)$ is related to the portion of the spectrum of the index of refraction which is greater than $\kappa = \omega/U_t$. This means that the spectrum $W_\chi(\omega)$ reveals the characteristics of the refractive index spectrum for κ greater than ω/U_t and thus it is possible to obtain information about the spectral shape of $\Phi_n(\kappa)$ from the temporal spectra.

Let us calculate the amplitude spectrum $W_\chi(\omega)$ using the Kolmogorov spectrum $\Phi_n(\kappa) = 0.033C_n{}^2\kappa^{-11/3}$. Equation (19-15) can be expressed using the confluent hypergeometric function $\psi(a, c; z)$ (Ishimaru, 1972). We have

$$\begin{aligned} W_\chi(\omega) &= \frac{8\pi^2}{U_t}(0.033C_n{}^2)k^{2/3}L^{7/3}\frac{\sqrt{\pi}}{2}\left(\frac{\omega}{\omega_t}\right)^{-8/3}[A - B] \\ A &= \frac{\Gamma(\frac{4}{3})}{\Gamma(\frac{11}{6})}, \qquad B = \mathrm{Im}\left\{\left(\frac{\omega_t}{\omega}\right)^2 \exp\left(-i\frac{\omega^2}{\omega_t{}^2}\right)\psi\left[\tfrac{1}{2}, \tfrac{4}{3}; -i\left(\frac{\omega}{\omega_t}\right)^2\right]\right\} \end{aligned} \tag{19-16}$$

where $\omega_t = U_t[k/L]^{1/2}$.

The frequency ω_t is important. When $\omega < \omega_t$, $W_\chi(\omega)$ is substantially constant, but when $\omega > \omega_t$, $W_\chi(\omega)$ decays as $\omega^{-8/3}$. These two asymptotic values are given by

$$W_\chi{}^0(\omega) \to 0.85 \frac{C_n{}^2 k^{-2/3} L^{7/3}}{U_t}$$

$$= 2.765 \frac{\sigma_\chi{}^2}{\omega_t} \qquad (\text{as } \omega \to 0) \tag{19-17}$$

$$W_\chi{}^\infty(\omega) \to 2.19 \frac{C_n{}^2 k^{2/3} L^{7/3}}{U_t} \left(\frac{\omega}{\omega_t}\right)^{-8/3}$$

$$= 7.13 \frac{\sigma_\chi{}^2}{\omega_t} \left(\frac{\omega}{\omega_t}\right)^{-8/3} \qquad (\text{as } \omega \to \infty) \tag{19-18}$$

where $\sigma_\chi{}^2 = 0.307 C_n{}^2 k^{7/6} L^{11/6}$ is the variance of the log-amplitude fluctuation. Note that these two asymptotes meet at the frequency $\omega = 1.43\omega_t$. The shape of the spectrum $W_\chi(\omega)$ is shown in Fig. 19-2.

The phase spectrum $W_S(\omega)$ is given by the same equation (19-16) except that the minus sign before the second term is replaced by the plus sign. The asymptotic form of the phase spectrum is given by

$$W_S(\omega) \to 2W_\chi{}^\infty(\omega) \qquad (\text{as } \omega \to 0) \tag{19-19}$$

$$W_S(\omega) \to W_\chi{}^\infty(\omega) \qquad (\text{as } \omega \to \infty). \tag{19-20}$$

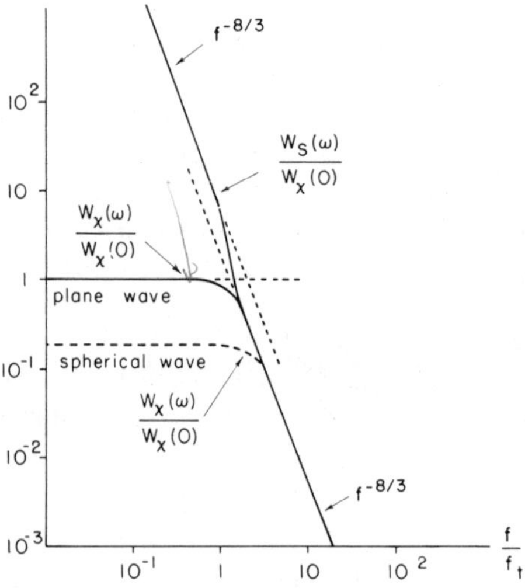

FIG. 19-2 Temporal frequency spectra of the log-amplitude fluctuation $W_\chi(\omega)$ and the phase fluctuation $W_S(\omega)$ normalized to $W_\chi(0)$. Both plane and spherical waves are shown.

These two asymptotes with $f^{-8/3}$ should be valid as long as ω/U_t in (19-15) is between $2\pi/L_0$ and $2\pi/l_0$. In terms of f/f_t, this requires

$$\sqrt{\lambda L}/L_0 < f/f_t < \sqrt{\lambda L}/l_0 \tag{19-21}$$

and, therefore, the departure of the actual spectrum from the asymptotes indicates the effects of the outer and inner scales of turbulence.

19-3 TEMPORAL SPECTRA DUE TO AVERAGE AND FLUCTUATING WIND VELOCITIES

We now take into account the average and the fluctuation of wind velocity. In this case, it is necessary to take into account the average over the velocity fluctuation $\mathbf{V}_f$. The spectral density $\Phi_n(\boldsymbol{\kappa})$ for a time-invariant case must then be replaced by the time-varying spectral density $\Phi_n(\boldsymbol{\kappa}, \tau)$. This is discussed in Section A-7 of Appendix A. We have

$$\Phi_n(\boldsymbol{\kappa}, \tau) = \Phi_n(\boldsymbol{\kappa})\chi(-\boldsymbol{\kappa}\tau) \tag{19-22}$$

where $\chi(-\boldsymbol{\kappa}\tau)$ is the characteristic function of the velocity fluctuation $\mathbf{V}_f$ and is given by

$$\chi(-\boldsymbol{\kappa}\tau) = \int \exp(-i\boldsymbol{\kappa} \cdot \mathbf{V}_f \tau) p(\mathbf{V}_f)\, d\mathbf{V}_f \tag{19-23}$$

and $p(\mathbf{V}_f)$ is the probability distribution function of $\mathbf{V}_f$.

The temporal correlation function $B_\chi(L, \tau)$ is therefore given by

$$B_\chi(L, \tau) = 2\pi^2 k^2 L \int_0^\infty \kappa\, d\kappa\, J_0(\kappa U_t \tau) f_\chi(\kappa) \Phi_n(\kappa, \tau) \tag{19-24}$$

where U_t is the magnitude of the transverse wind velocity, and the longitudinal component of the average velocity $\mathbf{U}$ has no effect on $B_\chi(L, \tau)$.

The temporal spectrum $W_\chi(\omega)$ when $p(\mathbf{V}_f)$ is Gaussian with the variance σ_v^2 is given by

$$W_\chi(\omega) = 8\pi^2 k^2 L \int_0^\infty \kappa\, d\kappa\, f_\chi(\kappa) \Phi_n(\kappa) I(\kappa, \omega) \tag{19-25}$$

$$I(\kappa, \omega) = \tfrac{1}{2} \int_{-\infty}^{\infty} J_0(\kappa U_t \tau) \exp[-(\kappa^2 \sigma_v^2 \tau^2/2) - i\omega\tau]\, d\tau. \tag{19-26}$$

In general, (19-25) cannot be expressed in a closed form. However, we can obtain the asymptotic form as $\omega \to 0$. As $\omega \to 0$, (19-26) can be evaluated:

$$I(\kappa, 0) = \frac{1}{2} \frac{(2\pi)^{1/2}}{\kappa \sigma_v} \exp\left(-\frac{U_t^2}{4\sigma_v^2}\right) I_0\left(\frac{U_t^2}{4\sigma_v^2}\right) \tag{19-27}$$

Substituting this into (19-25) and using the formula

$$\int_0^\infty \left[1 - \frac{\sin(A\kappa^2)}{A\kappa^2}\right]\kappa^{n+1}\, d\kappa = -\frac{\Gamma(n/2)\sin(n\pi/4)}{2A^{n/2+1}}, \tag{19-28}$$

valid for $-2 > n > -6$, we obtain

$$W_\chi(0) = 1.067 \frac{C_n{}^2 k^{2/3} L^{7/3}}{\sigma_v} \exp\left(-\frac{U_t{}^2}{4\sigma_v{}^2}\right) I_0\left(\frac{U_t{}^2}{4\sigma_v{}^2}\right)$$

$$= 4.913 \frac{\sigma_\chi{}^2}{\omega_f} \exp\left(-\frac{U_t{}^2}{4\sigma_v{}^2}\right) I_0\left(\frac{U_t{}^2}{4\sigma_v{}^2}\right) \tag{19-29}$$

where $\omega_f = \sqrt{2}\,\sigma_v\sqrt{k/L}$. Note that as $\sigma_v \to 0$, (19-29) reduces to $W_\chi{}^0(\omega)$ in (19-17). The asymptotic form of (19-25) as $\omega \to \infty$ can be obtained by noting that as $\omega \to \infty$, the major contribution to W_χ in (19-26) comes from the neighborhood of $\tau = 0$ and therefore we can approximate $J_0(\kappa U_t \tau)$ by $\exp[-(\kappa U_t \tau/2)^2]$. Thus we get

$$I(\kappa, \omega) \approx \frac{2\sqrt{\pi}}{\kappa[U_t{}^2 + 2\sigma_v{}^2]^{1/2}} \exp\left[-\frac{\omega^2}{\kappa^2(U_t{}^2 + 2\sigma_v{}^2)}\right]. \tag{19-30}$$

We note that (19-30) peaks in the neighborhood of $\kappa^2(U_t{}^2 + 2\sigma_v{}^2) \approx 2\omega^2$ and therefore as $\omega \to \infty$, we need only consider κ large. Thus we can approximate $f_\chi(\kappa) \approx 1$. With these considerations, we have

$$W_\chi{}^\infty(\omega) = 6.723(\sigma_\chi{}^2/\omega_{tf})(\omega/\omega_{tf})^{-8/3} \tag{19-31}$$

where

$$\omega_{tf} = (U_t{}^2 + 2\sigma_v{}^2)^{1/2}(k/L)^{1/2}. \tag{19-32}$$

Because of an approximation involved in (19-30), (19-31) does not exactly reduce to (19-18) as $\sigma_v \to 0$. It is expected that if the exact expression is used in place of (19-30), 7.13 instead of 6.723 should appear in (19-31).

19-4 TEMPORAL FREQUENCY SPECTRA OF A SPHERICAL WAVE

The general expression (19-15) for the temporal frequency spectrum for the log-amplitude fluctuation is applicable to the spherical wave case if the following filter function $f_\chi(\kappa)$ for the spherical wave is used:

$$f_\chi(\kappa) = \frac{1}{L}\int_0^L d\eta \left[1 - \cos\left(\frac{\eta}{L}\frac{(L-\eta)}{k}\kappa^2\right)\right]. \tag{19-33}$$

It is not too difficult to obtain the asymptotic form of $W_\chi(\omega)$ as $\omega \to 0$ and $\omega \to \infty$. First, let us consider $\omega \to 0$. In this case, we get

$$W_\chi(0) = \frac{2\pi k^2 L}{U_t} \int_0^\infty f_\chi(\kappa)\Phi_n(\kappa)\, d\kappa. \tag{19-34}$$

We first integrate (19-34) with respect to κ. For the integration with respect to η, we note that the difference between the spherical and plane waves is that the integrand contains $(\eta/L)(L-\eta)$ for the spherical wave whereas it contains $L-\eta$ for the plane wave. Therefore the difference is that of the following two integrals:

$$I_{sp} = \int_0^L \left[\frac{\eta}{L}(L-\eta)\right]^{(n-1)/2} d\eta$$

$$= L^{(n-1)/2} \frac{\{\Gamma[(n+1)/2]\}^2}{\Gamma(n+1)} \qquad \text{for spherical waves,} \tag{19-35a}$$

$$I_{pl} = \int_0^L (L-\eta)^{(n-1)/2}\, d\eta = L^{(n-1)/2}\left(\frac{2}{n+1}\right) \qquad \text{for plane waves} \tag{19-35b}$$

where $n = \frac{11}{3}$ for the Kolmogorov spectrum.

The spectrum $W_{\chi sp}(0)$ for the spherical wave is then obtained by multiplying the plane wave spectrum $W_{\chi pl}(0)$ by the ratio of (19-35a) and (19-35b):

$$W_{\chi sp}(0) = W_{\chi pl}(0)\left\{\frac{n+1}{2}\frac{[\Gamma(n+1)/2]^2}{\Gamma(n+1)}\right\}, \tag{19-36}$$

which becomes, for the Kolmogorov spectrum, $n = \frac{11}{3}$,

$$W_{\chi sp}(0) = 0.622(\sigma_{\chi pl}^2/\omega_t) = 1.540(\sigma_{\chi sp}^2/\omega_t) \tag{19-37}$$

where $\sigma_{\chi pl}^2$ and $\sigma_{\chi sp}^2$ are the variances for plane and spherical waves, respectively, and are given by

$$\sigma_{\chi pl}^2 = 0.307 C_n{}^2 k^{7/6} L^{11/6}, \qquad \sigma_{\chi sp}^2 = 0.124 C_n{}^2 k^{7/6} L^{11/6}. \tag{19-38}$$

For $\omega \to \infty$, we note that since $(K'^2 + \omega^2/U_t{}^2)^{1/2} \to \infty, f_\chi$ can be approximated by 1, and thus the spectrum $W_\chi(\omega)$ as $\omega \to \infty$ becomes identical to that of the plane wave:

$$W_{\chi sp}^\infty(\omega) = W_{\chi pl}^\infty(\omega) = 7.13 \frac{\sigma_{\chi pl}^2}{\omega_t}\left(\frac{\omega}{\omega_t}\right)^{-8/3} = 17.65 \frac{\sigma_{\chi sp}^2}{\omega_t}\left(\frac{\omega}{\omega_t}\right)^{-8/3}. \tag{19-39}$$

The phase spectrum $W_S(\omega)$ for a spherical wave is obtained by noting that $f_S(\kappa) \to 2$ as $\kappa \to 0$ and $f_S(\kappa) \to 1$ as $\kappa \to \infty$. From this, we conclude that

$$W_{S\text{sp}}(\omega) \to 2W_{\chi\text{sp}}^{\infty}(\omega) = 35.31(\sigma_{\chi\text{sp}}^2/\omega_t)(\omega/\omega_t)^{-8/3} \qquad (\text{as } \omega \to 0) \qquad (19\text{-}40)$$

$$W_{S\text{sp}}(\omega) \to W_{\chi\text{sp}}(\omega) = 17.65(\sigma_{\chi\text{sp}}^2/\omega_t)(\omega/\omega_t)^{-8/3} \qquad (\text{as } \omega \to \infty). \qquad (19\text{-}41)$$

These spectra for the spherical waves are shown in Fig. 19-2.

19-5 TWO-FREQUENCY CORRELATION FUNCTION

Up to this point, we have discussed the temporal correlations and frequency spectra of a monochromatic wave propagated through a moving random medium. In this section, we consider the cross correlation and cross spectra between waves at two different operating frequencies. This is important for several applications. For example, in 1974, the United States sent the Mariner 10 spacecraft to Venus. As it flew by Venus, it sent X and S band signals through the Venus atmosphere (Woo *et al.*, 1974) back to the earth. The fluctuation characteristics of these signals revealed important information on the small-scale turbulence in the Venusian atmosphere contributing to our understanding of the atmosphere's circulation and dynamics (Gurvich, 1969a; Golitsyn, 1970).

Let us consider two waves at two operating frequencies f_1 and f_2, respectively ($k_1 = 2\pi f_1/c$ and $k_2 = 2\pi f_2/c$). We then write the fields with the operating frequency f_1 at $\mathbf{r}_1$ and t_1 and with f_2 at $\mathbf{r}_2$ and t_2, respectively [see (19-3)]:

$$U(k_1, \mathbf{r}_1, t_1) = U_0(k_1, \mathbf{r}_1) \exp[\psi_1(k_1, \mathbf{r}_1, t_1)] \qquad (19\text{-}42)$$

$$U(k_2, \mathbf{r}_2, t_2) = U_0(k_2, \mathbf{r}_2) \exp[\psi_1(k_2, \mathbf{r}_2, t_2)]. \qquad (19\text{-}43)$$

The log-amplitude fluctuation χ and the phase fluctuation S_1 are given by

$$\psi_1(k_1, \mathbf{r}_1, t_1) = \chi(k_1, \mathbf{r}_1, t_1) + iS_1(k_1, \mathbf{r}_1, t_1) \qquad (19\text{-}44)$$

$$\psi_1(k_2, \mathbf{r}_2, t_2) = \chi(k_2, \mathbf{r}_2, t_2) + iS_1(k_2, \mathbf{r}_2, t_2). \qquad (19\text{-}45)$$

We now consider the amplitude correlation function B_χ, the phase correlation function B_S, and the cross correlation $B_{\chi S}$ between the amplitude and phase fluctuations. They are given by

$$B_\chi(k_1, \mathbf{r}_1, t_1; k_2, \mathbf{r}_2, t_2) = \langle \chi(k_1, \mathbf{r}_1, t_1)\chi(k_2, \mathbf{r}_2, t_2)\rangle \qquad (19\text{-}46\text{a})$$

$$B_S(k_1, \mathbf{r}_1, t_1; k_2, \mathbf{r}_2, t_2) = \langle S_1(k_1, \mathbf{r}_1, t_1)S_1(k_2, \mathbf{r}_2, t_2)\rangle \qquad (19\text{-}46\text{b})$$

$$B_{\chi S}(k_1, \mathbf{r}_1, t_1; k_2, \mathbf{r}_2, t_2) = \langle \chi(k_1, \mathbf{r}_1, t_1)S_1(k_2, \mathbf{r}_2, t_2)\rangle. \qquad (19\text{-}46\text{c})$$

At a given location $\mathbf{r} = \mathbf{r}_1 = \mathbf{r}_2$, these equations represent the temporal correlation functions. In most practical problems, the fluctuation characteristics can be assumed to be stationary and therefore the correlation functions depend only on the time difference $\tau = t_1 - t_2$. Under these assumptions (19-46a)–(19-46c) become

$$B_\chi(k_1, k_2, \mathbf{r}, \tau), \quad B_S(k_1, k_2, \mathbf{r}, \tau), \quad \text{and} \quad B_{\chi S}(k_1, k_2, \mathbf{r}, \tau),$$

respectively.

The temporal Fourier transform of these correlation functions is the temporal frequency spectra:

$$W_\chi(k_1, k_2, \mathbf{r}, \omega) = 2\int_{-\infty}^{\infty} B_\chi(k_1, k_2, \mathbf{r}, \tau)e^{-i\omega\tau}\, d\tau \tag{19-47a}$$

$$W_S(k_1, k_2, \mathbf{r}, \omega) = 2\int_{-\infty}^{\infty} B_S(k_1, k_2, \mathbf{r}, \tau)e^{-i\omega\tau}\, d\tau \tag{19-47b}$$

$$W_{\chi S}(k_1, k_2, \mathbf{r}, \omega) = 2\int_{-\infty}^{\infty} B_{\chi S}(k_1, k_2, \mathbf{r}, \tau)e^{-i\omega\tau}\, d\tau. \tag{19-47c}$$

The inverse transform is given by

$$B_\chi(k_1, k_2, \mathbf{r}, \tau) = \frac{1}{4\pi}\int_{-\infty}^{\infty} W_\chi(k_1, k_2, \mathbf{r}, \omega)e^{i\omega\tau}\, d\omega. \tag{19-48a}$$

We can write similar expressions for B_S and $B_{\chi S}$. Note that if we let $\tau = 0$ in (19-47), we obtain

$$B_\chi(k_1, k_2, \mathbf{r}, 0) = \frac{1}{4\pi}\int_{-\infty}^{\infty} W_\chi(k_1, k_2, \mathbf{r}, \omega)\, d\omega, \tag{19-48b}$$

indicating that the integral of the spectrum over all frequencies gives the correlation function at two different operating frequencies k_1 and k_2 at the same time.

General expressions for the correlation functions B and the spectra W have been obtained (Ishimaru, 1972). Here, we give the following expressions:

$$B_\chi(k_1, k_2, \mathbf{r}, \tau) = 2\pi^2 k_1 k_2 \int_0^\infty \kappa\, d\kappa \int_0^L d\eta\, g_\chi(\kappa, \eta) J_0(\kappa V\tau)\Phi_n(\kappa) \tag{19-49a}$$

$$B_S(k_1, k_2, \mathbf{r}, \tau) = 2\pi^2 k_1 k_2 \int_0^\infty \kappa\, d\kappa \int_0^\infty d\eta\, g_S(\kappa, \eta) J_0(\kappa V\tau)\Phi_n(\kappa) \tag{19-49b}$$

$$B_{\chi S}(k_1, k_2, \mathbf{r}, \tau) = 2\pi^2 k_1 k_2 \int_0^\infty \kappa\, d\kappa \int_0^\infty d\eta\, g_{\chi S}(\kappa, \eta) J_0(\kappa V\tau)\Phi_n(\kappa) \tag{19-49c}$$

where

$$g_\chi(\kappa, \eta) = \mathrm{Re}[h_1 h_2^* - h_1 h_2], \qquad g_S(\kappa, \eta) = \mathrm{Re}[h_1 h_2^* + h_1 h_2]$$

$$g_{\chi S}(\kappa, \eta) = -\mathrm{Im}[h_1 h_2^* + h_1 h_2]$$

$$h_1 = \exp\left[-i\frac{\gamma(L-\eta)}{2k_1}\kappa^2\right], \qquad h_2 = \exp\left[-i\frac{\gamma(L-\eta)}{2k_2}\kappa^2\right].$$

In these, $\gamma = 1$ for the plane wave case and $\gamma = \eta/L$ for the spherical wave case, $\mathbf{r}$ is at $x = L$ and $y = z = 0$, and V is the magnitude of the transverse wind velocity.

The temporal frequency spectra are obtained by using (19-47):

$$W_\chi(k_1, k_2, \mathbf{r}, \omega) = 8\pi^2 k_1 k_2 \int_{\omega/V}^{\infty} \kappa \, d\kappa \int_0^L d\eta \; g_\chi(\kappa, \eta)\Phi_n(\kappa)[(kV)^2 - \omega^2]^{-1/2}. \tag{19-50}$$

If the wind velocity V and the spectrum Φ_n are functions of position η, then (19-50) is still valid if V is modified to $V(\eta)$ and $\Phi_n(\kappa)$ is modified to $\Phi_n(\kappa, \eta)$. The spectra W_S and $W_{\chi S}$ are obtained by (19-50) using g_S and $g_{\chi S}$ in the integrand. General formulations for beam waves are given by Ishimaru (1972).

In general, the cross correlation $B_\chi(k_1, k_2, \mathbf{r}, \tau)$ may not be an even function of τ and therefore the cross spectrum $W_\chi(k_1, k_2, \mathbf{r}, \omega)$ may not be real. Therefore we write

$$W_\chi(k_1, k_2, \mathbf{r}, \tau) = C_\chi(k_1, k_2, \mathbf{r}, \omega) - iQ_\chi(k_1, k_2, \mathbf{r}, \omega) \tag{19-51}$$

where C_χ and Q_χ are called the cospectrum and quadrature spectrum, respectively. The coherence is then defined by

$$\mathrm{coh}_\chi(k_1, k_2, \omega) = \frac{C_\chi^2(k_1, k_2, \omega) + Q_\chi^2(k_1, k_2, \omega)}{W_\chi(k_1, k_1, \omega)W_\chi(k_2, k_2, \omega)}. \tag{19-52}$$

Let us consider a plane wave case. We have

$$g_\chi(\kappa, \eta) = \cos\left[\frac{(L-\eta)\kappa^2}{2}\left(\frac{1}{k_1} - \frac{1}{k_2}\right)\right] - \cos\left[\frac{(L-\eta)\kappa^2}{2}\left(\frac{1}{k_1} + \frac{1}{k_2}\right)\right]. \tag{19-53}$$

Substituting this into (19-50), we get the cross spectrum $W_\chi(k_1, k_2, \omega)$. We can obtain the asymptotic form of coh_χ in (19-52) when the spectral density $\Phi_n(\kappa)$ is given by the Kolmogorov spectrum:

$$\Phi_n(\kappa) = 0.033 C_n^2 \kappa^{-11/3}. \tag{19-54}$$

Using (19-53), (19-54), and (19-50) in (19-52), we obtain

$$\mathrm{coh}_\chi(k_1, k_2, 0) = \left(\frac{k_2}{k_1}\right)^{(n-1)/2}\left[\left(\frac{1 + k_1/k_2}{2}\right)^{(n-1)/2} - \left(\frac{1 - k_1/k_2}{2}\right)^{(n-1)/2}\right]^2 \tag{19-55}$$

where $n = \frac{11}{3}$.

The phase coherence $\mathrm{coh}_S(k_1, k_2, \omega)$ can be calculated using (19-50) and (19-52) and replacing g_χ by g_S. In the limit as $\omega \to 0$, we get $\mathrm{coh}_S(k_1, k_2, 0) \to 1$. Figure 19-3 shows a general shape of the coherence.

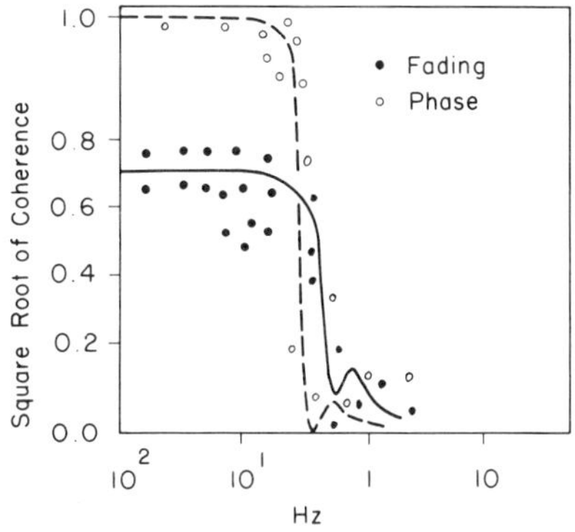

FIG. 19-3

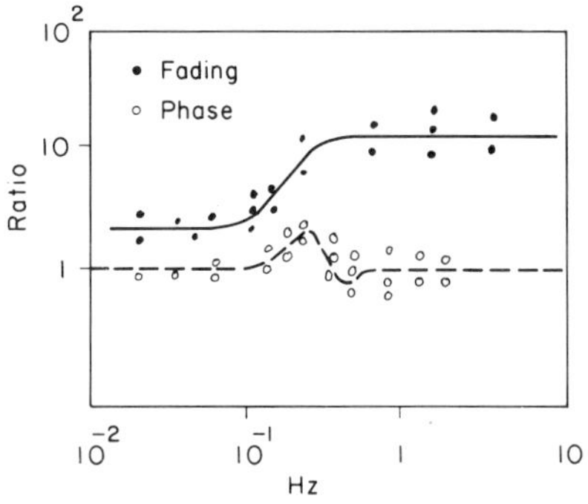

FIG. 19-4

FIG. 19-3 Theoretical and experimental curves of the log-amplitude (fading) coherence $\mathrm{coh}_\chi(k_1, k_2, \omega)$ and the phase coherence $\mathrm{coh}_S(k_1, k_2, \omega)$. $k_1 = 2\pi f_1/c$ and $k_2 = 2\pi f_2/c$, and $f_1 = 34.52$ GHz and $f_2 = 9.6$ GHz.

FIG. 19-4 Theoretical and experimental curves of the ratio of the log-amplitude (fading) specta $W_\chi(k_1, k_1, \omega)/W_\chi(k_2, k_2, \omega)$ and the ratio of the phase spectra $w_S(k_1, k_1, \omega)/w_S(k_2, k_2, \omega)$ where $w_S(k, k, \omega) = W_S(k, k, \omega)/(kL)^2$.

This is compared with the experimental data obtained in Hawaii using 9.6 and 34.52 GHz and the path length of 64.25 km, showing excellent agreement (Janes *et al.*, 1970; Janes and Thompson, 1973; Ishimaru, 1972). In Fig. 19-4, we also show the ratio of the log-amplitude spectra $W_\chi(k, k, \omega)$ and the phase spectra $W_S(k, k, \omega)/(k^2L^2)$ at 34.52 and 9.6 GHz, with the experimental data. Other studies on frequency spectra were reported by Mandics *et al.* (1973), Mandics and Lee (1969), and Clifford (1971).

19-6 CROSSED BEAMS

Consider two transmitters located at $\boldsymbol{\rho}_{t1}$ and $\boldsymbol{\rho}_{t2}$ at $x = 0$ (see Fig. 19-5). They are operating at two different frequencies (k_1 and k_2). At $x = L$, there

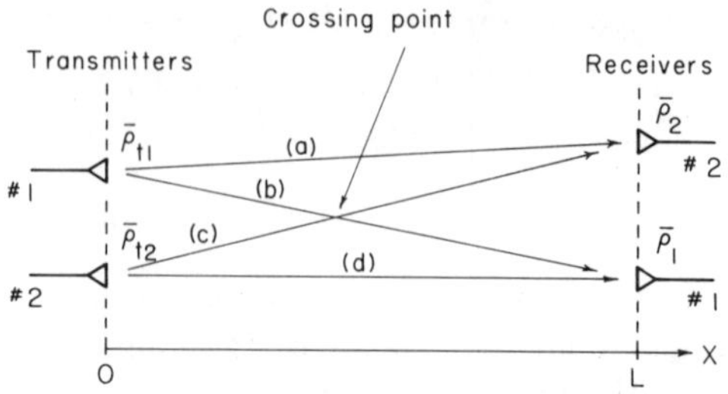

FIG. 19-5 Two beams crossing at a midpoint.

are two detectors at $\boldsymbol{\rho}_1$ and $\boldsymbol{\rho}_2$ which receive the signal at k_1 and k_2, respectively. Since they respond only to k_1 and k_2, the correlation between the fluctuations at $\boldsymbol{\rho}_1$ and $\boldsymbol{\rho}_2$ is determined only by the crossed beams (b) and (c) in Fig. 19-5. If $k_1 = k_2$, however, the detector at $\boldsymbol{\rho}_1$ receives waves through the path (b) and (d). Similarly, the wave at $\boldsymbol{\rho}_2$ consists of waves through (a) and (c). Therefore, the correlation between the fields at $\boldsymbol{\rho}_1$ and $\boldsymbol{\rho}_2$ consists of the correlations between (a) and (b), between (a) and (d), between (b) and (c), and between (c) and (d). However, if the distance $|\boldsymbol{\rho}_1 - \boldsymbol{\rho}_2|$ is much greater than the correlation distance (approximately $\sqrt{\lambda L}$), then the correlations between different paths are negligible except the crossed path (b) and (c). The crossed path provides a relatively strong correlation because of the common random medium at the crossing point. This correlation between the crossed beams has been used to probe the characteristics of the turbulence at the crossing point (Wang *et al.*, 1974; Fisher and Krause, 1967). In this section, we examine the correlation between the path (b) and (c). The correlation between other paths can be analyzed in a similar manner.

Consider the field at $x = L$ and $\boldsymbol{\rho}_1$ due to the transmitter operating at k_1 and located at $x = 0$ and $\boldsymbol{\rho}_{t1}$. In this case, we can use (18-3) and (19-49) and write

$$\psi_1(k_1, \mathbf{r}_1, t_1) = \int dv' \, h(k_1, \mathbf{r}_1, \mathbf{r}_1', t_1) n_1(\mathbf{r}_1', 0) \tag{19-56}$$

where

$$h(k_1, \mathbf{r}_1, \mathbf{r}_1', t_1) = \frac{k_1^{\,2}}{2\pi} \frac{1}{\gamma(L - x')} \exp\left[i \frac{k_1}{2} \frac{|\boldsymbol{\rho}_1' - \boldsymbol{\rho}_{t1} + \mathbf{V} t_1 - \gamma(\boldsymbol{\rho}_1 - \boldsymbol{\rho}_{t1})|^2}{\gamma(L - x')} \right],$$

$\mathbf{r}_1 = (L, \boldsymbol{\rho}_1)$, $\mathbf{r}_1' = (x', \boldsymbol{\rho}_1')$, and $\mathbf{V}$ is the transverse wind velocity. Note that the only difference between this case and the case of $\boldsymbol{\rho}_{t1} = \boldsymbol{\rho}_{t2} = 0$ is that $\gamma \boldsymbol{\rho}_1$

for the latter is replaced by $\gamma(\boldsymbol{\rho}_1 - \boldsymbol{\rho}_{t1}) + \boldsymbol{\rho}_{t1}$. The correlation function B_χ at $x = L$ is given, therefore, by

$$B_\chi(k_1, k_2, \boldsymbol{\rho}_1, \boldsymbol{\rho}_2, \tau) = \langle \chi(k_1, \boldsymbol{\rho}_1, t_1)\chi(k_2, \boldsymbol{\rho}_2, t_2)\rangle$$

$$= 2\pi^2 k_1 k_2 \int_0^\infty \kappa\, d\kappa \int_0^L d\eta\, g_\chi(\kappa, \eta) J_0(\kappa P)\Phi_n(\kappa) \qquad (19\text{-}57)$$

where $P = |\boldsymbol{\rho}_t + \gamma(\boldsymbol{\rho} - \boldsymbol{\rho}_t) + \mathbf{V}\tau|$, $\boldsymbol{\rho}_t = \boldsymbol{\rho}_{t1} - \boldsymbol{\rho}_{t2}$, $\boldsymbol{\rho} = \boldsymbol{\rho}_1 - \boldsymbol{\rho}_2$, $\gamma = \eta/L$, and g_χ is given in (19-49). We can generalize (19-57) to include the variations of the wind velocity $\mathbf{V}$ and the spectral density Φ_n along the propagation path by letting $\mathbf{V}$ and Φ_n be functions of the position: $\mathbf{V} = \mathbf{V}(\eta)$ and $\Phi_n = \Phi_n(\eta, \kappa)$. In this manner, we can obtain $\mathbf{V}(\eta)$ and $\Phi_n(\eta, \kappa)$ by measuring the correlation function B_χ.

Suppose that two transmitters are located at $\boldsymbol{\rho}_{t1} = (0, 0, 0)$ and $\boldsymbol{\rho}_{t2} = (0, d_t, 0)$ and two receivers are located at $\boldsymbol{\rho}_1 = (L, d_r, 0)$ and $\boldsymbol{\rho}_2 = (L, 0, 0)$. The crossing point is located at

$$\eta = [d_t/(d_t + d_r)]L. \qquad (19\text{-}58)$$

Note that when $\tau = 0$, at this crossing point, P becomes zero. This means that at the crossing point the integrand $J_0(\kappa P)$ becomes peaked and that the random medium in the neighborhood of the crossing point contributes most to the correlation of the received signal. This phenomenon has been used for remote sensing of the medium at the crossing point.

19-7 WAVE FLUCTUATIONS IN AN INHOMOGENEOUS RANDOM MEDIUM

In dealing with wave propagation in a random medium, we normally assume that the structure constant C_n may vary as a function of distance along the propagation path, but it is homogeneous in the direction transverse to the propagation path. This is usually permissible since in general, only the transverse dimension of the medium on the order of the Fresnel size $(\lambda L)^{1/2}$ contributes to the wave fluctuation, and the medium is usually considered homogeneous within this transverse dimension. However, there may be some cases in which the transverse dimension may be comparable to the Fresnel size, and in such a case we need to take into account the effect of the transverse medium variation on the wave fluctuation. This problem is discussed in the following section.

Another assumption usually made is that the wave propagation path is straight and the bending of the ray is negligible. This is valid in most practical situations. However, there are some cases in which the ray bending is significant. Examples are the radio occultation technique of probing a planetary atmosphere and acoustic fluctuation in inhomogeneous ocean water.

The bending of the ray affects the phase and amplitude so that the equivalent Fresnel size changes. The effect of this Fresnel shrinkage on the planetary radio occultation experiments has been pointed out by Fjeldbo and Eshleman (1969).

19-8 WAVE FLUCTUATIONS IN A LOCALIZED SMOOTHLY VARYING RANDOM MEDIUM

In analyzing wave propagation in a random medium, it is commonly assumed that the medium is homogeneous. In some situations, such as vertical propagation in the atmosphere, the random medium is obviously not homogeneous, but varies as a function of height. This situation can be easily handled by letting C_n^2 be a function of height.

There are other situations in which the random medium varies smoothly in the direction transverse to the direction of wave propagation. An example is a wave propagation through a rocket plume which is finite and localized. Another example is a radio link between a spacecraft and the earth when the propagation path passes through a thin layer of turbulent atmosphere of a planet. In general, if the medium is homogeneous over the distance of the Fresnel size $(\lambda L)^{1/2}$ in the transverse direction (y, z), we can safely neglect the variation of the random medium in y and z, and can take into account the variation of the structure constant in the propagation direction (x) by letting $C_n^2 = C_n^2(x)$.

If the medium has a transverse variation over the distance of the Fresnel size $(\lambda L)^{1/2}$, we need to devise a method to include the effect of inhomogeneities. In this section we discuss a technique originally proposed by Silverman (1957b) and extended by Ishimaru (1973; Woo and Ishimaru, 1973).

We assume that the correlation function $B_n(\mathbf{r}_1, \mathbf{r}_2)$ of the refractive index fluctuation n_1 is a product of a slowly varying variance $B_n{}^g(\mathbf{r}_c)$ and a correlation coefficient $B_n{}^l(\mathbf{r}_d)$. The variance $B_n{}^g(\mathbf{r}_c)$ is a function of the center coordinate $\mathbf{r}_c = \frac{1}{2}(\mathbf{r}_1 + \mathbf{r}_2)$ only and represents the "global" variation of the random medium (superscript g means "global"). The correlation coefficient $B_n{}^l(\mathbf{r}_d)$ is a function of the difference coordinate $\mathbf{r}_d = \mathbf{r}_1 - \mathbf{r}_2$ only and represents the "local" behavior of the medium:

$$B_n(\mathbf{r}_1, \mathbf{r}_2) = B_n{}^g(\mathbf{r}_c) B_n{}^l(\mathbf{r}_d). \tag{19-59}$$

The corresponding spectrum $\Phi_n(\mathbf{K}_1, \mathbf{K}_2)$ is given by†

$$\Phi_n(\mathbf{K}_1, \mathbf{K}_2) = \frac{1}{(2\pi)^6} \iint B_n(\mathbf{r}_1, \mathbf{r}_2) \exp(i\mathbf{K}_1 \cdot \mathbf{r}_1 - i\mathbf{K}_2 \cdot \mathbf{r}_2)\, d\mathbf{r}_1\, d\mathbf{r}_2. \tag{19-60}$$

† For a more rigorous spectral representation, see Ishimaru (1973).

Noting that

$$\mathbf{K}_1 \cdot \mathbf{r}_1 - \mathbf{K}_2 \cdot \mathbf{r}_2 = \mathbf{K}_d \cdot \mathbf{r}_c + \mathbf{K}_c \cdot \mathbf{r}_d \tag{19-61}$$

where $\mathbf{K}_d = \mathbf{K}_1 - \mathbf{K}_2$ and $\mathbf{K}_c = \frac{1}{2}(\mathbf{K}_1 + \mathbf{K}_2)$, we get

$$\Phi_n(\mathbf{K}_1, \mathbf{K}_2) = \Phi_n{}^g(\mathbf{K}_d)\Phi_n{}^l(\mathbf{K}_c)$$

$$\Phi_n{}^g(\mathbf{K}_d) = \frac{1}{(2\pi)^3}\int B_n{}^g(\mathbf{r}_c)\exp(i\mathbf{K}_d \cdot \mathbf{r}_c)\,d\mathbf{r}_c \tag{19-62}$$

$$\Phi_n{}^l(\mathbf{K}_c) = \frac{1}{(2\pi)^3}\int B_n{}^l(\mathbf{r}_d)\exp(i\mathbf{K}_c \cdot \mathbf{r}_d)\,d\mathbf{r}_d.$$

We also need the relationship between the two-dimensional global and local spectra, $F_n{}^g(x_c, \boldsymbol{\kappa}_d)$ and $F_n{}^l(x_d, \boldsymbol{\kappa}_c)$, and the three-dimensional global and local spectra, $\Phi_n{}^g(\mathbf{K}_d)$ and $\Phi_n{}^l(\mathbf{K}_c)$. We write

$$B_n{}^g(x_c, \boldsymbol{\rho}_c) = \int \exp(i\boldsymbol{\kappa}_d \cdot \boldsymbol{\rho}_c)F_n{}^g(x_c, \boldsymbol{\kappa}_d)\,d\boldsymbol{\kappa}_d$$

$$B_n{}^l(x_d, \boldsymbol{\rho}_d) = \int \exp(i\boldsymbol{\kappa}_c \cdot \boldsymbol{\rho}_d)F_n{}^l(x_d, \boldsymbol{\kappa}_c)\,d\boldsymbol{\kappa}_c \tag{19-63}$$

$$F_n{}^g(x_c, \boldsymbol{\kappa}_d) = \int \exp(iK_{d1}x_c)\Phi_n{}^g(K_{d1}, \boldsymbol{\kappa}_d)\,dK_{d1}$$

$$F_n{}^l(x_d, \boldsymbol{\kappa}_c) = \int \exp(iK_{c1}x_d)\Phi_n{}^l(K_{c1}, \boldsymbol{\kappa}_c)\,dK_{c1}$$

where $\mathbf{r}_c = x_c\hat{\mathbf{x}} + \boldsymbol{\rho}_c$, $\mathbf{r}_d = x_d\hat{\mathbf{x}} + \boldsymbol{\rho}_d$, $\mathbf{K}_c = K_{c1}\hat{\mathbf{x}} + \boldsymbol{\kappa}_c$, and $\mathbf{K}_d = K_{d1}\hat{\mathbf{x}} + \boldsymbol{\kappa}_d$. Since $F_n{}^g$ and $F_n{}^l$ are one-dimensional Fourier transforms of $\Phi_n{}^g$ and $\Phi_n{}^l$, we can also express $\Phi_n{}^g$ and $\Phi_n{}^l$ as one-dimensional inverse Fourier transforms of $F_n{}^g$ and $F_n{}^l$. In particular, we have the following useful relationship;

$$\Phi_n(0, \boldsymbol{\kappa}_c) = \frac{1}{2\pi}\int F_n{}^l(x_d, \boldsymbol{\kappa}_c)\,dx_d. \tag{19-64}$$

Equations (19-60)–(19-64) are the fundamental relationships between the correlation function (19-59) and the spectrum Φ_n. Note that if the medium is homogeneous, $B_n{}^g(\mathbf{r}_c)$ is a constant and we get $\Phi_n{}^g = (B_n{}^g)\,\delta(\mathbf{K}_d)$.

Keeping this formulation in mind, we now examine the wave propagation in such a medium. For a plane wave, the Rytov solution of the field $U(\mathbf{r})$

is given by

$$U(\mathbf{r}) = U_0(\mathbf{r}) \exp[\chi(\mathbf{r}) + iS_1(\mathbf{r})]$$

$$\chi(L, \boldsymbol{\rho}) = \int_0^L dx' \iint \exp(i\boldsymbol{\kappa} \cdot \boldsymbol{\rho}) H_r(L - x', \boldsymbol{\kappa}) \, d\nu(x', \boldsymbol{\kappa}) \tag{19-65}$$

$$S_1(L, \boldsymbol{\rho}) = \int_0^L dx' \iint \exp(i\boldsymbol{\kappa} \cdot \boldsymbol{\rho}) H_i(L - x', \boldsymbol{\kappa}) \, d\nu(x', \boldsymbol{\kappa})$$

where

$$H_r(L - x', \boldsymbol{\kappa}) = k \sin \frac{(L - x')}{2k} \kappa^2, \qquad H_i(L - x', \boldsymbol{\kappa}) = k \cos \frac{(L - x')}{2k} \kappa^2.$$

Now consider the amplitude correlation function

$$B_\chi(L, \boldsymbol{\rho}_1, \boldsymbol{\rho}_2) = \langle \chi(L, \boldsymbol{\rho}_1) \chi(L, \boldsymbol{\rho}_2) \rangle. \tag{19-66}$$

Using (19-65), we obtain

$$B_\chi(L, \boldsymbol{\rho}_1, \boldsymbol{\rho}_2) = \int_0^L dx_c \int d\boldsymbol{\kappa}_d \int d\boldsymbol{\kappa}_c \exp(i\boldsymbol{\kappa}_d \cdot \boldsymbol{\rho}_c + i\boldsymbol{\kappa}_c \cdot \boldsymbol{\rho}_d)$$
$$\times H_{r1} H_{r2} F_n{}^g(x_c, \boldsymbol{\kappa}_d) 2\pi \Phi_n{}^l(\boldsymbol{\kappa}_c) \tag{19-67}$$

where

$$H_{r1} = H_r(L - x_c, \boldsymbol{\kappa}_c + \tfrac{1}{2}\boldsymbol{\kappa}_d), \qquad H_{r2} = H_r(L - x_c, \boldsymbol{\kappa}_c - \tfrac{1}{2}\boldsymbol{\kappa}_d),$$

and $F_n{}^g(x_c, \boldsymbol{\kappa}_d)$ is the two-dimensional spectrum given in (19-63).

Making use of (19-63), we can write (19-67) as a convolution integral:

$$B_\chi(L, \boldsymbol{\rho}_c, \boldsymbol{\rho}_d) = \frac{1}{2\pi} \int_0^L dx_c \int d\boldsymbol{\rho}_c' \, B_n{}^g(x_c, \boldsymbol{\rho}_c') G_\chi(x_c, \boldsymbol{\rho}_c - \boldsymbol{\rho}_c', \boldsymbol{\rho}_d)$$

where

$$G_\chi(x_c, \boldsymbol{\rho}_c - \boldsymbol{\rho}_c', \boldsymbol{\rho}_d) = \int d\boldsymbol{\kappa}_d \int d\boldsymbol{\kappa}_c \exp[i\boldsymbol{\kappa}_d \cdot (\boldsymbol{\rho}_c - \boldsymbol{\rho}_c') + i\boldsymbol{\kappa}_c \cdot \boldsymbol{\rho}_d]$$
$$\times H_{r1} H_{r2} \Phi_n(\boldsymbol{\kappa}_c). \tag{19-68}$$

This is the fundamental equation for the correlation function B_χ when the medium has a smoothly varying variance $B_n{}^g(x_c, \boldsymbol{\rho}_c')$. We also get from (19-67)

$$H_{r1} H_{r2} = \frac{k^2}{2} \left\{ \cos\left[\frac{(L - x_c)}{k} \boldsymbol{\kappa}_c \cdot \boldsymbol{\kappa}_d\right] - \cos\left[\frac{(L - x_c)}{k}\left(\kappa_c{}^2 + \frac{\kappa_d{}^2}{4}\right)\right]\right\}. \tag{19-69}$$

For the Kolmogorov spectrum, we use

$$B_n{}^g(\mathbf{r}_c) = 0.033 C_n{}^2(\mathbf{r}_c), \qquad \Phi_n(\boldsymbol{\kappa}_c) = (\kappa_c{}^2 + 1/L_0{}^2)^{-11/6}. \tag{19-70}$$

The slowly varying variance is represented by $B_n{}^g(\mathbf{r}_c)$.

For example, if a plane wave is incident on a random medium with a transverse dimension of b, then the variance $\sigma_\chi{}^2$ is proportional to $L^{5/6}$ if the propagation distance $L \ll L_0{}^2/\lambda$. This is the same as the usual homogeneous medium case. But in the range $L_0{}^2/\lambda \ll L \ll bL_0/\lambda$, the variance $\sigma_\chi{}^2$ is almost constant and in the range $L \gg bL_0/\lambda$, $\sigma_\chi{}^2$ decreases as L^{-2}. If a plane wave is incident on a random medium whose structure constant varies according to

$$C_n{}^2(x_c, \boldsymbol{\rho}_c) = C_{n0}^2 \exp(-x_c{}^2/a^2 - \rho_c{}^2/b^2), \tag{19-71}$$

then the variance at $(L, 0, 0)$ is given by (see Fig. 19-6)

$$\sigma_\chi{}^2(L, 0) = 0.563 k^{7/6} L^{5/6} \sqrt{\pi}\, a C_{n0}^2 \tag{19-72a}$$

in the region $L \ll L_0{}^2/\lambda$;

$$\sigma_\chi{}^2(L, 0) = 0.391 k^2 L_0^{5/3} \sqrt{\pi}\, a C_{n0}^2 \tag{19-72b}$$

in the region $L_0{}^2/\lambda \ll L \ll bL_0/\lambda$; and

$$\sigma_\chi{}^2(L, 0) = 0.104 K^4 L_0^{11/3} L^{-2} \pi^{3/2} a b^2 C_{n0}^2 \tag{19-72c}$$

in the region $bL_0/\lambda \ll L$. This result has been applied to the remote probing of the Venus atmosphere by the Mariner's occultation experiment (for details, see Ishimaru, 1973; Woo and Ishimaru, 1974). The frequency spectrum of fluctuations due to the motion of spacecraft is also investigated.

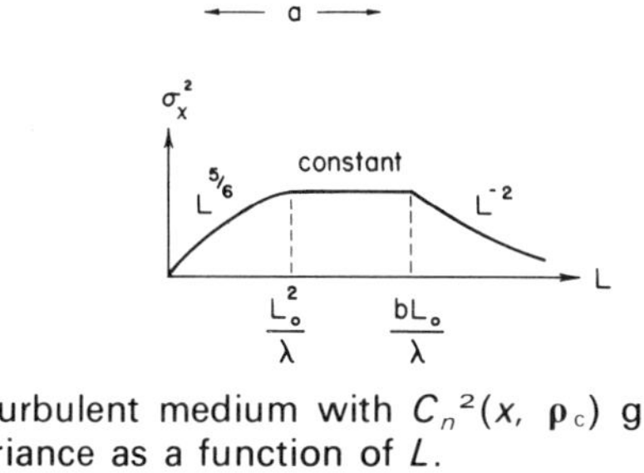

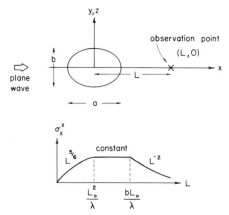

FIG. 19-6 Turbulent medium with $C_n{}^2(x, \rho_c)$ given in (19-71) and the log-amplitude variance as a function of L.

For a spherical wave case (Ishimaru, 1973) we need to replace $\boldsymbol{\rho}$ and $L - x_c$ for a plane wave case by $(x_c/L)\boldsymbol{\rho}$ and $(x_c/L)(L - x_c)$, respectively. Thus we have

$$B_\chi(L, \boldsymbol{\rho}_0, \boldsymbol{\rho}_d) = \frac{1}{2\pi}\int_0^L dx_c \int d\boldsymbol{\rho}_c' \, B_n{}^g(x_c, \boldsymbol{\rho}_c')G_\chi$$

$$G_\chi = \int d\boldsymbol{\kappa}_d \int d\boldsymbol{\kappa}_c \exp\left[i\boldsymbol{\kappa}_d \cdot \left(\frac{x_c}{L}\boldsymbol{\rho}_c - \boldsymbol{\rho}_c'\right) + i\boldsymbol{\kappa}_c \cdot \frac{x_c}{L}\boldsymbol{\rho}_d\right] H_{r1}H_{r2}\Phi_n{}^l(\boldsymbol{\kappa}_c)$$

(19-73)

$$H_{r1}H_{r2} = \frac{k^2}{2}\left\{\cos\left[\left(\frac{x_c}{L}\right)\frac{(L - x_c)}{k}\boldsymbol{\kappa}_c \cdot \boldsymbol{\kappa}_d\right] - \cos\left[\left(\frac{x_c}{L}\right)\frac{(L - x_c)}{k}\left(\kappa_c{}^2 + \frac{\kappa_d{}^2}{4}\right)\right]\right\}.$$

CHAPTER 20 □ STRONG FLUCTUATION THEORY

In Chapters 16–19 we discussed the weak fluctuation theory. Generally, the weak fluctuation theory is valid only when the log-amplitude variance is less than about 0.2–0.5. For the Kolmogorov spectrum $\Phi_n(\kappa) = 0.033C_n^2\kappa^{-11/3}$, this condition is given by

$$\sigma_\chi^2 = 0.307C_n^2k^{7/6}L^{11/6} < 0.2\text{–}0.5. \tag{20-1}$$

When the limit is exceeded, the fluctuation is considered "strong." For example, in the earth's atmosphere, C_n ranges from 10^{-9} $\mathrm{m}^{-1/3}$ for weak turbulence to 10^{-7} $\mathrm{m}^{-1/3}$ for strong turbulence. Therefore, for microwave propagation in the earth's atmosphere, the weak fluctuation theory is valid for almost any distance encountered in practice. For optical propagation, however, this limit is exceeded within a few kilometers. On the other hand, even for microwave frequencies, when the wave propagates over a great distance, such as in the ionosphere or solar corona, the weak fluctuation theory is no longer valid.

Several techniques have been proposed to deal with the strong fluctuation problem. They are the diagram method (de Wolf, 1968; Tatarski, 1971; Frisch, 1968; Barabanenkov *et al.*, 1971), the integral equation method (Brown, 1971, 1972), including the Dyson and the Bethe–Salpeter equations, the extended Huygens–Fresnel principle (Lutomirski and Yura, 1971a; Yura, 1972, 1973; Kon, 1970; Clifford *et al.*, 1975; Feizulin and Kravtsov, 1967; Banakh *et al.*, 1974), and the parabolic equation method (Tatarski, 1971; Ho and Beran, 1968; Ho, 1969; Beran and Whitman, 1971; Dowling and Livingston, 1973; Furutsu, 1972; Shishov, 1968; Whitman and Beran, 1970).

It has been shown (Strohbehn, 1968; Yura, 1972). that for the second moment, all of these techniques are equivalent to each other under certain assumptions. For this reason, in this chapter, we discuss the second moment using the parabolic equation method.

For the fourth order moments, the Huygens–Fresnel method yields relatively simple solutions (Clifford *et al.*, 1974; and Banakh *et al.*, 1974). However, the equivalence between this and other methods has not been established. It may be noted that the parabolic equations have been obtained for all higher moments, and at present they are considered more fundamental than other formulations. Their exact solutions, however, have not been obtained except for the second moment, and this constitutes one of the most important unsolved problems in strong fluctuation theory.

20-1 PARABOLIC EQUATION

Let us consider a wave $u(\mathbf{r})$ propagating in a random medium characterized by the relative dielectric constant $\varepsilon_r(\mathbf{r})$:

$$\varepsilon_r(\mathbf{r}) = \langle \varepsilon_r \rangle [1 + \varepsilon_1(\mathbf{r})]. \tag{20-2}$$

Using the average wave number

$$k^2 = \omega^2 \mu_0 \varepsilon_0 \langle \varepsilon_r \rangle = k_0{}^2 \langle \varepsilon_r \rangle, \tag{20-3}$$

we write a wave equation for a scalar field $u(\mathbf{r})$:

$$[\nabla^2 + k^2(1 + \varepsilon_1(\mathbf{r}))]u(\mathbf{r}) = 0. \tag{20-4}$$

The first step in obtaining the parabolic equation is to recognize that as a wave $u(\mathbf{r})$ propagates in the x direction, its phase progresses substantially as ikx. Therefore if we write

$$u(\mathbf{r}) = U(\mathbf{r})e^{ikx}, \tag{20-5}$$

then $U(\mathbf{r})$ should be a slowly varying function of x. Substituting (20-5) into (20-4) we obtain the following exact equation for $U(\mathbf{r})$:

$$2ik\frac{\partial U(\mathbf{r})}{\partial x} + \nabla^2 U(\mathbf{r}) + k^2\varepsilon_1(\mathbf{r})U(\mathbf{r}) = 0. \tag{20-6a}$$

Since $U(\mathbf{r})$ is a slowly varying function of x and varies only over the distance of the scale size l of the random medium, we note that

$$|k\,\partial U/\partial x| \gg |\partial^2 U/\partial x^2| \qquad \text{as long as} \quad l \gg \lambda. \tag{20-6b}$$

Therefore we can replace ∇^2 in (20-6) by the transverse Laplacian $\nabla_t{}^2 = \partial^2/\partial y^2 + \partial^2/\partial z^2$, and we obtain the following parabolic equation for $U(\mathbf{r})$:

$$2ik\frac{\partial U(\mathbf{r})}{\partial x} + \nabla_t{}^2 U(\mathbf{r}) + k^2\varepsilon_1(\mathbf{r})U(\mathbf{r}) = 0. \tag{20-7}$$

This is the starting point of our analysis. In the subsequent sections, we shall examine the average value of the field and higher moments.

20-2 ASSUMPTION FOR THE REFRACTIVE INDEX FLUCTUATIONS

In our analysis of strong fluctuations, we make use of the following three important assumptions:

(a) The parabolic equation, which has already been stated.

(b) The fluctuation $\varepsilon_1(\mathbf{r})$ is a Gaussian random field, and therefore its characteristics are completely described by the correlation functions

$$B_\varepsilon(\mathbf{r}-\mathbf{r}') = \langle \varepsilon_1(\mathbf{r})\varepsilon_1(\mathbf{r}')\rangle. \tag{20-8}$$

(c) $\varepsilon_1(\mathbf{r})$ is delta correlated in the direction of propagation (the x direction):

$$\langle \varepsilon_1(x, \boldsymbol{\rho})\varepsilon_1(x', \boldsymbol{\rho}')\rangle = \delta(x-x')A(\boldsymbol{\rho}-\boldsymbol{\rho}'). \tag{20-9}$$

The assumption expressed in (20-9) is based on the fact that, although the correlation of the dielectric constant in the transverse direction $\boldsymbol{\rho}$ has a direct bearing on the transverse correlation of the field, the correlation of the dielectric constant in the direction of the wave propagation has little effect on the fluctuation characteristics of the wave. In the Rytov solution, we have in effect used this assumption when we noted that H_r at x' and x'' can be equated to H_r at $\frac{1}{2}(x' + x'')$.

The relationship between the function $A(\boldsymbol{\rho}-\boldsymbol{\rho}')$ and the spectrum of the refractive index fluctuation $\Phi_n(\boldsymbol{\kappa})$ is given by the following: We write the correlation function as a Fourier transform of the spectral density

$$B_\varepsilon(\mathbf{r}-\mathbf{r}') = 4B_n(\mathbf{r}-\mathbf{r}') = \iiint \Phi_\varepsilon(\mathbf{K}) \exp[i\mathbf{K}\cdot(\mathbf{r}-\mathbf{r}')]\, d\mathbf{K} \tag{20-10}$$

where $\phi_\varepsilon(\mathbf{K}) = 4\Phi_n(\mathbf{K})$ and $d\mathbf{K} = dK_1\, dK_2\, dK_3 = dK_1\, d\boldsymbol{\kappa}$. Taking the inverse transform of (20-10) and using (20-9), we obtain

$$\Phi_\varepsilon(\boldsymbol{\kappa}) = \frac{1}{(2\pi)^3}\iint A(\boldsymbol{\rho}-\boldsymbol{\rho}')\exp[-i\boldsymbol{\kappa}\cdot(\boldsymbol{\rho}-\boldsymbol{\rho}')]\, d(\boldsymbol{\rho}-\boldsymbol{\rho}'). \tag{20-11}$$

This is a two-dimensional Fourier transform. Taking the inverse, we get

$$A(\boldsymbol{\rho}) = 2\pi \int \Phi_\varepsilon(\boldsymbol{\kappa})\, e^{i\boldsymbol{\kappa}\cdot\boldsymbol{\rho}}\, d\boldsymbol{\kappa}. \tag{20-12}$$

If the turbulence is isotropic, then we get

$$A(\rho) = (2\pi)^2 \int_0^\infty J_0(\kappa\rho)\Phi_\varepsilon(\kappa)\kappa\, d\kappa. \tag{20-13}$$

$A(\rho)$ is also related to the correlation function $B_\varepsilon(x, \rho)$ through

$$A(\rho) = \int_{-\infty}^{\infty} B_\varepsilon(x, \rho)\, dx = 4 \int_{-\infty}^{\infty} B_n(x, \rho)\, dx. \tag{20-14}$$

20-3 EQUATION FOR THE AVERAGE FIELD AND GENERAL SOLUTION

Let us take the ensemble average of the parabolic equation (20-7):

$$2ik \frac{\partial}{\partial x} \langle U(\mathbf{r})\rangle + \nabla_t{}^2 \langle U(\mathbf{r})\rangle + k^2 \langle \varepsilon_1(\mathbf{r}) U(\mathbf{r})\rangle = 0. \tag{20-15}$$

If we can express the last term in terms of the average field in the following manner:

$$\langle \varepsilon_1(\mathbf{r}) U(\mathbf{r})\rangle = g(\mathbf{r}) \langle \mathbf{U}(\mathbf{r})\rangle, \tag{20-16}$$

then we have a differential equation for $\langle U(\mathbf{r})\rangle$.

In order to obtain $g(\mathbf{r})$ in (20-16), we first note that $U(\mathbf{r})$ is a functional† on $\varepsilon_1(\mathbf{r})$, and make use of the following relationship valid for a Gaussian random field $\varepsilon_1(\mathbf{r})$ and a functional $U(\mathbf{r})$ on $\varepsilon_1(\mathbf{r})$ (Appendix 20B):

$$\langle \varepsilon_1(\mathbf{r}) U(\mathbf{r})\rangle = \int dV' \langle \varepsilon_1(\mathbf{r}) \varepsilon_1(\mathbf{r}')\rangle \left\langle \frac{\delta U(\mathbf{r})}{\delta \varepsilon_1(\mathbf{r}')} \right\rangle \tag{20-17}$$

where $\delta U / \delta \varepsilon_1$ is called the functional derivative or variational derivative (see Appendix 20A). Equation (20-17) is sometimes called the Navikov–Furutsu formula. Using the delta-correlated assumption (20-9), Eq. (20-17) becomes

$$\langle \varepsilon_1(\mathbf{r}) U(\mathbf{r})\rangle = \iint d\boldsymbol{\rho}' A(\boldsymbol{\rho} - \boldsymbol{\rho}') \left\langle \frac{\delta U(x, \boldsymbol{\rho})}{\delta \varepsilon(x, \boldsymbol{\rho}')} \right\rangle. \tag{20-18}$$

In Appendix 20C, it will be shown that

$$\frac{\delta U(x, \boldsymbol{\rho})}{\delta \varepsilon_1(x', \boldsymbol{\rho}')} = \frac{ik}{4} \delta(\boldsymbol{\rho} - \boldsymbol{\rho}') U(x, \boldsymbol{\rho}') \tag{20-19a}$$

$$\frac{\delta U^*(x, \boldsymbol{\rho})}{\delta \varepsilon_1(x', \boldsymbol{\rho}')} = -\frac{ik}{4} \delta(\boldsymbol{\rho} - \boldsymbol{\rho}') U^*(x, \boldsymbol{\rho}'). \tag{20-19b}$$

Substituting (20-19a) into (20-18), we obtain

$$\langle \varepsilon_1(\mathbf{r}) U(\mathbf{r})\rangle = \frac{ik}{4} A(0) \langle U(x, \boldsymbol{\rho})\rangle. \tag{20-20}$$

† A functional is a quantity that depends on a function $\varepsilon(\mathbf{r})$ (see Appendix 20A). In contrast, a function is a quantity that depends on a variable.

Therefore, (20-15) becomes

$$\left[2ik\frac{\partial}{\partial x} + \nabla_t^2 + \frac{ik^3}{4}A(0)\right]\langle U(x, \boldsymbol{\rho})\rangle = 0. \tag{20-21}$$

Together with the boundary condition at $x = 0$,

$$\langle U(0, \boldsymbol{\rho})\rangle = U_0(\boldsymbol{\rho}), \tag{20-22}$$

(20-21) completely determines the coherent field $\langle U(x, \boldsymbol{\rho})\rangle$.

The solution to (20-21) can be easily obtained by writing

$$\langle U(x, \boldsymbol{\rho})\rangle = f(x, \boldsymbol{\rho}) \exp(-\alpha_0 x) \tag{20-23}$$

where

$$\alpha_0 = k^2 A(0)/8 = 2\pi^2 k^2 \int_0^\infty \Phi_n(\kappa)\kappa \, d\kappa. \tag{20-24}$$

Substituting (20-23) into (20-21), we obtain

$$[2ik\, \partial/\partial x + \nabla_t^2] f(x, \boldsymbol{\rho}) = 0. \tag{20-25}$$

However, (20-25) is the parabolic differential equation for the field in free space ($\varepsilon_1 = 0$) and therefore $f(x, \boldsymbol{\rho})$ is the field in free space in the absence of the randomness. Letting the free space field be $f(x, \boldsymbol{\rho}) = U_0(x, \boldsymbol{\rho})$, we get the final solution:

$$\langle U(x, \boldsymbol{\rho})\rangle = U_0(x, \boldsymbol{\rho}) \exp(-\alpha_0 x). \tag{20-26}$$

The coherent intensity is therefore given by

$$|\langle U(x, \boldsymbol{\rho})\rangle|^2 = |U_0(x, \boldsymbol{\rho})|^2 \exp(-2\alpha_0 x). \tag{20-27}$$

We note that $2\alpha_0$ given by (20-24) is equal to the total scattering cross section per unit volume of the turbulence given in Eq. (16-56) of Chapter 16.

In general, α_0 is related to the correlation function $B_n(\mathbf{r})$ of the refractive index fluctuation through (20-14):

$$\alpha_0 = k^2 \int_0^\infty B_n(x)\, dx = k^2 \sigma_n^2 L_n \tag{20-28}$$

where $\sigma_n^2 = \langle n_1^2\rangle$ is the variance and L_n is called the integral scale of the random medium:

$$L_n = \frac{\int_0^\infty B_n(x)\, dx}{B_n(0)}. \tag{20-29}$$

For the Kolmogorov spectrum given by

$$\Phi_n(\kappa) = 0.033 C_n^2 (\kappa^2 + 1/L_0^2)^{-11/6} \exp(-\kappa^2/\kappa_m^2), \tag{20-30}$$

where $\kappa_m = 5.92/l_0$, we get

$$\alpha_0 = 0.033\pi^2 C_n{}^2 k^2 L_0^{5/3} \psi[1, \tfrac{1}{6}; (\kappa_m L_0)^{-2}] \tag{20-31}$$

where $\psi(a, b; z)$ is a confluent hypergeometric function independent of the Kummer function. Since $\kappa_m L_0 = 5.92 L_0/l_0$, $(\kappa_m L_0)^{-2}$ is much smaller than unity, and therefore we use

$$\psi(a, b; z) \to \frac{\Gamma(1-b)}{\Gamma(1+a-b)} \qquad \text{for} \quad \operatorname{Re} b < 1, \quad |z| \ll 1,$$

and obtain

$$\alpha_0 = 0.391 C_n{}^2 k^2 L_0^{5/3}. \tag{20-32}$$

20-4 PARABOLIC EQUATION FOR THE MUTUAL COHERENCE FUNCTION

We now consider the second moment

$$\Gamma(x, \boldsymbol{\rho}_1, \boldsymbol{\rho}_2) = \langle U(x, \boldsymbol{\rho}_1) U^*(x, \boldsymbol{\rho}_2) \rangle \tag{20-33}$$

which is also called the mutual coherence function (Gurvich and Tatarski, 1975; Tatarski, 1971).

To obtain the differential equation for $\Gamma(x, \boldsymbol{\rho}_1, \boldsymbol{\rho}_2)$, we start with the parabolic equation (20-7):

$$2ik \frac{\partial}{\partial x} U(x, \boldsymbol{\rho}_1) + \nabla_t{}^2 U(x, \boldsymbol{\rho}_1) + k^2 \varepsilon_1(x, \boldsymbol{\rho}_1) U(x, \boldsymbol{\rho}_1) = 0. \tag{20-34}$$

We multiply this by $U^*(x, \boldsymbol{\rho}_2)$, and write

$$2ik \frac{\partial U_1}{\partial x} U_2{}^* + \nabla_{t1}^2 U_1 U_2{}^* + k^2 \varepsilon_1(x, \boldsymbol{\rho}_1) U_1 U_2{}^* = 0 \tag{20-35}$$

where $U_1 = U(x, \boldsymbol{\rho}_1)$, $U_2 = U(x, \boldsymbol{\rho}_2)$, and ∇_{t1}^2 is a Laplacian with respect to $\boldsymbol{\rho}_1$. We take the conjugate of (20-34) with $\boldsymbol{\rho}_1$ replaced by $\boldsymbol{\rho}_2$ and multiply it by U_1:

$$-2ik \frac{\partial U_2{}^*}{\partial x} U_1 + \nabla_{t2}^2 U_2{}^* U_1 + k^2 \varepsilon_1(x, \boldsymbol{\rho}_2) U_2{}^* U_1 = 0. \tag{20-36}$$

Subtracting (20-36) from (20-35) and averaging, we obtain

$$2ik \frac{\partial}{\partial x} \Gamma(x, \boldsymbol{\rho}_1, \boldsymbol{\rho}_2) + (\nabla_{t1}^2 - \nabla_{t2}^2) \Gamma(x, \boldsymbol{\rho}_1, \boldsymbol{\rho}_2)$$

$$+ k^2 \langle [\varepsilon_1(x, \boldsymbol{\rho}_1) - \varepsilon_1(x, \boldsymbol{\rho}_2)] U(x, \boldsymbol{\rho}_1) U^*(x, \boldsymbol{\rho}_2) \rangle = 0. \tag{20-37}$$

The last term in (20-37) can be shown to be proportional to $\Gamma(x, \boldsymbol{\rho}_1, \boldsymbol{\rho}_2)$. To do this, we make use of (20-18) and write

$$\langle \varepsilon_1(x, \boldsymbol{\rho}_1) Z(x, \boldsymbol{\rho}_1, \boldsymbol{\rho}_2) \rangle = \int d\rho_1' A(\boldsymbol{\rho}_1 - \boldsymbol{\rho}_1') \left\langle \frac{\delta Z(x, \boldsymbol{\rho}_1, \boldsymbol{\rho}_2)}{\delta \varepsilon_1(x, \boldsymbol{\rho}_1')} \right\rangle \tag{20-38}$$

where $Z(x, \boldsymbol{\rho}_1, \boldsymbol{\rho}_2) = U(x, \boldsymbol{\rho}_1) U^*(x, \boldsymbol{\rho}_2)$. Making use of (20-19a) and (20-19b), we obtain

$$\begin{aligned} \frac{\delta Z(x, \boldsymbol{\rho}_1, \boldsymbol{\rho}_2)}{\delta \varepsilon_1(x, \boldsymbol{\rho}')} &= \frac{\delta U(x, \boldsymbol{\rho}_1)}{\delta \varepsilon_1(x, \boldsymbol{\rho}')} U^*(x, \boldsymbol{\rho}_2) + U(x, \boldsymbol{\rho}_1) \frac{\delta U^*(x, \boldsymbol{\rho}_2)}{\delta \varepsilon_1(x, \boldsymbol{\rho}')} \\ &= \frac{ik}{4} [\delta(\boldsymbol{\rho}_1 - \boldsymbol{\rho}') U(x, \boldsymbol{\rho}') U^*(x, \boldsymbol{\rho}_2) \\ &\quad - \delta(\boldsymbol{\rho}_2 - \boldsymbol{\rho}') U(x, \boldsymbol{\rho}_1) U^*(x, \boldsymbol{\rho}')]. \end{aligned} \tag{20-39}$$

Averaging and substituting (20-39) into (20-38), we get

$$\langle \varepsilon_1(x, \boldsymbol{\rho}_1) Z(x, \boldsymbol{\rho}_1, \boldsymbol{\rho}_2) \rangle = (ik/4)[A(0) - A(\boldsymbol{\rho}_1 - \boldsymbol{\rho}_2)] \Gamma(x, \boldsymbol{\rho}_1, \boldsymbol{\rho}_2). \tag{20-40}$$

By interchanging $\boldsymbol{\rho}_1$ and $\boldsymbol{\rho}_2$ in (20-38) and taking the conjugate, we get

$$\begin{aligned} \langle \varepsilon_1(x, \boldsymbol{\rho}_2) Z^*(x, \boldsymbol{\rho}_2, \boldsymbol{\rho}_1) \rangle &= \langle \varepsilon_1(x, \boldsymbol{\rho}_2) Z(x, \boldsymbol{\rho}_1, \boldsymbol{\rho}_2) \rangle \\ &= -\frac{ik}{4} [A(0) - A(\boldsymbol{\rho}_1 - \boldsymbol{\rho}_2)] \Gamma(x, \boldsymbol{\rho}_1, \boldsymbol{\rho}_2). \end{aligned} \tag{20-41}$$

Substituting (20-40) and (20-41) into (20-37), we obtain the differential equation for the coherence function:

$$\left\{ 2ik \frac{\partial}{\partial x} + (\nabla_{t1}^2 - \nabla_{t2}^2) + \frac{ik^3}{2} [A(0) - A(\boldsymbol{\rho}_1 - \boldsymbol{\rho}_2)] \right\} \Gamma(x, \boldsymbol{\rho}_1, \boldsymbol{\rho}_2) = 0. \tag{20-42}$$

In a similar manner, we obtain the differential equation for $\langle U(x, \boldsymbol{\rho}_1) U(x, \boldsymbol{\rho}_2) \rangle$:

$$\left\{ 2ik \frac{\partial}{\partial x} + (\nabla_{t1}^2 + \nabla_{t2}^2) + \frac{ik^3}{2} [A(0) + A(\boldsymbol{\rho}_1 - \boldsymbol{\rho}_2)] \right\} \langle U(x, \boldsymbol{\rho}_1) U(x, \boldsymbol{\rho}_2) \rangle = 0. \tag{20-43}$$

Equations (20-42) and (20-43) constitute the basic differential equations for the second moment of the field.

There is a close relationship between the mutual coherence function Γ in a random continuum and the specific intensity I in a random distribution of scatterers. It has been shown (Ishimaru, 1975) that the small angle approximation of the equation of transfer given in (13-5) is identical to the parabolic equation (20-42) with the following correspondence:

$$\Gamma(x, \boldsymbol{\rho}_c, \boldsymbol{\rho}_d) = \int I(x, \boldsymbol{\rho}_c, \mathbf{s}) \exp(ik\mathbf{s} \cdot \boldsymbol{\rho}_d)\, d\mathbf{s}$$

$$2\pi k^4 \Phi_n(\boldsymbol{\kappa}) = \rho_n |f(\mathbf{s})|^2 = \frac{\rho_n \sigma_t}{4\pi} p(\mathbf{s}), \qquad \boldsymbol{\kappa} = k\mathbf{s}. \tag{20-44}$$

20-5 SOLUTIONS FOR THE MUTUAL COHERENCE FUNCTION

In this section, we first obtain the mutual coherence function for a plane wave, and then we derive an expression for the mutual coherence function for a general case.

For a plane wave, $\Gamma(x, \boldsymbol{\rho}_1, \boldsymbol{\rho}_2)$ must be a function of x and $\rho = |\boldsymbol{\rho}_1 - \boldsymbol{\rho}_2|$ only. If the wave is perfectly coherent at $x = 0$, then

$$\Gamma(0, \boldsymbol{\rho}_1 - \boldsymbol{\rho}_2) = 1, \tag{20-45}$$

and therefore we get the solution to (20-42):

$$\Gamma(x, \boldsymbol{\rho}_1, \boldsymbol{\rho}_2) = \exp\left\{-\frac{k^2}{4}[A(0) - A(\boldsymbol{\rho}_1 - \boldsymbol{\rho}_2)]x\right\}$$

$$= \exp\left\{-k^2 x \int_{-\infty}^{\infty} [B_n(\eta, 0) - B_n(\eta, \rho)]\, d\eta\right\}. \tag{20-46a}$$

Using (20-13) we can express (20-46a) in the form

$$\Gamma(x, \boldsymbol{\rho}) = \exp\{-\tfrac{1}{2} D(x, \boldsymbol{\rho})\}$$

$$D(x, \boldsymbol{\rho}) = 8\pi^2 k^2 x \int_0^{\infty} [1 - J_0(\kappa\rho)] \Phi_n(\kappa) \kappa\, d\kappa. \tag{20-46b}$$

We note that the structure functions for the log-amplitude fluctuation and the phase fluctuations in the Rytov approximation are (see Chapter 17)

$$D_\chi(x, \rho) = 2[B_\chi(x, 0) - B_\chi(x, \rho)]$$

$$= 4\pi^2 k^2 x \int_0^{\infty} [1 - J_0(\kappa\rho)] \left[1 - \frac{\sin(\kappa^2 x/k)}{\kappa^2 x/k}\right] \Phi_n(\kappa) \kappa\, d\kappa \tag{20-47}$$

$$D_S(x, \rho) = 2[B_S(x, 0) - B_S(x, \rho)]$$

$$= 4\pi^2 k^2 x \int_0^{\infty} [1 - J_0(\kappa\rho)] \left[1 + \frac{\sin(\kappa^2 x/k)}{\kappa^2 x/k}\right] \Phi_n(\kappa) \kappa\, d\kappa. \tag{20-48}$$

Therefore, we conclude that $D(x, \boldsymbol{\rho})$ in (20-46b) is identical to

$$D(x, \rho) = D_\chi(x, \rho) + D_S(x, \rho). \tag{20-49}$$

$D(x, \boldsymbol{\rho})$ is called the wave structure function.

Let us next consider a general solution of (20-42) which satisfies a given boundary condition at $x = 0$:

$$\Gamma(0, \boldsymbol{\rho}_1, \boldsymbol{\rho}_2) = \Gamma_0(\boldsymbol{\rho}_1, \boldsymbol{\rho}_2). \tag{20-50}$$

Recognizing that the mutual coherence function $\Gamma(x, \boldsymbol{\rho}_1, \boldsymbol{\rho}_2)$ can be better described in terms of the difference coordinate $\boldsymbol{\rho}_d = \boldsymbol{\rho}_1 - \boldsymbol{\rho}_2$ and the average coordinate $\boldsymbol{\rho}_c = \frac{1}{2}(\boldsymbol{\rho}_1 + \boldsymbol{\rho}_2)$, we first transform (20-42) using $\boldsymbol{\rho}_d$ and $\boldsymbol{\rho}_c$. We note that†

$$\nabla_{t1}^2 - \nabla_{t2}^2 = 2\left(\frac{\partial^2}{\partial y_c\, \partial z_d} + \frac{\partial^2}{\partial z_c\, \partial z_d}\right),$$

with $\boldsymbol{\rho}_c = y_c \hat{\mathbf{y}} + z_c \hat{\mathbf{z}}$ and $\boldsymbol{\rho}_d = y_d \hat{\mathbf{y}} + z_d \hat{\mathbf{z}}$, and therefore we write (20-42) as follows:

$$\left\{2ik\frac{\partial}{\partial x} + 2\left(\frac{\partial^2}{\partial y_c\, \partial y_d} + \frac{\partial^2}{\partial z_c\, \partial z_d}\right) + \frac{ik^3}{2}[A(0) - A(\rho_d)]\right\}\Gamma(x, \boldsymbol{\rho}_c, \boldsymbol{\rho}_d) = 0. \tag{20-51}$$

Compared with the plane wave case, the second term is the one which characterizes a general case. We now consider the following double Fourier transform with respect to $\boldsymbol{\rho}_c$ and $\boldsymbol{\rho}_d$:

$$M(x, \boldsymbol{\kappa}_d, \boldsymbol{\kappa}_c) = \frac{1}{(2\pi)^4}\iint \Gamma(x, \boldsymbol{\rho}_c, \boldsymbol{\rho}_d) \times \exp(-i\boldsymbol{\kappa}_d \cdot \boldsymbol{\rho}_c - i\boldsymbol{\kappa}_c \cdot \boldsymbol{\rho}_d)\, d\boldsymbol{\rho}_c\, d\boldsymbol{\rho}_d \tag{20-52}$$

$$\Gamma(x, \boldsymbol{\rho}_c, \boldsymbol{\rho}_d) = \iint M(x, \boldsymbol{\kappa}_d, \boldsymbol{\kappa}_d) \times \exp(i\boldsymbol{\kappa}_d \cdot \boldsymbol{\rho}_c + i\boldsymbol{\kappa}_c \cdot \boldsymbol{\rho}_d)\, d\boldsymbol{\kappa}_d\, d\boldsymbol{\kappa}_c. \tag{20-53}$$

Taking the double Fourier transform of (20-51), we get

$$2ik\frac{\partial}{\partial x}M(x, \boldsymbol{\kappa}_d, \boldsymbol{\kappa}_c) - 2\boldsymbol{\kappa}_d \cdot \boldsymbol{\kappa}_c M(x, \boldsymbol{\kappa}_d, \boldsymbol{\kappa}_c) + \frac{ik^3}{2}\int d\boldsymbol{\kappa}_c' F_A(\boldsymbol{\kappa}_c - \boldsymbol{\kappa}_c')M(x, \boldsymbol{\kappa}_d, \boldsymbol{\kappa}_c') = 0 \tag{20-54}$$

† Note that

$$\frac{\partial}{\partial y_1} = \frac{\partial}{\partial y_c}\frac{\partial y_c}{\partial y_1} + \frac{\partial}{\partial y_1}\frac{\partial y_d}{\partial y_1} = \frac{1}{2}\frac{\partial}{\partial y_c} + \frac{\partial}{\partial y_d} \quad \text{and} \quad \frac{\partial}{\partial y_2} = \frac{1}{2}\frac{\partial}{\partial y_c} - \frac{\partial}{\partial y_d}.$$

where

$$F_A(\boldsymbol{\kappa}_c) = \frac{1}{(2\pi)^2} \int [A(0) - A(\boldsymbol{\rho}_d)] \exp(-i\boldsymbol{\kappa}_c \cdot \boldsymbol{\rho}_d)\, d\boldsymbol{\rho}_d$$
$$= \delta(\boldsymbol{\kappa}_c) A(0) - (2\pi)\Phi_\varepsilon(\boldsymbol{\kappa}_c)$$

is a Fourier transform of $A(0) - A(\boldsymbol{\rho}_d)$. The last term in (20-54) was obtained by noting that it is a Fourier transform of a product of two functions $A(0) - A(\boldsymbol{\rho}_d)$ and $\Gamma(x, \boldsymbol{\rho}_c, \boldsymbol{\rho}_d)$.

In (20-54), we recognize that the second term contains $\boldsymbol{\kappa}_d$. This term can be eliminated by writing

$$M(x, \boldsymbol{\kappa}_d, \boldsymbol{\kappa}_c) = M_0(x, \boldsymbol{\kappa}_d, \boldsymbol{\kappa}_c) \exp[-i(\boldsymbol{\kappa}_d \cdot \boldsymbol{\kappa}_c/k)x]. \tag{20-55}$$

We then get a simpler differential equation for M_0:

$$2ik \frac{\partial}{\partial x} M_0(x, \boldsymbol{\kappa}_d, \boldsymbol{\kappa}_c) + \frac{ik^3}{2} \int d\boldsymbol{\kappa}_c' F_A(\boldsymbol{\kappa}_c - \boldsymbol{\kappa}_c')$$
$$\times \exp\left[\frac{i\boldsymbol{\kappa}_d \cdot (\boldsymbol{\kappa}_c - \boldsymbol{\kappa}_c')}{k} x\right] M_0(x, \boldsymbol{\kappa}_d, \boldsymbol{\kappa}_c') = 0. \tag{20-56}$$

We now take the inverse Fourier transform of (20-56) with respect to $\boldsymbol{\kappa}_c$:

$$\Gamma_0(x, \boldsymbol{\kappa}_d, \boldsymbol{\rho}_d) = \int M_0(x, \boldsymbol{\kappa}_d, \boldsymbol{\kappa}_c) \exp(i\boldsymbol{\kappa}_c \cdot \boldsymbol{\rho}_d)\, d\boldsymbol{\kappa}_c. \tag{20-57}$$

Then we get

$$2ik \frac{\partial}{\partial x} \Gamma_0(x, \boldsymbol{\kappa}_d, \boldsymbol{\rho}_d) + \frac{ik^3}{2} \left[A(0) - A\left(\boldsymbol{\rho}_d + \frac{\boldsymbol{\kappa}_d x}{k}\right)\right] \Gamma_0(x, \boldsymbol{\kappa}_d, \boldsymbol{\rho}_d) = 0, \tag{20-58}$$

which can be solved to yield

$$\Gamma_0(x, \boldsymbol{\kappa}_d, \boldsymbol{\rho}_d) = \Gamma_0(0, \boldsymbol{\kappa}_d, \boldsymbol{\rho}_d) \exp\left\{-\int_0^x \frac{k^2}{4} \left[A(0) - A\left(\boldsymbol{\rho}_d + \frac{\boldsymbol{\kappa}_d x'}{k}\right)\right] dx'\right\}. \tag{20-59}$$

The final expression for $\Gamma(x, \boldsymbol{\rho}_c, \boldsymbol{\rho}_d)$ is obtained by taking the inverse transform of (20-57) and using (20-55) and (20-53):

$$\Gamma(x, \boldsymbol{\rho}_c, \boldsymbol{\rho}_d) = \iint M_0(x, \boldsymbol{\kappa}_d, \boldsymbol{\kappa}_c)$$
$$\times \exp\left(-i \frac{\boldsymbol{\kappa}_d \cdot \boldsymbol{\kappa}_c}{k} x + i\boldsymbol{\kappa}_d \cdot \boldsymbol{\rho}_c + i\boldsymbol{\kappa}_c \cdot \boldsymbol{\rho}_d\right) d\boldsymbol{\kappa}_d\, d\boldsymbol{\kappa}_c. \tag{20-60}$$

Using (20-57), this becomes

$$\Gamma(x, \boldsymbol{\rho}_c, \boldsymbol{\rho}_d) = \int \Gamma_0\left(x, \boldsymbol{\kappa}_d, \boldsymbol{\rho}_d - \frac{\boldsymbol{\kappa}_d \mathbf{x}}{k}\right) \exp(i\boldsymbol{\kappa}_d \cdot \boldsymbol{\rho}_c)\, d\boldsymbol{\kappa}_d$$

$$= \int \Gamma_0\left(0, \boldsymbol{\kappa}_d, \boldsymbol{\rho}_d - \frac{\boldsymbol{\kappa}_d x}{k}\right) \exp\left\{i\boldsymbol{\kappa}_d \cdot \boldsymbol{\rho}_c - \int_0^x \frac{k^2}{4}\left[A(0) - A\left(\boldsymbol{\rho}_d - \frac{\boldsymbol{\kappa}_d}{k}(x - x')\right)\right] dx'\right\} d\boldsymbol{\kappa}_d \tag{20-61}$$

where $\Gamma_0(0, \boldsymbol{\kappa}_d, \boldsymbol{\rho}_d)$ is given by the value of Γ at $x = 0$:

$$\Gamma(0, \boldsymbol{\rho}_1, \boldsymbol{\rho}_2) = \Gamma_0(0, \boldsymbol{\rho}_c, \boldsymbol{\rho}_d)$$

$$\Gamma_0(0, \boldsymbol{\kappa}_d, \boldsymbol{\rho}_d) = \frac{1}{(2\pi)^2} \int \Gamma_0(0, \boldsymbol{\rho}_c, \boldsymbol{\rho}_d) \exp(-i\boldsymbol{\kappa}_d \cdot \boldsymbol{\rho}_c)\, d\boldsymbol{\rho}_c. \tag{20-62}$$

Equation (20-61) together with (20-62) constitutes the general solution of (20-51).

Alternatively, we can use $\boldsymbol{\rho}_d' = \boldsymbol{\rho}_d - \boldsymbol{\kappa}_d x/k$ instead of $\boldsymbol{\kappa}_d$, and express (20-61) in the form

$$\Gamma(x, \boldsymbol{\rho}_c, \boldsymbol{\rho}_d) = \left(\frac{k}{2\pi x}\right)^2 \int d\boldsymbol{\rho}_c' \int d\boldsymbol{\rho}_d' \Gamma(0, \boldsymbol{\rho}_c', \boldsymbol{\rho}_d') \exp\left[i\frac{k}{x}(\boldsymbol{\rho}_d - \boldsymbol{\rho}_d') \cdot (\boldsymbol{\rho}_c - \boldsymbol{\rho}_c') - H\right] \tag{20-63}$$

where

$$H = \frac{k^2}{4}\int_0^x \left\{A(0) - A\left[\boldsymbol{\rho}_d' + (\boldsymbol{\rho}_d - \boldsymbol{\rho}_d')\frac{x'}{x}\right]\right\} dx'$$

$$= 4\pi^2 k^2 \int_0^x dx' \int_0^\infty [1 - J_0(\kappa\rho)]\Phi_n(\kappa)\kappa\, d\kappa$$

$$\rho = \left|\boldsymbol{\rho}_d' + (\boldsymbol{\rho}_d - \boldsymbol{\rho}_d')\frac{x'}{x}\right|.$$

We note that $\Gamma(0, \boldsymbol{\rho}_c', \boldsymbol{\rho}_d')$ is the mutual coherence function at $x = 0$ between two points $\boldsymbol{\rho}_1'$ and $\boldsymbol{\rho}_2'$ where $\boldsymbol{\rho}_1' = \boldsymbol{\rho}_c' + \frac{1}{2}\boldsymbol{\rho}_d'$ and $\boldsymbol{\rho}_2' = \boldsymbol{\rho}_c' - \frac{1}{2}\boldsymbol{\rho}_d'$, and $\Gamma(x, \boldsymbol{\rho}_c, \boldsymbol{\rho}_d)$ is the mutual coherence function at x between $\boldsymbol{\rho}_1 = \boldsymbol{\rho}_c + \frac{1}{2}\boldsymbol{\rho}_d$ and $\boldsymbol{\rho}_2 = \boldsymbol{\rho}_c - \frac{1}{2}\boldsymbol{\rho}_d$. We note also that $\exp[i(k/x)(\boldsymbol{\rho}_d - \boldsymbol{\rho}_d') \cdot (\boldsymbol{\rho}_c - \boldsymbol{\rho}_c')]$ is the parabolic approximation of $\exp[ikr_1 - ikr_2]$ where r_1 and r_2 are the distances between the source points $(0, \boldsymbol{\rho}_1')$ and $(0, \boldsymbol{\rho}_2')$ and the observation points $(x, \boldsymbol{\rho}_1)$ and $(x, \boldsymbol{\rho}_2)$, respectively. The function H is related to the wave structure function D for spherical waves. Consider two spherical waves at $(x, \boldsymbol{\rho}_1)$

and $(x, \boldsymbol{\rho}_2)$ originating at $(0, \boldsymbol{\rho}_1')$ and $(0, \boldsymbol{\rho}_2')$, respectively. Let the log-amplitude and phase fluctuations of these two spherical waves be χ_1 and χ_2 and S_1 and S_2, respectively. Then the function H in (20-63) is given by†

$$H = \tfrac{1}{2}D, \qquad D = D_\chi + D_S, \qquad D_\chi = \langle |\chi_1 - \chi_2|^2 \rangle, \qquad D_S = \langle |S_1 - S_2|^2 \rangle. \tag{20-64}$$

20-6 EXAMPLES OF MUTUAL COHERENCE FUNCTIONS

In this section, we consider a few examples.

(a) *Spherical wave* A spherical wave is given by

$$u_0(\mathbf{r}) = e^{ikr}/r. \tag{20-65}$$

The parabolic approximation of (20-65) is given by

$$u_0(\mathbf{r}) = U_0(x, \boldsymbol{\rho}) \exp(ikx), \qquad U_0(x, \boldsymbol{\rho}) = \frac{1}{x} \exp\left(i \frac{k|\boldsymbol{\rho}|^2}{2x}\right). \tag{20-66}$$

The mutual coherence function of a spherical wave at $x = 0$ can be expressed using delta functions:

$$\Gamma(0, \boldsymbol{\rho}_c', \boldsymbol{\rho}_d') = \left(\frac{2\pi}{k}\right)^2 \delta(\boldsymbol{\rho}_c')\, \delta(\boldsymbol{\rho}_d'). \tag{20-67}$$

We note that in free space ($H = 0$), if we substitute (20-67) into (20-63), we obtain the correct mutual coherence function:

$$\Gamma(x, \boldsymbol{\rho}_c, \boldsymbol{\rho}_d) = \frac{1}{x^2} \exp\left(i \frac{k|\boldsymbol{\rho}_1|^2}{2x} - i \frac{k|\boldsymbol{\rho}_2|^2}{2x}\right) = \frac{1}{x^2} \exp\left(i \frac{k}{x} \boldsymbol{\rho}_c \cdot \boldsymbol{\rho}_d\right). \tag{20-68}$$

In a random medium, substituting (20-67) into (20-63), we get the mutual coherence function for a spherical wave:

$$\Gamma(x, \boldsymbol{\rho}_c, \boldsymbol{\rho}_d) = \frac{1}{x^2} \exp\left(i \frac{k}{x} \boldsymbol{\rho}_c \cdot \boldsymbol{\rho}_d - H\right)$$

$$H = \frac{k^2}{4} \int_0^x \left[A(0) - A\left(\boldsymbol{\rho}_d \frac{x'}{x}\right)\right] dx' \tag{20-69}$$

$$= 4\pi^2 k^2 \int_0^x dx' \int_0^\infty \left[1 - J_0\left(\kappa \rho_d \frac{x'}{x}\right)\right] \Phi_n(\kappa) \kappa \, d\kappa.$$

† This is the same as that of two crossing spherical waves discussed in Section 19-6. Note that $\boldsymbol{\rho}_d' + (\boldsymbol{\rho}_d - \boldsymbol{\rho}_d')(x'/x)$ in H is identical to $\boldsymbol{\rho}_t + \gamma(\boldsymbol{\rho} - \boldsymbol{\rho}_t)$ in (19-57).

(b) *Beam wave* At $x = 0$, a beam wave is given by (see Section 18-4)

$$U_0(\rho) = \exp(-\tfrac{1}{2}k\alpha\rho^2), \qquad \alpha = \alpha_r + i\alpha_i = \frac{\lambda}{\pi W_0^2} + \frac{1}{R_0}. \tag{20-70}$$

Therefore the mutual coherence function at $x = 0$ is given by

$$\begin{aligned}\Gamma(0, \boldsymbol{\rho}_c', \boldsymbol{\rho}_d') &= \exp\left[-\frac{k\alpha}{2}\rho_1'^2 - \frac{k\alpha^*}{2}\rho_2'^2\right] \\ &= \exp\{-k[\alpha_r(\rho_c'^2 + \tfrac{1}{4}\rho_d'^2) + i\alpha_i \boldsymbol{\rho}_c' \cdot \boldsymbol{\rho}_d']\}.\end{aligned} \tag{20-71}$$

Using (20-62), we write

$$\Gamma_0(0, \boldsymbol{\kappa}_d, \boldsymbol{\rho}_d) = \frac{1}{4\pi k\alpha_r}\exp\left[-k\alpha_r\frac{\rho_d^2}{4} - \frac{|\boldsymbol{\kappa}_d + k\alpha_i\boldsymbol{\rho}_d|^2}{4k\alpha_r}\right]. \tag{20-72}$$

Substituting this into (20-61), we obtain

$$\Gamma(x, \boldsymbol{\rho}_c, \boldsymbol{\rho}_d) = \frac{W_0^2}{8\pi}\int d\boldsymbol{\kappa}_d \exp[-a\rho_d^2 - b\kappa_d^2 + c\boldsymbol{\rho}_d \cdot \boldsymbol{\kappa}_d + i\boldsymbol{\kappa}_d \cdot \boldsymbol{\rho}_c - H] \tag{20-73}$$

where

$$a = \frac{1}{2W_0^2}\left(1 + \frac{\alpha_i^2}{\alpha_r^2}\right), \qquad b = \frac{W^2}{8} = \frac{W_0^2}{8}[(\alpha_r x)^2 + (1 - \alpha_i x)^2]$$

$$c = \tfrac{1}{2}\left[\alpha_r x - \frac{\alpha_i}{\alpha_r}(1 - \alpha_i x)\right], \qquad H = \frac{k^2}{4}\int_0^x \left[A(0) - A\left(\boldsymbol{\rho}_d + \frac{\boldsymbol{\kappa}_d x'}{k}\right)\right] dx'.$$

(c) *Incoherent source* If the source is completely incoherent at $x = 0$, we have (Beran and Parrent, 1964, p. 67)

$$\Gamma(0, \boldsymbol{\rho}_c', \boldsymbol{\rho}_d') = \frac{4\pi^2}{k^2} I(\boldsymbol{\rho}_c')\,\delta(\boldsymbol{\rho}_d') \tag{20-74}$$

where $I(\boldsymbol{\rho}_c)$ is the intensity at $\boldsymbol{\rho}_c$. Substituting (20-74) into (20-63), we obtain

$$\Gamma(x, \boldsymbol{\rho}_c, \boldsymbol{\rho}_d) = \frac{1}{x^2}\int d\boldsymbol{\rho}_c' I(\boldsymbol{\rho}_c') \exp\left[i\frac{k}{x}\boldsymbol{\rho}_d \cdot (\boldsymbol{\rho}_c - \boldsymbol{\rho}_c') - H\right]$$

$$H = \frac{k^2}{4}\int_0^x \left\{A(0) - A\left(\boldsymbol{\rho}_d\frac{x'}{x}\right)\right\} dx'. \tag{20-75}$$

20-7 MUTUAL COHERENCE FUNCTION IN A TURBULENT MEDIUM

In a turbulent medium, the spectrum of the refractive index fluctuation may be approximated by the modified von Karman spectrum:

$$\Phi_n(\kappa) = 0.033C_n{}^2(\kappa^2 + 1/L_0{}^2)^{-11/6} \exp(-\kappa^2/\kappa_m{}^2) \qquad (20\text{-}76)$$

where $\kappa_m = 5.92/l_0$. Let us consider H using this spectrum:

$$H = \frac{k^2}{4} \int_0^x \{A(0) - A(\boldsymbol{\rho})\}\, dx'$$

$$= 4\pi^2 k^2 \int_0^x dx' \int_0^\infty [1 - J_0(\kappa\rho)]\Phi_n(\kappa)\kappa\, d\kappa. \qquad (20\text{-}77)$$

It can be shown (Ishimaru, 1977b) that approximately

$$A(0) - A(\rho) = 6.56C_n{}^2 \frac{\rho^2}{l_0^{1/3}} \quad \text{for} \quad \rho < l_0 \qquad (20\text{-}78a)$$

$$= 5.83C_n{}^2\rho^{5/3} \quad \text{for} \quad l_0 < \rho < L_0 \qquad (20\text{-}78b)$$

$$= 3.127C_n{}^2 L_0^{5/3} \quad \text{for} \quad L_0 < \rho, \qquad (20\text{-}78c)$$

and each of these three approximations may be used for the following three propagation distances, respectively:

$$x \gg x_i \qquad (20\text{-}79a)$$

$$x_i \gg x \gg x_c \qquad (20\text{-}79b)$$

$$x_c \gg x \qquad (20\text{-}79c)$$

where $x_i = [0.39C_n{}^2k^2l_0^{5/3}]^{-1}$ and $x_c = [0.39C_n{}^2k^2L_0^{5/3}]^{-1}$. For the beam wave at distance $x \gg x_i$, we use (20-78a) and obtain (Ishimaru, 1977b)

$$\Gamma(x, \boldsymbol{\rho}_c, \boldsymbol{\rho}_d) = (W_0/W)^2 \exp[-2\rho_c{}^2/W^2 - p\rho_d{}^2 + i2q\boldsymbol{\rho}_d \cdot \boldsymbol{\rho}_c] \qquad (20\text{-}80a)$$

where

$$W^2 = W_0{}^2[(\alpha_r x)^2 + (1 - \alpha_i x)^2] + 4.38C_n{}^2 l_0^{-1/3} x^3$$

$$p = \frac{1}{2W_0{}^2}\left(1 + \frac{\alpha_i{}^2}{\alpha_r{}^2}\right) + \beta x - \frac{2}{W^2}F^2$$

$$q = \frac{2}{W^2}F, \qquad F = \frac{\alpha_r x}{2}\left(1 + \frac{\alpha_i{}^2}{\alpha_r{}^2}\right) - \frac{\alpha_i}{2\alpha_r} + \beta\frac{x^2}{k}$$

$$\beta = 1.64k^2C_n{}^2 l_0^{-1/3}.$$

For the beam wave at $x_i \gg x \gg x_c$, we use (20-73) with the following expression for H:

$$H = 1.46 \int_0^x k^2 C_n{}^2 \left| \boldsymbol{\rho}_d - \frac{\mathbf{K}_d}{k} x' \right|^{5/3} dx'. \tag{20-80b}$$

This is the case usually encountered in practice. Figure 20-1 shows an example of the average intensity on the beam axis as a function of distance. It shows that in turbulence, the intensity decreases considerably faster than in free space, that the intensity for a focused beam approaches that for a collimated beam, and the focusing effect disappears within a short distance.

Figure 20-2 shows the beam broadening and the correlation distance of a collimated beam. At 0.5 km, the beam spread is negligible, but the correla-

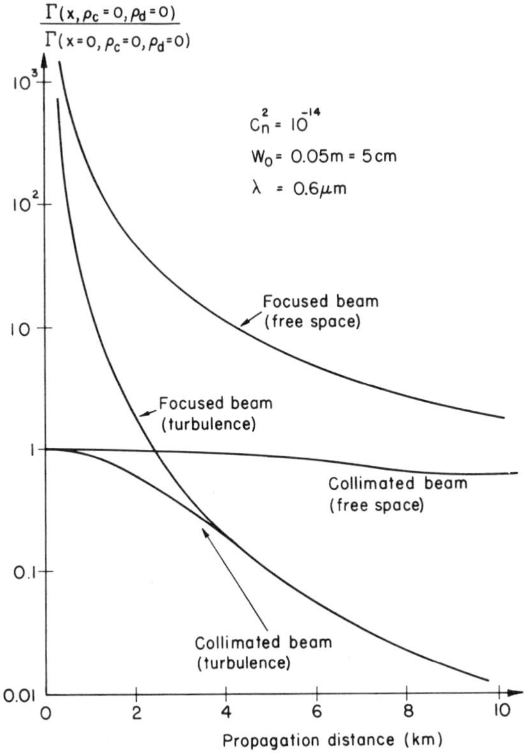

FIG. 20-1 Average intensity $\langle I(x, \rho_c = 0) \rangle = \Gamma(x, \rho_c = 0, \rho_d = 0)$ of collimated and focused beams on the beam axis as functions of distance.

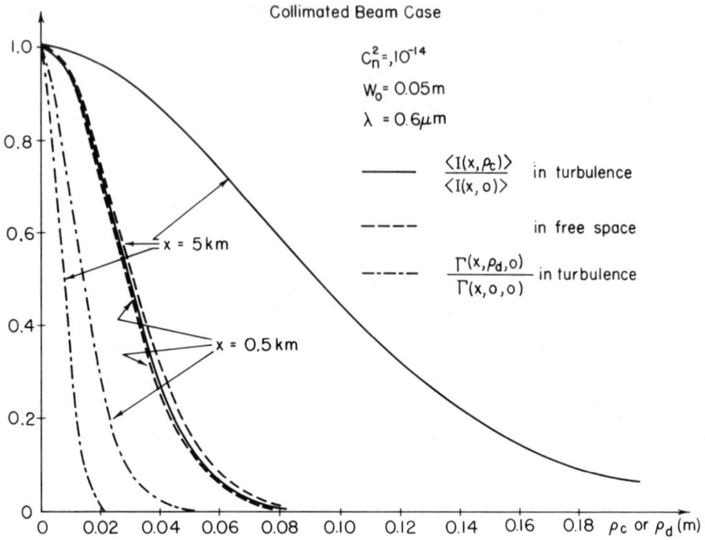

FIG. 20-2 Normalized average intensity $\langle I(x, \rho_c)\rangle$ as functions of transverse distance ρ_c and normalized coherence function $\Gamma(x, \rho_d, \rho_c = 0)$ on the beam axis as functions of separation ρ_d.

tion distance is reduced to $\sqrt{\lambda L}$. At 5 km, the beam spread is considerable and the correlation distance is much smaller than $\sqrt{\lambda L}$.

In the region $x_i \gg x \gg x_c$, the mutual coherence function for a plane wave is given by

$$\Gamma(x, \rho_d) = \exp(-1.46k^2 x C_n{}^2 \rho_d^{5/3}). \tag{20-81a}$$

For a spherical wave, we obtain

$$\Gamma(x, \boldsymbol{\rho}_c, \boldsymbol{\rho}_d) = (1/x^2)\exp(i(k/x)\boldsymbol{\rho}_c \cdot \boldsymbol{\rho}_d - 0.547k^2 x C_n{}^2 \rho_d^{5/3}). \tag{20-81b}$$

20-8 TEMPORAL FREQUENCY SPECTRA

In Chapter 19 we discussed the temporal fluctuations and spectra of a wave when the random medium is moving with the wind. It was noted that the wind velocity parallel to the direction of wave propagation has little effect on the temporal fluctuation, whereas the wind velocity transverse to the propagation path has a direct effect on the temporal fluctuations. In this section, we consider the transverse wind velocity $\mathbf{V}$ and assume Taylor's "frozen-in" hypothesis that eddies do not change their shape within the time required for these eddies to move over the distance of the eddy size. Under this assumption, the parabolic equation (20-42) becomes

$$\left\{2ik\frac{\partial}{\partial x} + (\nabla_{t2}^2 - \nabla_{t2}^2) + \frac{ik^3}{2}[A(0) - A(\boldsymbol{\rho}_1 - \boldsymbol{\rho}_2 - \mathbf{V}\tau)]\right\}\Gamma(x, \boldsymbol{\rho}_1, \boldsymbol{\rho}_2, \tau) = 0 \tag{20-82}$$

where

$$\Gamma(x, \boldsymbol{\rho}_1, \boldsymbol{\rho}_2, \tau) = \langle U(x, \boldsymbol{\rho}_1, t + \tau)U^*(x, \boldsymbol{\rho}_2, t)\rangle.$$

The solution to (20-82) is identical to (20-61) or (20-63) except that $A(\boldsymbol{\rho})$ is replaced by $A(\boldsymbol{\rho} - \mathbf{V}\tau)$.

For a spherical wave, the temporal fluctuation is obtained from (20-69):

$$\Gamma(x, \boldsymbol{\rho}_c, \boldsymbol{\rho}_d, \tau) = (1/x^2)\exp(i(k/x)\boldsymbol{\rho}_c \cdot \boldsymbol{\rho}_d - H)$$

$$H = \frac{k^2}{4}\int_0^x \left[A(0) - A\left(\boldsymbol{\rho}_d\frac{x'}{x} - \mathbf{V}\tau\right)\right]dx' \tag{20-83}$$

$$= 4\pi^2k^2\int_0^x dx'\int_0^\infty \left[1 - J_0\left(\kappa\left|\boldsymbol{\rho}_d\frac{x'}{x} - \mathbf{V}\tau\right|\right)\right]\Phi_n(\kappa)\kappa\, d\kappa.$$

On the x axis, $\boldsymbol{\rho}_c = \boldsymbol{\rho}_d = 0$ and we have

$$\Gamma(x, 0, 0, \tau) = (1/x^2)\exp(-H)$$

$$H = 4\pi^2k^2\int_0^x dx'\int_0^\infty [1 - J_0(\kappa V\tau)]\Phi_n(\kappa)\kappa\, d\kappa. \tag{20-84}$$

The mutual coherence function (20-84) can be written as the sum of the coherent part Γ_c and the incoherent part Γ_i. The coherent part Γ_c is the coherent intensity given in (20-27) and independent of τ:

$$\Gamma_c(x, 0, 0, \tau) = \frac{1}{x^2}\exp(-2\alpha_0 x), \qquad 2\alpha_0 = 4\pi^2k^2\int_0^x dx'\int_0^\infty \Phi_n(\kappa)\kappa\, d\kappa. \tag{20-85}$$

Therefore, we write

$$\Gamma(x, 0, 0, \tau) = \Gamma_c + \Gamma_i, \qquad \Gamma_i(x, 0, 0, \tau) = x^{-2}[\exp(-H) - \exp(-2\alpha_0 x)]. \tag{20-86a}$$

Note that as $\tau \to \infty$, $\Gamma_i \to 0$.

If the turbulence is strong and the distance x is large so that the coherent part Γ_c disappears, then we have approximately

$$\Gamma(x, 0, 0, \tau) \approx \frac{1}{x^2}\exp\left[-1.46k^2(V\tau)^{5/3}\int_0^x C_n{}^2(x')\, dx'\right]. \tag{20-86b}$$

Let us next consider the temporal frequency spectrum of the spherical wave. Taking the Fourier transform of (20-86), we get

$$W_f(x, \omega) = 2\int_{-\infty}^{\infty} \Gamma(x, 0, 0, \tau)e^{-i\omega\tau}\, d\tau = W_{fc}(x, \omega) + W_{fi}(x, \omega)$$

$$W_{fc}(x, \omega) = \frac{4\pi}{x^2}\exp(-2\alpha_0 x)\,\delta(\omega) \tag{20-87}$$

$$W_{fi}(x, \omega) = \frac{2}{x^2}\int_{-\infty}^{\infty} [\exp(-H) - \exp(-2\alpha_0 x)]e^{-i\omega\tau}\, d\tau.$$

Note that ω in (20-87) is the frequency deviation from the operating frequency ω_0 ($k = \omega_0/c$). We note also that the coherent intensity Γ_c gives rise to a delta function spectrum at $\omega = 0$, and this means that the coherent field is a monochromatic wave at ω_0. The incoherent part Γ_i has the spectrum broadening given by W_{fi} in (20-87).

For a beam wave, we obtain the temporal fluctuation from (20-73) by modifying H as follows:

$$H = \frac{k^2}{4}\int_0^x \left[A(0) - A\left(\boldsymbol{\rho}_d + \frac{\boldsymbol{\kappa}_d x'}{k} - \mathbf{V}\tau\right)\right] dx'. \tag{20-88}$$

On the beam axis ($\boldsymbol{\rho}_c = \boldsymbol{\rho}_d = 0$), we have

$$\Gamma(x, 0, 0, \tau) = \Gamma_c + \Gamma_i$$

$$\Gamma_c = \frac{W_0^{\,2}}{8\pi}\int d\boldsymbol{\kappa}_d \exp[-b\kappa_d^{\,2} - 2\alpha_0 x]$$

$$= \frac{W_0^{\,2}}{W^2}\exp(-2\alpha_0 x) \tag{20-89}$$

$$\Gamma_i = \frac{W_0^{\,2}}{4\pi}\int d\boldsymbol{\kappa}_d \exp(-b\kappa_d^{\,2})[\exp(-H) - \exp(-2\alpha_0 x)].$$

The temporal frequency spectra for Γ_c and Γ_i are given by

$$W_{fc}(x, \omega) = \frac{W_0^{\,2}}{W^2}\exp(-2\alpha_0 x)\,\delta(\omega), \qquad W_{fi}(x, \omega) = 2\int_{-\infty}^{\infty} \Gamma_i e^{-i\omega\tau}\, d\tau. \tag{20-90}$$

20-9 TWO-FREQUENCY CORRELATION FUNCTION

The mutual coherence function Γ (Hong and Ishimaru, 1976; Ishimaru and Hong, 1975a; Lee, 1974; Lee and Jokipii, 1975; Liu, *et al.*, 1974;

Shishov, 1974; Uscinski, 1974) between two different operating frequencies $k_1 = \omega_1/c$ and $k_2 = \omega_2/c$ is defined by

$$
\begin{aligned}
&\langle u(x, \boldsymbol{\rho}_1, t+\tau, k_1) u^*(x, \boldsymbol{\rho}_2, t, k_2)\rangle \\
&\quad = \langle U(x, \boldsymbol{\rho}_1, t+\tau, k_1) U^*(x, \boldsymbol{\rho}_2, t, k_2)\rangle \exp[-i\omega_1(t+\tau) + i\omega_2 t] \\
&\quad = \Gamma(x, \boldsymbol{\rho}_1, \boldsymbol{\rho}_2, \tau, k_1, k_2) \exp[-i\omega_1(t+\tau) + i\omega_2 t]. \qquad (20\text{-}91)
\end{aligned}
$$

Employing the procedure given in (20-4), the parabolic differential equation for Γ is given by

$$
\left[2i\frac{\partial}{\partial x} + \frac{1}{k_1}\nabla_{t1}^2 - \frac{1}{k_2}\nabla_{t2}^2 + \frac{i}{4}\{k_1{}^2 A_1(0) + k_2{}^2 A_2(0) - k_1 k_2[A_1(\boldsymbol{\rho}_d - \mathbf{V}\tau) + A_2(\boldsymbol{\rho}_d - \mathbf{V}\tau)]\}\right]\Gamma = 0 \qquad (20\text{-}92)
$$

where $A_1(\boldsymbol{\rho})$ is given in terms of the spectral density Φ_{n1} of the refractive index fluctuation at k_1:

$$
A_1(\boldsymbol{\rho}) = 16\pi^2 \int_0^\infty J_0(\kappa\rho)\Phi_{n1}(\kappa)\kappa\, d\kappa, \qquad (20\text{-}93)
$$

and A_2 is given by the same equation with Φ_{n1} replaced by Φ_{n2}, the spectral density at k_2.

The two-frequency correlation function Γ is important for a broadband pulse propagation problem. It is also important in analyzing the effect of interstellar turbulence on the shape of a pulse emitted by a pulsar. A general solution of (20-92) has not been obtained to date. For a plane wave case, some simplifications are possible and we present this result in the following section.

20-10 PLANE WAVE SOLUTION FOR THE TWO-FREQUENCY MUTUAL COHERENCE FUNCTION

For a plane wave case, we have $\nabla_{t1} = \nabla_{t2} = \nabla_t$ in (20-92). For a turbulent medium, we have approximately [see (20-78b)]

$$
A(0) - A(\rho) = 5.83 C_n{}^2 \rho^{5/3}. \qquad (20\text{-}94)
$$

Now we let, using $k_d = k_1 - k_2$,

$$
\Gamma = \Gamma_1 \exp[-(k_d{}^2/8)A(0)z]. \qquad (20\text{-}95)
$$

Using (20-94), we get

$$
\left[\frac{\partial}{\partial z} + ia\nabla_t{}^2 + bc\rho^{\nu}\right]\Gamma_1 = 0 \qquad (20\text{-}96)
$$

where

$$a = \frac{k_d/2}{[k_c^2 - (k_d/2)^2]} \approx \frac{k_d/2}{k^2}, \qquad b = \tfrac{1}{4}[k_c^2 - (k_d/2)^2] \approx k^2/4$$

$$k_c = \tfrac{1}{2}(k_1 + k_2) \approx k, \qquad k_d = k_1 - k_2, \qquad c = 5.83C_n^2, \qquad \nu = 5/3.$$

The approximations for a and b are valid in most practical situations. In the preceding, k is the wave number for the carrier frequency or the center frequency.

Next, we normalize (20-96) by using [see (15-119)] $z/L = z'$ and $\rho/\rho_0 = \rho'$, where L is the length of the turbulent medium $(0 \le z' \le 1)$ and ρ_0 is the correlation distance. We then obtain

$$\left[\frac{\partial}{\partial z'} + i\left(\frac{k_d}{k_{coh1}}\right)\nabla'^2 + \rho'^{\nu}\right]\Gamma_1 = 0 \tag{20-97}$$

where $\rho_0 = (5.83C_n^2Lk^2/4)^{-3/5}$ and $k_{coh1} = 0.31k/\sigma_\chi^{12/5}$. The correlation distance ρ_0 is the distance ρ at which the mutual coherence function (20-46b) decreases to e^{-1} of the value at $\rho = 0$. Note that for a turbulent medium, (20-46b) becomes

$$\Gamma(x, \boldsymbol{\rho}) = \exp[-(\rho/\rho_0)^{5/3}]. \tag{20-98}$$

The other quantity k_{coh1} is the wave number for the coherence bandwidth and σ_χ^2 is the variance of the log-amplitude fluctuation of a plane wave given by

$$\sigma_\chi^2 = 0.307C_n^2k^{7/6}L^{11/6}. \tag{20-99}$$

We can write the coherence bandwidth f_{coh1}

$$f_{coh1} = 0.31f/\sigma_\chi^{12/5} = 1.28fC_n^{-12/5}k^{-7/5}L^{-11/5}. \tag{20-100}$$

From this, it is seen that for a nonionized turbulence, the coherence bandwidth is proportional to $f^{-2/5}$.

Let us consider an ionized plasma. The relative dielectric constant ε_r is given by

$$\varepsilon_r = 1 - (\omega_p/\omega)^2 \tag{20-101}$$

where ω_p is the plasma frequency defined by

$$\omega_p^2 = e^2N_e/m\varepsilon_0, \tag{20-102}$$

N_e is the electron number density (m^{-3}), e and m the charge and mass of an electron, and ε_0 the free space permittivity. We then have the fluctuation of the relative dielectric constant:

$$\varepsilon_r = \langle\varepsilon_r\rangle(1+\varepsilon_1), \qquad \langle\varepsilon_r\rangle = 1 - \left\langle\left(\frac{\omega_p}{\omega}\right)^2\right\rangle, \qquad \varepsilon_1 = -\frac{e^2}{(m\varepsilon_0\langle\varepsilon_r\rangle)}\frac{\delta N_e}{\omega^2}. \tag{20-103}$$

This shows that since $\langle\varepsilon_r\rangle$ is only slightly dependent on ω, the refractive index fluctuation n_1 $(=\varepsilon_1/2)$ is approximately proportional to ω^{-2}. Since the structure constant $C_n{}^2$ is proportional to $\langle n_1{}^2\rangle$, we have

$$C_n{}^2 = \frac{1}{4}\left[\frac{e^2}{m\varepsilon_0\langle\varepsilon_r\rangle}\right]^2\frac{1}{\omega^4}C_N{}^2 \tag{20-104}$$

where C_N is the structure constant for the electron density fluctuation. Substituting (20-104) into (20-100), we obtain

$$f_{\text{coh}1} \sim C_N^{-12/5} f^{22/5} L^{-11/5}. \tag{20-105}$$

Since the pulse broadening is inversely proportional to the coherence bandwidth, we have

$$\text{pulse broadening} \sim \lambda^{22/5}. \tag{20-106}$$

It has been shown that the width of a pulse from a pulsar has a wavelength dependence closely represented by (20-106). Figure 20-3 shows a numerical solution of (20-97).

Let us rewrite (20-95) as

$$\Gamma = \Gamma_1(k_d/k_{\text{coh}1}, \rho/\rho_0)\exp(-k_d{}^2/k_{\text{coh}2}^2) \tag{20-107}$$

where $k_{\text{coh}2} = k/(\alpha_0 L)^{1/2}$, and α_0 is given in (20-24). From this, we note that when the optical distance $2\alpha_0 L$ is comparable to unity, $k_{\text{coh}2} < k_{\text{coh}1}$, and therefore $k_{\text{coh}2}$ determines the pulse broadening. On the other hand, at a large distance, $k_{\text{coh}2} > k_{\text{coh}1}$, and $k_{\text{coh}1}$ determines the pulse characteristics. Note that $k_{\text{coh}1}$ is proportional to $L^{-11/5}$ whereas $k_{\text{coh}2}$ is proportional to $L^{-1/2}$.

The effects of interstellar inhomogeneities on the shape of the radio pulses from pulsars have been studied extensively (Erukhimov, 1972; Williamson, 1972, 1973; Lang, 1971; Ables *et al.*, 1970; Lee and Jokipii, 1975a,b; Shishov, 1974; Uscinski, 1974; for other references on pulse propagation in a random medium, see Plonus *et al.*, 1972; Gardner and Plonus, 1975; Knollman, 1965; Mintzer, 1953; Shirokova, 1963; El-Khamy and McIntosh, 1973; Brown and Clifford, 1973; Bucher and Lerner, 1973).

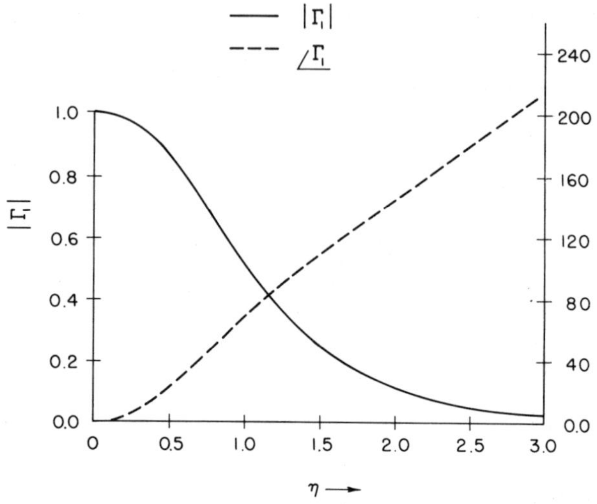

FIG. 20-3 Numerical solution for Γ_1 when $\rho = 0$ as a function of $\eta = (k_d/k_{coh})^{5/11}$.

20-11 PULSE SHAPE

It has been shown (Hong and Ishimaru, 1976; Sreenivasiah *et al.*, 1976) that the output intensity $I(t)$ for an input intensity pulse $I_i(t)$ is given by the convolution integral [see (15-85)]

$$I(t) = \int G(t - t')I_i(t')\,dt' \tag{20-108}$$

where $G(t)$ is the intensity response to a delta function input and is given by a Fourier transform of the two-frequency correlation function Γ:

$$G(t) = \frac{1}{2\pi}\int d\omega_d \Gamma(\omega_d) \exp(-i\omega_d t). \tag{20-109}$$

Since Γ is given by (20-107), $G(t)$ can be expressed as a convolution integral

$$G(t) = \int I_1(t - t')I_2(t')\,dt' \tag{20-110}$$

where

$$I_1(t) = \frac{1}{2\pi}\int \Gamma_1(\omega_d) \exp\left[i\omega_d\left(t - \frac{z}{c_0}\right)\right] d\omega_d$$

$$I_2(t) = \frac{1}{2\pi}\int \exp\left(-\frac{k_d{}^2}{k_{coh2}^2}\right) \exp\left[i\omega_d\left(t - \frac{z}{c_0}\right)\right] d\omega_d .$$

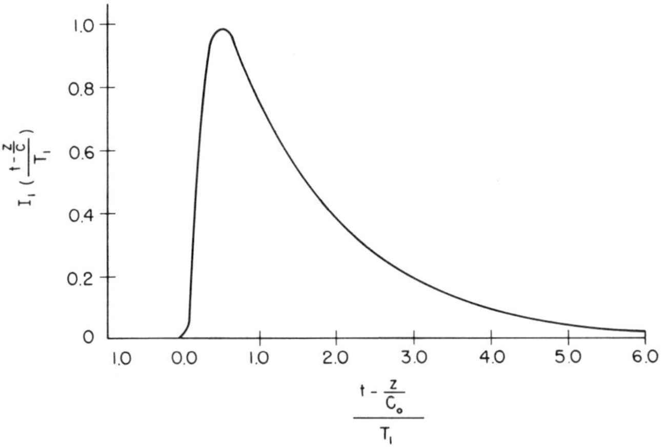

FIG. 20-4 The pulse shape I_1 due to a delta function input.

Figure 20-4 shows a numerical calculation of $I_1(t - z/c_0)$ as a function of a normalized time $(t - z/c_0)/T_1$ where $T_1 = [\omega_{\text{coh}1}]^{-1}$. The integral for $I_2(t - z/c_0)$ can be easily carried out to yield a Gaussian pulse shape. The effects of I_1 and I_2 on pulse broadening have been discussed at the end of the preceding section.

20-12 ANGULAR AND TEMPORAL FREQUENCY SPECTRA†

In (20-46b), we showed the mutual coherence function for a plane wave case. Using (20-78b) for turbulence, we rewrite (20-46b) as

$$\Gamma(x, \boldsymbol{\rho}) = \exp[-(\rho/\rho_0)^{5/3}] \tag{20-111}$$

where $\rho_0 = [1.46 C_n{}^2 k^2 x]^{-3/5}$.

The angular spectrum $\Gamma(x, \theta)$ of the intensity is given by a Fourier transform of the mutual coherence function $\Gamma(x, \boldsymbol{\rho})$. Therefore we get

$$\begin{aligned}\Gamma(x, \theta) &= \frac{k^2}{(2\pi)^2} \int \Gamma(x, \boldsymbol{\rho}) \exp(-ik\boldsymbol{\rho} \cdot \boldsymbol{\theta})\, d\boldsymbol{\rho} \\ &= \frac{k^2}{2\pi} \int_0^\infty \exp\left[-\left(\frac{\rho}{\rho_0}\right)^{5/3}\right] J_0(k\rho\theta)\rho\, d\rho \\ &= \frac{1}{2\pi\theta_c{}^2} \int_0^\infty \exp(-t^{5/3}) J_0\left(\frac{\theta}{\theta_c} t\right) t\, dt \end{aligned} \tag{20-112}$$

where $\theta_c = (k\rho_0)^{-1}$. This is plotted in Fig. 20-5.

† In Section 15-2 the angular spectrum is related to the specific intensity and is expressed as $I(z, \theta)$.

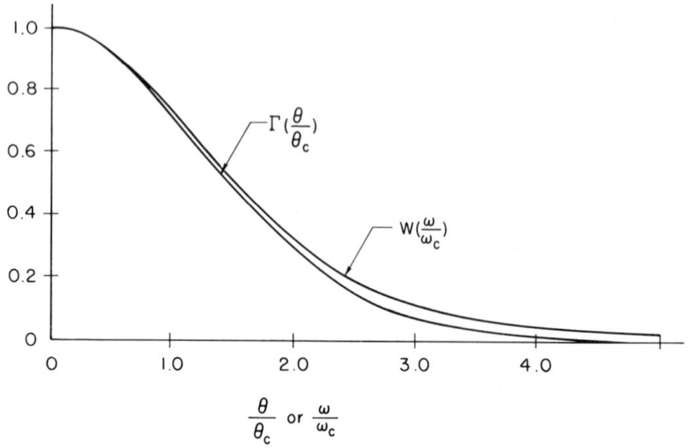

FIG. 20-5 Angular spectrum $\Gamma(\theta/\theta_c)$ and temporal frequency spectrum $W(\omega/\omega_c)$.

Let us next calculate the temporal frequency spectrum $W(\omega)$. Letting $\mathbf{V}$ be the wind velocity transverse to the propagation path, we get

$$W(\omega) = 2 \int \Gamma(x, \mathbf{V}\tau) \exp(-i\omega\tau)\, d\tau$$

$$= \frac{2}{\omega_c} \int \exp(-t^{5/3}) \exp\left(-i\frac{\omega}{\omega_c}t\right) dt \tag{20-113}$$

where $\omega_c = V/\rho_0$. This is also shown in Fig. 20-5. Note that the frequency spread of the frequency spectrum is given by

$$\omega_c = V/\rho_0 = V[1.46 C_n{}^2 k^2 x]^{3/5}. \tag{20-114}$$

If the structure constant $C_n{}^2$ varies along the propagation path, we then have

$$\omega_c = V\left[\int_0^L 1.46 C_n{}^2(x) k^2\, dx\right]^{3/5}. \tag{20-115}$$

As an example, consider the frequency spectrum of the fluctuation of a microwave transmitted from a spacecraft through the turbulent plasma surrounding the sun. The scale size (outer scale) of this turbulence is estimated (Woo *et al.*, 1976b)† to be at least 10^6 km. Since the Fresnel size $(\lambda L)^{1/2}$ at 1 A.U.‡ is approximately 150 km at 2 GHz, we can make use of formula

† Also see Lotova (1975) for a summary of recent studies in the Soviet Union.
‡ A.U. = astronomical unit = mean earth-sun distance = 149,598,000 km.

(20-115). If we assume that the structure constant is a function of distance from the sun, then we have

$$C_n^2(r) = C_n^2[(r_0^2 + x^2)^{1/2}] \tag{20-116}$$

where r_0 is the distance from the sun to the closest approach of the propagation path and is given by $r_0 = (1 \text{ A.U.}) \sin \varepsilon$. The angle ε is called the elongation of the source (see Fig. 20-6). The velocity V should be the transverse component V_t of the solar wind. The frequency spectrum (20-113) and (20-115) for a microwave propagated through the turbulent plasma surrounding the sun has been experimentally verified (Woo *et al.*, 1976a,b).

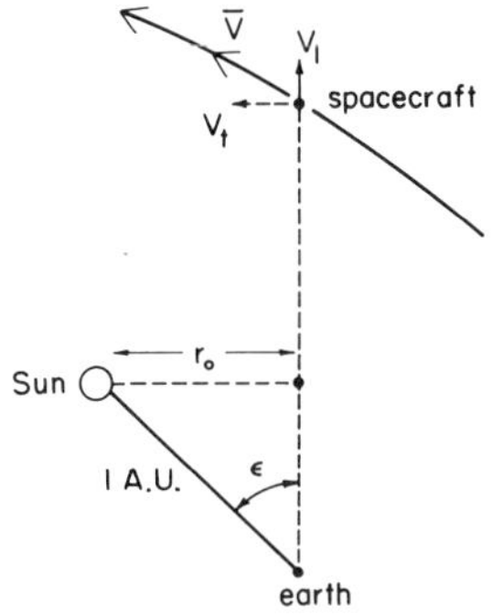

FIG. 20-6 Geometry showing the sun, the earth, a spacecraft, and the elongation ε.

20-13 FOURTH ORDER MOMENTS

The parabolic equation for the higher order moment has been obtained by means of a variety of techniques (Tatarski, 1971; Brown, 1972; Furutsu, 1972; Shishov, 1968; Beran and Ho, 1969; Gurvich and Tatarski, 1975). It has the following form:

$$\left[2ik\frac{\partial}{\partial x} + (\Delta_1 + \cdots + \Delta_n - \Delta_1' - \cdots - \Delta_m') + \frac{ik^3}{4} F_{nm}\right]\Gamma_{nm} = 0 \tag{20-117}$$

where

$$\Gamma_{nm} = \langle U(x, \boldsymbol{\rho}_1) \cdots U(x, \boldsymbol{\rho}_n) U^*(x, \boldsymbol{\rho}_1') \cdots U^*(x, \boldsymbol{\rho}_m')\rangle$$

and $\Delta_1, \ldots, \Delta_n, \Delta_1', \ldots, \Delta_m'$ are the transverse Laplacian with respect to $\boldsymbol{\rho}_1, \ldots, \boldsymbol{\rho}_n, \boldsymbol{\rho}_1', \ldots, \boldsymbol{\rho}_m'$, respectively. F_{nm} is given by

$$F_{nm} = \sum_{i=1}^{n} \sum_{j=1}^{n} A(\boldsymbol{\rho}_i - \boldsymbol{\rho}_j) - 2\sum_{i=1}^{n} \sum_{k=1}^{m} A(\boldsymbol{\rho}_i - \boldsymbol{\rho}_k') + \sum_{k=1}^{m} \sum_{l=1}^{m} A(\boldsymbol{\rho}_k' - \boldsymbol{\rho}_l'). \tag{20-118}$$

To date, however, no general solution has been obtained even for the fourth moment, except for approximate solutions (Tatarski, 1971; Fante, 1975).

Let us consider the fourth order moment Γ_4:

$$\Gamma_4 = \langle U(x, \boldsymbol{\rho}_1)U(x, \boldsymbol{\rho}_2)U^*(x, \boldsymbol{\rho}_1{}')U^*(x, \boldsymbol{\rho}_2{}')\rangle. \tag{20-119}$$

We make use of a set of new variables:

$$\mathbf{R} = \tfrac{1}{4}(\boldsymbol{\rho}_1 + \boldsymbol{\rho}_2 + \boldsymbol{\rho}_1{}' + \boldsymbol{\rho}_2{}'), \qquad \boldsymbol{\rho} = \boldsymbol{\rho}_1 + \boldsymbol{\rho}_2 - \boldsymbol{\rho}_1{}' - \boldsymbol{\rho}_2{}'$$
$$\mathbf{r}_1 = \tfrac{1}{2}(\boldsymbol{\rho}_1 - \boldsymbol{\rho}_2 + \boldsymbol{\rho}_1{}' - \boldsymbol{\rho}_2{}'), \qquad \mathbf{r}_2 = \tfrac{1}{2}(\boldsymbol{\rho}_1 - \boldsymbol{\rho}_2 - \boldsymbol{\rho}_1{}' + \boldsymbol{\rho}_2{}'). \tag{20-120}$$

Equation (20-117) can then be written as

$$\left[\frac{\partial}{\partial x} - \frac{i}{k}(\nabla_R \cdot \nabla_\rho + \nabla_{r1} \cdot \nabla_{r2}) + Q\right]\Gamma_4 = 0 \tag{20-121}$$

where

$$Q = \frac{\pi k^2}{4}\left[H\left(\mathbf{r}_1 + \frac{\boldsymbol{\rho}}{2}\right) + H\left(\mathbf{r}_1 - \frac{\boldsymbol{\rho}}{2}\right) + H\left(\mathbf{r}_2 + \frac{\boldsymbol{\rho}}{2}\right)\right.$$
$$\left. + H\left(\mathbf{r}_2 - \frac{\boldsymbol{\rho}}{2}\right) - H(\mathbf{r}_1 + \mathbf{r}_2) - H(\mathbf{r}_1 - \mathbf{r}_2)\right]$$

$$H(\boldsymbol{\rho}) = \frac{1}{\pi}[A(0) - A(\boldsymbol{\rho})].$$

Next, consider a plane wave propagation in a random medium. The wave should have no dependence on the center coordinate $\mathbf{R}$, and therefore $\nabla_R = 0$. Since ∇_ρ is also eliminated from the equation, $\boldsymbol{\rho}$ is only a parameter

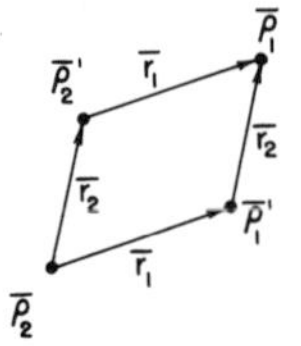

FIG. 20-7 Locations of $\boldsymbol{\rho}_1$, $\boldsymbol{\rho}_2$, $\boldsymbol{\rho}_1{}'$, and $\boldsymbol{\rho}_2{}'$.

and can be set to zero. In this case, $\boldsymbol{\rho}_1, \boldsymbol{\rho}_2, \boldsymbol{\rho}_1{}', \boldsymbol{\rho}_2{}'$ are located at the vertices of a parallelogram (Fig. 20-7). We then write (20-121) in the form

$$\left[\frac{\partial}{\partial x} - \frac{i}{k}\nabla_{r1} \cdot \nabla_{r2} + Q\right]\Gamma_4 = 0 \tag{20-122}$$

where $Q = D'(\mathbf{r}_1) + D'(\mathbf{r}_2) - \frac{1}{2}D'(\mathbf{r}_1 + \mathbf{r}_2) - \frac{1}{2}D'(\mathbf{r}_1 - \mathbf{r}_2)$. The function $D'(\mathbf{r})$

is the derivative of the wave structure function given in (20-49) with respect to x:

$$D'(\mathbf{r}) = \frac{\partial}{\partial x} D(x, \mathbf{r}) = 8\pi^2 k^2 \int_0^\infty [1 - J_0(\kappa r)]\Phi_n(\kappa)\kappa \, d\kappa = 2.92 C_n{}^2 k^2 r^{5/3}. \tag{20-123}$$

The last expression of (20-123) is for a turbulent medium in the range given in (20-78b).

For a plane wave incident at $x = 0$, the boundary condition for Γ_4 is $\Gamma_4 = 1$ at $x = 0$. Equation (20-122) and this boundary condition constitute the mathematical description of the problem.

Let us examine a few general characteristics of the fourth order moment $\Gamma_4(x, \mathbf{r}_1, \mathbf{r}_2)$ in (20-122). First of all, since $D'(\mathbf{r})$ is an even function of $|\mathbf{r}|$, Q in (20-122) is unchanged if $\mathbf{r}_1$ and $\mathbf{r}_2$ are interchanged. Therefore we have

$$\Gamma_4(x, \mathbf{r}_1, \mathbf{r}_2) = \Gamma_4(x, \mathbf{r}_2, \mathbf{r}_1). \tag{20-124}$$

The intensity correlation function is given by

$$\langle I(x, \boldsymbol{\rho}_1) I(x, \boldsymbol{\rho}_2)\rangle = \langle U(x, \boldsymbol{\rho}_1) U^*(x, \boldsymbol{\rho}_1) U(x, \boldsymbol{\rho}_2) U^*(x, \boldsymbol{\rho}_2)\rangle. \tag{20-125}$$

This is identical to the fourth order moment Γ_4 when $\boldsymbol{\rho}_1 = \boldsymbol{\rho}_1{}'$ and $\boldsymbol{\rho}_2 = \boldsymbol{\rho}_2{}'$, and therefore we have

$$\Gamma_4(x, \mathbf{r}_1, 0) = \langle I(x, \boldsymbol{\rho}_2 + \mathbf{r}_1) I(x, \boldsymbol{\rho}_2)\rangle. \tag{20-126}$$

The intensity covariance function B_I is then given by

$$B_I = \langle I(x, \boldsymbol{\rho}_1) I(x, \boldsymbol{\rho}_2)\rangle - \langle I\rangle^2 = \Gamma_4(x, \mathbf{r}_1, 0) - \Gamma_2(x, 0)^2 \tag{20-127}$$

where $\Gamma_2(x, \boldsymbol{\rho})$ is the second order moment given in (20-46b).

We also note that as $|\mathbf{r}_2| \to \infty$, the points $\boldsymbol{\rho}_1$ and $\boldsymbol{\rho}_2{}'$ are separated from $\boldsymbol{\rho}_1{}'$ and $\boldsymbol{\rho}_2$ at infinite distance and the correlation between these two groups of points disappears. Therefore we have in the limit $|\mathbf{r}_2| \to \infty$,

$$\begin{aligned}\Gamma_4 &= \langle U(x, \boldsymbol{\rho}_1) U^*(x, \boldsymbol{\rho}_2{}')\rangle \langle U(x, \boldsymbol{\rho}_2) U^*(x, \boldsymbol{\rho}_1{}')\rangle \\ &= \Gamma_2(x, \mathbf{r}_1)\Gamma_2(x, \mathbf{r}_1)^* = |\Gamma_2(x, \mathbf{r}_1)|^2. \end{aligned} \tag{20-128}$$

It can also be shown (Tatarski, 1971, p. 426) that the intensity covariance B_I satisfies

$$\int_{-\infty}^{\infty} B_I(x, \mathbf{r}_1) \, d\mathbf{r}_1 = 0. \tag{20-129}$$

This is a consequence of the law of energy conservation for a plane wave.

20-14 THIN SCREEN THEORY

Equation (20-122) for the fourth order moment has been studied extensively for the two limiting situations. One is called the thin screen or phase screen theory, where it is assumed that the random medium is concentrated within a "thin screen" and that this thin screen causes only the "phase" modulation of the wave propagating through it. The other is called the extended medium where the medium is assumed to be homogeneously extended throughout the space.

The phase screen theory has been used extensively for problems in astronomy (Prokhorov *et al.*, 1975; Rumsey, 1975; Mercier, 1962; Salpeter, 1967; Taylor, 1972). In this theory, we assume that the strength of the random medium can be strong enough to cause strong fluctuation, but the thickness of the screen is thin so that at the exit plane, only the phase modulation results. This phase modulated wave propagates through free space, resulting in the fluctuation of the intensity at a distance L (see Fig. 20-8).

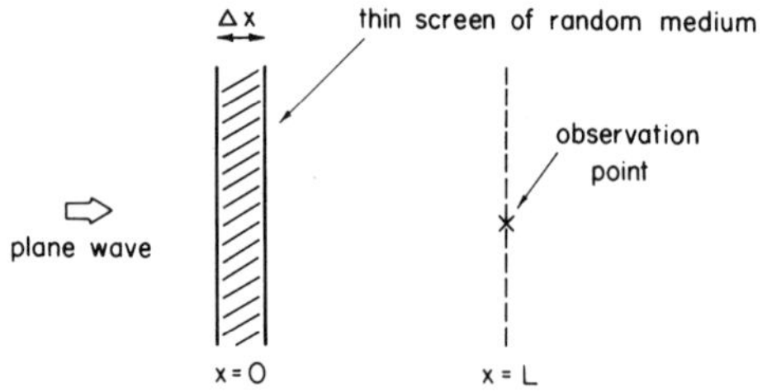

FIG. 20-8 Geometry for the thin screen (phase screen) theory.

Let us designate Γ_4 by Γ and write Γ as

$$\Gamma = \exp(\psi). \tag{20-130}$$

Substituting this into (20-122), we obtain

$$\frac{\partial \psi}{\partial x} - \frac{i}{k}[\nabla_{r2}\psi \cdot \nabla_{r1}\psi + \nabla_{r2} \cdot \nabla_{r1}\psi] + Q = 0. \tag{20-131}$$

Integrating this, we get

$$\psi = -Qx + \frac{i}{k}\int_0^x [\nabla_{r2}\psi \cdot \nabla_{r1}\psi + \nabla_{r2} \cdot \nabla_{r1}\psi]\, dx. \tag{20-132}$$

For a thin screen with thickness Δx, we expand ψ in a Taylor's series:

$$\psi = \sum_{n=1}^{\infty} a_n (\Delta x)^n, \tag{20-133}$$

and obtain a recurrence formula for a_n. It is clear that the second term of (20-132) is on the order of $(\Delta x)^2$ or higher and for a thin screen, we have approximately

$$\psi = -Q\,\Delta x. \tag{20-134}$$

The fourth order moment Γ at the exit plane ($x = 0$) is therefore given by

$$\begin{aligned} \Gamma_0 &= \exp(\psi) \\ \psi &= -[D(\mathbf{r}_1) + D(\mathbf{r}_2) - \tfrac{1}{2}D(\mathbf{r}_1 + \mathbf{r}_2) - \tfrac{1}{2}D(\mathbf{r}_1 - \mathbf{r}_2)] \\ D(\mathbf{r}) &= 2.92 C_n{}^2 k^2 r^{5/3}(\Delta x). \end{aligned} \tag{20-135}$$

Next we consider the propagation of the fourth moment in free space from $x = 0$ to $x = L$. In free space, (20-122) becomes

$$\left(\frac{\partial}{\partial x} - \frac{i}{k}\nabla_{r1}\cdot\nabla_{r2}\right)\Gamma = 0. \tag{20-136}$$

We take a Fourier transform with respect to $\mathbf{r}_1$ and $\mathbf{r}_2$. We let

$$M(x, \boldsymbol{\kappa}_1, \boldsymbol{\kappa}_2) = \frac{1}{(2\pi)^4}\int d\mathbf{r}_1\, d\mathbf{r}_2\, \Gamma \exp(-i\boldsymbol{\kappa}_1\cdot\mathbf{r}_1 - i\boldsymbol{\kappa}_2\cdot\mathbf{r}_2), \tag{20-137}$$

and obtain

$$\left(\frac{\partial}{\partial x} + \frac{i}{k}\boldsymbol{\kappa}_1\cdot\boldsymbol{\kappa}_2\right)M(x, \boldsymbol{\kappa}_1, \boldsymbol{\kappa}_2) = 0. \tag{20-138}$$

Solving this, we get

$$M(L, \boldsymbol{\kappa}_1, \boldsymbol{\kappa}_2) = M(0, \boldsymbol{\kappa}_1, \boldsymbol{\kappa}_2)\exp[-(i/k)\boldsymbol{\kappa}_1\cdot\boldsymbol{\kappa}_2 L] \tag{20-139}$$

where $M(0, \boldsymbol{\kappa}_1, \boldsymbol{\kappa}_2)$ is the Fourier transform of Γ at $x = 0$ (Γ_0) given in (20-135):

$$M(0, \boldsymbol{\kappa}_1, \boldsymbol{\kappa}_2) = \frac{1}{(2\pi)^4}\int d\mathbf{r}_1\, d\mathbf{r}_2 \exp(\psi)\exp(-i\boldsymbol{\kappa}_1\cdot\mathbf{r}_1 - i\boldsymbol{\kappa}_2\cdot\mathbf{r}_2). \tag{20-140}$$

Taking the inverse Fourier transform of (20-139) with respect to $\boldsymbol{\kappa}_1$ and $\boldsymbol{\kappa}_2$, we get the fourth order moment $\Gamma(L, \mathbf{r}_1, \mathbf{r}_2)$. We can perform integration with respect to $\boldsymbol{\kappa}_1$ and $\boldsymbol{\kappa}_2$, and obtain

$$\begin{aligned} &\Gamma(L, \mathbf{r}_1, \mathbf{r}_2) \\ &= \left(\frac{k}{2\pi L}\right)^2 \int d\mathbf{r}_1{}'\, d\mathbf{r}_2{}' \exp\left[i\frac{k}{L}(\mathbf{r}_1 - \mathbf{r}_1{}')\cdot(\mathbf{r}_2 - \mathbf{r}_2{}') + \psi(\mathbf{r}_1{}', \mathbf{r}_2{}')\right]. \end{aligned} \tag{20-141}$$

Alternatively, we can perform integration with respect to $\boldsymbol{\kappa}_2$ and $\mathbf{r}_2'$ and let $\boldsymbol{\kappa}_1 = \boldsymbol{\kappa}$. We then get

$$\Gamma(L, \mathbf{r}_1, \mathbf{r}_2) = \int d\boldsymbol{\kappa} \exp(i\boldsymbol{\kappa} \cdot \mathbf{r}_1) S(L, \boldsymbol{\kappa}, \mathbf{r}_2)$$

$$S(L, \boldsymbol{\kappa}, \mathbf{r}_2) = \frac{1}{(2\pi)^2} \int d\mathbf{r}_1' \exp\left[-i\boldsymbol{\kappa} \cdot \mathbf{r}_1' + \psi\left(\mathbf{r}_1', \mathbf{r}_2 - \frac{\boldsymbol{\kappa} L}{k}\right)\right] \qquad (20\text{-}142)$$

where ψ is given in (20-135).

Since the intensity correlation is given by $\Gamma(L, \mathbf{r}_1, 0)$ in (20-126), the function $S(L, \boldsymbol{\kappa}, 0)$ is the spectral density of the intensity fluctuation. The expression for S given in (20-142) has been obtained in several investigations (Prokhorov *et al.*, 1975; Rumsey, 1975). Since S is a Fourier transform of the intensity correlation function, it also represents the angular distribution of the intensity arriving from different directions. Therefore $S(L, \boldsymbol{\kappa}, 0)$ is often called the angular spectrum of the intensity fluctuation. In this case, we use $\boldsymbol{\kappa} = k\boldsymbol{\theta}$, where $\boldsymbol{\theta}$ is a two-dimensional vector representing directional cosines. If the direction of the intensity of a wave arriving at the observation point is represented by a unit vector $\hat{\mathbf{i}} = l\hat{\mathbf{x}} + m\hat{\mathbf{y}} + n\hat{\mathbf{z}}$, then $\boldsymbol{\theta} = m\hat{\mathbf{y}} + n\hat{\mathbf{z}}$.

It is often convenient and instructive to express the final results in normalized form. This can be done in the following manner. From (20-135), we recognize that D has a common factor. Therefore we let

$$D(\mathbf{r}) = (\kappa_0 r)^{5/3} = t^{5/3} \qquad (20\text{-}143)$$

where $\kappa_0 = [2.92 C_n^2 k^2 \Delta x]^{3/5}$. Using the normalized wave number vector $\boldsymbol{\kappa}_n = \boldsymbol{\kappa}/\kappa_0$, we write

$$S(L, \boldsymbol{\kappa}_n, 0) = \frac{1}{(2\pi\kappa_0)^2} \int d\mathbf{t} \exp[-i\boldsymbol{\kappa}_n \cdot \mathbf{t} + \psi]$$

$$\psi = -[t^{5/3} + (\alpha\kappa_n)^{5/3} - \tfrac{1}{2}|\mathbf{t} + \alpha\boldsymbol{\kappa}_n|^{5/3} - \tfrac{1}{2}|\mathbf{t} - \alpha\boldsymbol{\kappa}_n|^{5/3}] \qquad (20\text{-}144)$$

where $\alpha = (\kappa_0^2 L)/k$. Note that for numerical calculation, we may use

$$\int d\mathbf{t} = \int_0^\infty t\, dt \int_0^{2\pi} d\phi, \qquad \boldsymbol{\kappa}_n \cdot \mathbf{t} = \kappa_n t \cos\phi$$
$$(20\text{-}145)$$
$$|\mathbf{t} + \alpha\boldsymbol{\kappa}_n|^{5/3} = [t^2 + \alpha^2\kappa_n^2 + 2\alpha t\kappa_n \cos\phi]^{5/6}.$$

We also note that as $\kappa_n \to 0$, the integral diverges and $S \to (1/\kappa_0^2)\,\delta(\boldsymbol{\kappa}_n) = \delta(\boldsymbol{\kappa})$ representing the spectrum of $\langle I \rangle^2 = 1$ [see (20-127) and (20-150b)].

In the foregoing, we note that the intensity spectrum S is expressed in terms of the normalized wave number κ_n and a parameter α. To see the meaning of this parameter α let us examine the Rytov solution for the thin screen. From (17-111), we have

$$\sigma_\chi{}^2 = 0.563k^{7/6} \int_0^L d\eta C_n{}^2(\eta)(L-\eta)^{5/6}. \tag{20-146}$$

For a thin screen, we write

$$C_n{}^2(\eta) = C_n{}^2(\Delta x)\,\delta(\eta). \tag{20-147}$$

We then get the Rytov solution for $\sigma_\chi{}^2$:

$$\sigma_\chi{}^2 = 0.563k^{7/6}C_n{}^2(\Delta x)L^{5/6}. \tag{20-148}$$

Now it can easily be verified that the parameter α is related to $\sigma_\chi{}^2$ through

$$\alpha = [5.2\sigma_\chi{}^2]^{6/5}. \tag{20-149}$$

We can therefore express the intensity spectrum S in terms of $\sigma_\chi{}^2$.

The general expressions for the intensity correlation $\Gamma(L, \mathbf{r}_1, 0)$ in (20-126), the intensity covariance B_I, and the variance $\sigma_I{}^2$ of the intensity fluctuation are given by

$$\Gamma(L, \mathbf{r}_1, 0) = \int d\boldsymbol{\kappa}\, \exp(i\boldsymbol{\kappa}\cdot\mathbf{r}_1)S(L, \boldsymbol{\kappa}, 0) = \kappa_0{}^2 \int d\boldsymbol{\kappa}_{\mathrm{n}}\, \exp(i\boldsymbol{\kappa}_{\mathrm{n}}\cdot\mathbf{r}_{\mathrm{n}})S(L, \boldsymbol{\kappa}_{\mathrm{n}}, 0)$$

where

$$\mathbf{r}_{\mathrm{n}} = \kappa_0\mathbf{r}_1. \tag{20-150a}$$

$$B_I(L, \mathbf{r}_1) = \kappa_0{}^2 \int d\boldsymbol{\kappa}_{\mathrm{n}}[\exp(i\boldsymbol{\kappa}_{\mathrm{n}}\cdot\mathbf{r}_n)S(L, \boldsymbol{\kappa}_{\mathrm{n}}, 0) - \langle I\rangle^2\,\delta(\boldsymbol{\kappa}_{\mathrm{n}})] \tag{20-150b}$$

$$\sigma_I{}^2 = \kappa_0{}^2 \int d\boldsymbol{\kappa}_{\mathrm{n}}[S(L, \boldsymbol{\kappa}_{\mathrm{n}}, 0) - \langle I\rangle^2\,\delta(\boldsymbol{\kappa}_{\mathrm{n}})]. \tag{20-150c}$$

Note that since the incident intensity is normalized to unity, we have $\langle I\rangle^2 = 1$. The scintillation index m^2 is given by

$$m^2 = \frac{\langle(I-\langle I\rangle)^2\rangle}{\langle I\rangle^2} = \sigma_I{}^2. \tag{20-151}$$

Next let us consider the temporal frequency spectrum $W(x, \omega)$ of the intensity fluctuation $I - \langle I\rangle$. Letting $\mathbf{V} = V\hat{\mathbf{y}}$ be the wind velocity transverse to the x axis, we get from (20-150b),

$$\begin{aligned} W(L, \omega) &= 2\int B_I(L, \mathbf{V}\tau)\exp(-i\omega\tau)\,d\tau \\ &= 4\pi\kappa_0{}^2 \int d\boldsymbol{\kappa}_{\mathrm{n}}[\delta(\kappa_0\mathbf{V}\cdot\boldsymbol{\kappa}_{\mathrm{n}} - \omega)S(L, \boldsymbol{\kappa}_{\mathrm{n}}, 0) - \delta(\boldsymbol{\kappa}_{\mathrm{n}})\,\delta(\omega)] \\ &= \frac{4\pi\kappa_0}{V}\int_{-\infty}^{\infty} dq\left[S\left(L, \frac{\omega}{\omega_0}\hat{\mathbf{y}} + q\hat{\mathbf{z}}, 0\right) - \delta(q)\,\delta(\omega)\right] \end{aligned} \tag{20-152}$$

where we used $\boldsymbol{\kappa}_n = (\omega/\omega_0)\hat{\mathbf{y}} + q\hat{\mathbf{z}}$, and

$$\omega_0 = \kappa_0 V = [2.92 C_n{}^2 k^2 \, \Delta x]^{3/5} V. \tag{20-153}$$

Note that the temporal frequency spectrum W is given in terms of the normalized frequency ω/ω_0 and α (or $\sigma_\chi{}^2$).

In calculating $W(L, \omega)$, we note that S contains $\delta(\boldsymbol{\kappa}_n)\,\delta(\omega)$ which is canceled by the second term inside the integral, and that S is a function of the magnitude $|\boldsymbol{\kappa}_n|$ only. Therefore we write for $\omega \neq 0$,

$$W(L, \omega) = \frac{8\pi\kappa_0}{V} \int_0^\infty dq S\left(L, \left[\left(\frac{\omega}{\omega_0}\right)^2 + q^2\right]^{1/2}, 0\right) \tag{20-154}$$

where S is given in (20-144) and (20-145). At $\omega = 0$, S contains $\delta(q)$ which is canceled as shown in (20-152) and therefore the integral (20-154) should start with $0+$ excluding $q = 0$. Alternatively, (20-154) can be evaluated for $\omega \neq 0$ and then a limit as $\omega \to 0$ can be taken.

20-15 APPROXIMATE SOLUTION FOR THE THIN SCREEN THEORY

Consider the power spectrum $S(L, \boldsymbol{\kappa}, 0)$ of the intensity given in (20-142):

$$S(L, \boldsymbol{\kappa}, 0) = \frac{1}{(2\pi)^2} \int d\mathbf{r}_1{}' \exp\left[-i\boldsymbol{\kappa} \cdot \mathbf{r}_1{}' + \psi\left(\mathbf{r}_1{}', -\frac{\boldsymbol{\kappa}L}{k}\right)\right] \tag{20-155}$$

where ψ is given in (20-135). Alternatively we can consider (20-144). We consider two extreme cases: $|\alpha\kappa_n| \gg t$ and $|\alpha\kappa_n| \ll t$ in (20-144). Since the range of $|t|$ which contributes mostly to the integral is $|t| \gtrsim |\kappa_n|^{-1}$, these two extreme cases correspond to $|\kappa| \gg (k/L)^{1/2}$ and $|\kappa| \ll (k/L)^{1/2}$.

For $|\kappa| \gg (k/L)^{1/2}$, we have $|\mathbf{r}_1| \ll |\mathbf{r}_2|$ in (20-135) and therefore $D(\mathbf{r}_2) \approx D(\mathbf{r}_1 + \mathbf{r}_2) \approx D(\mathbf{r}_1 - \mathbf{r}_2)$. Thus we get

$$\psi(\mathbf{r}_1{}', -\boldsymbol{\kappa}L/k) \approx -D(\mathbf{r}_1{}'). \tag{20-156}$$

The spectral density S is given by

$$S(L, \boldsymbol{\kappa}, 0) = \frac{1}{(2\pi)^2} \int d\mathbf{r}_1{}' \exp[-i\boldsymbol{\kappa} \cdot \mathbf{r}_1{}' - D(\mathbf{r}_1{}')]. \tag{20-157}$$

This has the same form as (20-112) and therefore we write S in terms of the normalized $\boldsymbol{\kappa}_n$:

$$S(L, \kappa_n, 0) = \frac{1}{2\pi\kappa_0{}^2} \int_0^\infty t \, dt \; J_0(\kappa_n t) \exp(-t^{5/3}). \tag{20-158}$$

This is applicable to the case $\alpha\kappa_n \gg \kappa_n^{-1}$. Since $\alpha \gg 1$ in the strong fluctuation region ($\sigma_\chi{}^2 \gg 1$), this equation is applicable to most regions of κ_n.

For $|\kappa| \ll (k/L)^{1/2}$, we have $|\mathbf{r}_1| \gg |\mathbf{r}_2|$ in (20-135) and therefore we expand ψ in a Taylor's series about $\mathbf{r}_2 = \boldsymbol{\kappa}L/k = 0$:

$$\psi = -[D(\mathbf{r}_2) - \tfrac{1}{2}(\mathbf{r}_2 \cdot \nabla_{r1})^2 D(\mathbf{r}_1) + \cdots]. \tag{20-159}$$

We further expand $\exp(\psi)$ as follows:

$$e^{\psi} = \exp[-D(\mathbf{r}_2)][1 + \tfrac{1}{2}(\mathbf{r}_2 \cdot \nabla_{r1})^2 D(\mathbf{r}_1) + \cdots]. \tag{20-160}$$

and note

$$D(\mathbf{r}_1) = 4\pi k^2(\Delta x) \int [1 - \exp(i\boldsymbol{\kappa} \cdot \mathbf{r}_1)]\Phi_n(\boldsymbol{\kappa})\, d\boldsymbol{\kappa}$$

$$\tfrac{1}{2}(\mathbf{r}_2 \cdot \nabla_{r1})^2 D(\mathbf{r}_1) = 2\pi k^2(\Delta x) \int (\mathbf{r}_2 \cdot \boldsymbol{\kappa})^2 \exp(i\boldsymbol{\kappa} \cdot \mathbf{r}_1)\Phi_n(\boldsymbol{\kappa})\, d\boldsymbol{\kappa}. \tag{20-161}$$

Using (20-160) and (20-161) in (20-142) and performing integration with respect to $\mathbf{r}_1'$, we finally obtain

$$S(L, \boldsymbol{\kappa}, 0) = \delta(\boldsymbol{\kappa}) + \kappa^4(L/k)^2\Phi_s(\kappa) \exp[-D(-\boldsymbol{\kappa}L/k)] \tag{20-162}$$

where $\Phi_s(\kappa)$ is the spectral density for the wave structure function $D(\mathbf{r}_1)$ and is proportional to the spectral density $\Phi_n(\kappa)$ of the refractive index fluctuation:

$$D(\mathbf{r}_1) = 2 \int [1 - \exp(i\boldsymbol{\kappa} \cdot \mathbf{r}_1)]\Phi_s(\boldsymbol{\kappa})\, d\boldsymbol{\kappa}, \qquad \Phi_s(\kappa) = 2\pi k^2(\Delta x)\Phi_n(\kappa). \tag{20-163}$$

Equations (20-157) and (20-162) are consistent with the results obtained by several workers (Prokhorov *et al.*, 1975; Rumsey, 1975; Fante, 1976).

Let us next consider the scintillation index m^2 in (20-151). For the strong fluctuation case $\alpha \gg 1$, (20-157) is applicable to most regions in κ_n except when κ_n is small where (20-162) is applicable. Integrating (20-162) in the region $\kappa_n \approx 0$, we get unity. Integrating (20-157) over κ_n, we also get unity:

$$\int S(L, \boldsymbol{\kappa}, 0)\, d\boldsymbol{\kappa} = \int d\mathbf{r}_1'\, \delta(\mathbf{r}_1') \exp[-D(\mathbf{r}_1')] = 1. \tag{20-164}$$

We therefore obtain $m^2 \to 1$ as $\sigma_\chi{}^2 \to \infty$.

An approximate expression for the temporal frequency spectrum $W(L, \omega)$ in (20-154) can be obtained by using (20-158) in (20-154). We perform integration with respect to q using Sonine's formula (Magnus and Oberhettinger, 1954, p. 29). We then get

$$W(L, \omega) = \frac{4}{\omega_0} \int_0^{\infty} dt \exp(-t^{5/3}) \cos\left[\left(\frac{\omega}{\omega_0}\right)t\right]. \tag{20-165}$$

Note that ω_0 is given by (20-153):

$$\omega_0 = \kappa_0 V = [2.92 C_n^2 k^2 \, \Delta x]^{3/5} V = [5.2\sigma_\chi^2]^{3/5} (k/L)^{1/2} V. \quad (20\text{-}166)$$

20-16 THIN SCREEN THEORY FOR SPHERICAL WAVES

In Chapter 18, we examined the spherical wave solution for a weak fluctuation case and concluded in (18-4) that the results for the plane wave case can be converted into the spherical wave case by the replacement

$$\boldsymbol{\rho} \to (x'/x)\boldsymbol{\rho}, \qquad x - x' \to (x'/x)(x - x'). \quad (20\text{-}167)$$

Since the medium for thin screen theory is concentrated at one plane, we see that x' and x correspond to L_1 and $L_1 + L_2$, respectively, where L_1 is the distance from the source to the screen and L_2 is the distance from the screen to the observation point. Also, since the propagation path L_1 and L_2 is in free space, we expect that (20-167) is valid for the thin screen theory (Prokhorov *et al.*, 1975, p. 802).

Therefore we conclude that the thin screen theory for the spherical wave is obtained from the plane wave solution by the replacement

$$\mathbf{r}_1 \text{ and } \mathbf{r}_2 \to \frac{L_1}{L_1 + L_2}\mathbf{r}_1 \quad \text{and} \quad \frac{L_1}{L_1 + L_2}\mathbf{r}_2$$
$$L \to \frac{L_1 L_2}{L_1 + L_2}, \qquad \boldsymbol{\kappa} \to \frac{L_1 + L_2}{L_1}\boldsymbol{\kappa}. \quad (20\text{-}168)$$

For example, S in (20-144) represents the spherical wave solution if the following parameters are used:

$$\alpha = \frac{\kappa_0^2 L_1 L_2}{(L_1 + L_2)k}, \qquad \boldsymbol{\kappa}_n = \frac{(L_1 + L_2)}{L_1}\frac{\boldsymbol{\kappa}}{\kappa_0}. \quad (20\text{-}169)$$

20-17 EXTENDED SOURCES

In the preceding sections, we discussed the intensity fluctuation of a plane wave as it propagates through a thin screen of a random medium. This is equivalent to a point source located at an infinite distance from the screen. In problems in astronomy, we often need to consider sources with a finite angular size. The effect of the finite source size is in general the reduction of scintillation. For example, stars twinkle but planets twinkle much less. In this section we develop a formula (Salpeter, 1967; Rumsey, 1975; Budden and Uscinski, 1970) to show this effect.

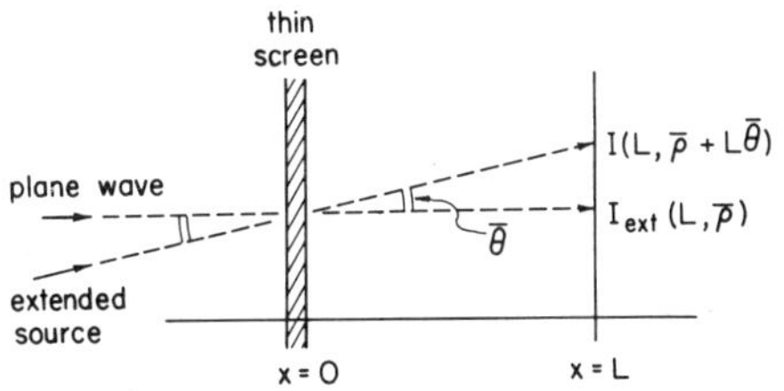

FIG. 20-9 Intensity from extended source at $(L, \boldsymbol{\rho})$ is equal to the intensity due to a plane wave at $(L, \boldsymbol{\rho} + L\boldsymbol{\theta})$.

Consider the intensity $I(L, \boldsymbol{\rho})$ at the observation point due to a plane wave with unit magnitude. The intensity $I_\theta(L, \boldsymbol{\rho})$ at the point $(L, \boldsymbol{\rho})$ due to the wave from an extended source in the direction $\boldsymbol{\theta}$ with unit magnitude, which has passed through the same random medium, should be approximately equal to the intensity $I(L, \boldsymbol{\rho} + L\boldsymbol{\theta})$ due to the plane wave at a point $(L, \boldsymbol{\rho} + L\boldsymbol{\theta})$ if $|\boldsymbol{\theta}| \ll 1$ (see Fig. 20-9). If we let $b(\boldsymbol{\theta})$ be the brightness distribution of the source in the direction $\boldsymbol{\theta}$, the intensity $I_{\text{ext}}(L, \boldsymbol{\rho})$ at $(L, \boldsymbol{\rho})$ due to the extended source is given by

$$I_{\text{ext}}(L, \boldsymbol{\rho}) = \int d\boldsymbol{\theta}\, b(\boldsymbol{\theta}) I_\theta(L, \boldsymbol{\rho}) = \int d\boldsymbol{\theta}\, b(\boldsymbol{\theta}) I(L, \boldsymbol{\rho} + L\boldsymbol{\theta}). \tag{20-170}$$

The intensity correlation is therefore given by

$$\Gamma_{\text{ext}}(L, \boldsymbol{\rho}_1 - \boldsymbol{\rho}_2) = \int d\boldsymbol{\theta}\, d\boldsymbol{\theta}'\, b(\boldsymbol{\theta}) b(\boldsymbol{\theta}') \Gamma(L, \boldsymbol{\rho}_1 - \boldsymbol{\rho}_2 + L(\boldsymbol{\theta} - \boldsymbol{\theta}')) \tag{20-171}$$

where

$$\Gamma_{\text{ext}}(L, \boldsymbol{\rho}_1 - \boldsymbol{\rho}_2) = \langle I_{\text{ext}}(L, \boldsymbol{\rho}_1) I_{\text{ext}}(L, \boldsymbol{\rho}_2) \rangle$$

$$\Gamma(L, \boldsymbol{\rho}_1 - \boldsymbol{\rho}_2 + L(\boldsymbol{\theta} - \boldsymbol{\theta}')) = \langle I(L, \boldsymbol{\rho}_1 + L\boldsymbol{\theta}) I(L, \boldsymbol{\rho}_2 + L\boldsymbol{\theta}') \rangle.$$

Now we consider the angular spectrum of the intensity fluctuation

$$S_{\text{ext}}(L, \boldsymbol{\kappa}) = \frac{1}{(2\pi)^2} \int \Gamma_{\text{ext}}(L, \boldsymbol{\rho}) \exp(-i\boldsymbol{\kappa} \cdot \boldsymbol{\rho})\, d\boldsymbol{\rho}. \tag{20-172}$$

Substituting (20-171) into (20-172), we can easily show that

$$S_{\text{ext}}(L, \boldsymbol{\kappa}) = |B(\boldsymbol{\kappa} L)|^2 S(L, \kappa) \tag{20-173}$$

where

$$B(\boldsymbol{\kappa} L) = \int d\boldsymbol{\theta}\, b(\boldsymbol{\theta}) \exp(i\boldsymbol{\kappa} \cdot \boldsymbol{\theta} L),$$

$$S(L, \kappa) = \frac{1}{(2\pi)^2} \int \Gamma(L, \boldsymbol{\rho}) \exp(-i\boldsymbol{\kappa} \cdot \boldsymbol{\rho})\, d\boldsymbol{\rho}.$$

This is the fundamental relationship relating the angular spectrum S_{ext} of the intensity fluctuation for the extended source to that of the plane wave spectrum† S and the brightness distribution of the source $b(\boldsymbol{\theta})$. This relationship was first derived independently by Cohen and Salpeter. It shows that if the source is extended, $B(\boldsymbol{\kappa}L)$ may be concentrated within a certain range of $|\boldsymbol{\kappa}|$ and therefore the spectrum S_{ext} may also be limited within a certain range of $\boldsymbol{\kappa}$. This means that the angular spectrum for an extended source is limited in a narrower angular range than for a point source, and therefore the temporal spectrum is also limited in a narrower frequency range, giving rise to "less" twinkling.

20-18 EXTENDED MEDIUM

Extensive studies have been reported on the fourth order moments in an extended medium. At present, available solutions include approximations obtained by (Tatarski, 1971, Section 72; Gurvich and Tatarski, 1975) and Fante (1975). Since these approximate solutions are well documented, only a brief summary is given here.

For a plane wave in an extended medium, we need to start with (20-122). Using the spectrum $S(x, \boldsymbol{\kappa}, \mathbf{r}_2)$ [see (20-142)],

$$\Gamma(x, \mathbf{r}_1, \mathbf{r}_2) = \int d\boldsymbol{\kappa} \exp(i\boldsymbol{\kappa} \cdot \mathbf{r}_1) S(x, \boldsymbol{\kappa}, \mathbf{r}_2) \tag{20-174}$$

and letting $\mathbf{r}_2 = \mathbf{r}$, (20-122) can be converted into the equation (Tatarski, 1971, p. 431)

$$\frac{\partial}{\partial x} S + \frac{\boldsymbol{\kappa}}{k} \cdot \nabla_r S + D'(\mathbf{r}) S = G(x, \boldsymbol{\kappa}, \mathbf{r}) \tag{20-175}$$

$$G(x, \boldsymbol{\kappa}, \mathbf{r}) = 4\pi k^2 \int [1 - \exp(i\boldsymbol{\kappa}' \cdot \mathbf{r})] \Phi_n(\boldsymbol{\kappa}') S(x, \boldsymbol{\kappa} - \boldsymbol{\kappa}', \mathbf{r})\, d\boldsymbol{\kappa}'$$

where we used (20-123):

$$D'(\mathbf{r}) = 4\pi k^2 \int [1 - \exp(i\boldsymbol{\kappa} \cdot \mathbf{r})] \Phi_n(\boldsymbol{\kappa})\, d\boldsymbol{\kappa}. \tag{20-176}$$

The boundary condition at $x = 0$ is $\Gamma = 1$ and therefore

$$S(0, \boldsymbol{\kappa}, \mathbf{r}) = \delta(\boldsymbol{\kappa}). \tag{20-177}$$

Tatarski showed that (20-175) can be converted into an integral equation for $S(x, \boldsymbol{\kappa}, \mathbf{r})$. This integral equation can be solved by an iteration technique.

† Note that $S(L, \kappa)$ is the intensity spectrum for a plane wave discussed in Section 20-14 [see (20-142) and (20-144)].

However, the complexity of the equation is such that only the first iteration can be expressed in reasonably simple form.

Tatarski used the incident spectrum $S = \delta(\boldsymbol{\kappa})$ in the integral to obtain the first iteration. This should be valid for small x, but it is not applicable for large x. Fante, on the other hand, recognized that as $x \to \infty$ [see (20-128)],

$$\Gamma(x, \mathbf{r}_1, \mathbf{r}_2) \to \begin{cases} |\Gamma_2(x, \mathbf{r}_1)|^2 & \text{as} \quad \mathbf{r}_2 \to \infty \\ |\Gamma_2(x, \mathbf{r}_2)|^2 & \text{as} \quad \mathbf{r}_1 \to \infty. \end{cases} \tag{20-178}$$

Therefore he used

$$\Gamma(x, \mathbf{r}_1, \mathbf{r}_2) = |\Gamma_2(x, \mathbf{r}_1)|^2 + |\Gamma_2(x, \mathbf{r}_2)|^2 \tag{20-179}$$

in the integral to compute the first iteration. This should give an excellent result for large distance.

For other iteration solutions and approximate expressions of angular spectra for various regions of $|\boldsymbol{\kappa}|$, the readers are referred to a survey paper by Prokhorov *et al.* (1975).

There is a heuristic technique based on the "extended Huygens–Fresnel" principle (Feizulin and Kravtsov, 1967; Yura, 1974a; Clifford *et al.*, 1974; Fante, 1976). This principle states that the field $U(x, \boldsymbol{\rho})$ at $(x, \boldsymbol{\rho})$ due to the field $U_0(\boldsymbol{\rho}')$ at $(0, \boldsymbol{\rho}')$ is given by the following extension of the Huygens–Fresnel formula:

$$U(x, \boldsymbol{\rho}) = \frac{k \exp(ikx)}{2\pi i x} \int d\boldsymbol{\rho}' U_0(\boldsymbol{\rho}') \exp\left[i \frac{k|\boldsymbol{\rho} - \boldsymbol{\rho}'|^2}{2x} + \psi(x, \boldsymbol{\rho}, \boldsymbol{\rho}') \right] \tag{20-180}$$

where $\psi(x, \boldsymbol{\rho}, \boldsymbol{\rho}') = \chi(x, \boldsymbol{\rho}, \boldsymbol{\rho}') + iS_1(x, \boldsymbol{\rho}, \boldsymbol{\rho}')$, and χ and S_1 are the log-amplitude and phase fluctuations, respectively, of a spherical wave from $(0, \boldsymbol{\rho}')$ to $(x, \boldsymbol{\rho})$.

Starting with (20-180), the second moments and the fourth moments have been calculated. The second moments have been shown to be identical to those obtained from the exact solution of the parabolic equation. The fourth moments are found to agree well with experimental data. However, the correspondence between the extended Fresnel–Huygens solution and the parabolic equation solution has not been established except for the thin screen case (Fante, 1976).

The analysis of the strong fluctuation problem can also be made using the diagram method based on an application of the Feynman diagram in quantum field theory. This technique clarifies the relationship among various scattering processes and is useful in understanding the essence of the multiple scattering theory. Readers are referred to some excellent review articles on this subject (Tatarski, 1971, p. 335; Frisch, 1968; Marcuvitz, 1974).

20-19 OPTICAL PROPAGATION IN A TURBULENT MEDIUM

In optical communication through the atmosphere, the random fluctuation of the refractive index affects the propagation characteristics of the light beam. It causes a spreading of the beam, a decrease in the temporal and spatial coherence, a beam wander, and scintillations of the received intensity. In this section we outline some of the important results Strohbehn (1977) (Fante, 1974, 1975; Prokhorov *et al.*, 1975; Gurvich and Tatarski, 1975; Beran, 1970, 1975; de Wolf, 1968, 1969, 1971, 1972, 1973, 1974; Dabberdt 1973; Collin, 1971, Bremmer, 1973; Tatarski and Gertsenshtein, 1963; Klyatskin, 1970; Yura, 1969, 1974; Collett *et al.*, 1973; Beran, 1970; Dabberdt and Johnson, 1973; Feyzulin, 1970; Ochs and Lawrence, 1969; Fante and Poirier, 1973; Dolin, 1968; Lutomirski and Buser, 1973; Lutomirski and Yura, 1969, 1971; Hufnagel and Stanley, 1964; Kerr *et al.*, 1970; Klyatskin and Tatarski, 1969, 1970; Lawrence, 1972; Livingston *et al.*, 1970; Mano, 1969; Molyneux, 1971; Sancer and Varvatsis, 1970; S'edin *et al.*, 1970; Torrieri and Taylor, 1972; Gebhardt, 1976; Lee, *et al.*, 1976).

(a) *Short- and long-term beam spread* If we place a photographic plate in a turbulent medium and record a light beam spot, we see the following. At a relatively short distance, the beam shape in the turbulence may be substantially the same as that in free space, but the beam spot wanders around due to the random motion of the turbulence. Therefore, if the exposure time is short compared with the wandering time Δt of the beam of the order of (beam size)/(wind velocity), the beam size should be substantially the same as that in free space. This beam radius ρ_s is called the short-term beam spread. If the exposure time is much longer than Δt, the wander of the center of gravity of the beam ρ_c in addition to the short-term spread ρ_s contributes to the recorded picture of the beam. This is called the long-term beam spread ρ_L and is related to ρ_s and ρ_c through

$$\langle \rho_L{}^2 \rangle = \langle \rho_s{}^2 \rangle + \langle \rho_c{}^2 \rangle. \tag{20-181}$$

Mathematically, the long-term spread is given by

$$\langle \rho_L{}^2 \rangle = \frac{\int \langle I(x, \boldsymbol{\rho}) \rangle \rho^2 \, d\boldsymbol{\rho}}{\int \langle I(x, \boldsymbol{\rho}) \rangle \, d\boldsymbol{\rho}} \tag{20-182}$$

where $\langle I(x, \boldsymbol{\rho}) \rangle = \Gamma(x, \boldsymbol{\rho}, 0)$ is given in (20-73). The center of gravity $\boldsymbol{\rho}_c$ is given by

$$\boldsymbol{\rho}_c = \frac{\int I(x, \boldsymbol{\rho}) \boldsymbol{\rho} \, d\boldsymbol{\rho}}{\int \langle I(x, \boldsymbol{\rho}) \rangle \, d\boldsymbol{\rho}} \tag{20-183}$$

Therefore we have

$$\langle \boldsymbol{\rho}_c{}^2 \rangle = \frac{\iint \langle I(x, \boldsymbol{\rho}_1) I(x, \boldsymbol{\rho}_2) \rangle (\boldsymbol{\rho}_1 \cdot \boldsymbol{\rho}_2)\, d\boldsymbol{\rho}_1\, d\boldsymbol{\rho}_2}{[\int \langle I(x, \boldsymbol{\rho}) \rangle\, d\boldsymbol{\rho}]^2}. \tag{20-184}$$

Some approximate calculations of this have been made. Note that W^2 in (20-80) is equivalent to $\langle \rho_L{}^2 \rangle$ for the case $x \gg x_i$:

$$\langle \rho_L{}^2 \rangle = \frac{W_0{}^2}{2} [(\alpha_r x)^2 + (1 - \alpha_i x)^2] + 2.2 C_n{}^2 l_0^{-1/3} x^3. \tag{20-185}$$

For $x \ll x_i$, it has been shown (Fante, 1975) that

$$\langle \rho_L{}^2 \rangle = \frac{W_0{}^2}{2} [(\alpha_r x)^2 + (1 - \alpha_i x)^2] + \frac{4x^2}{k^2 \rho_0{}^2} \tag{20-186}$$

where ρ_0 is given by

$$\rho_0 = [1.46 k^2 x C_n{}^2 (\tfrac{3}{8})]^{-3/5}. \tag{20-187}$$

For the short-term beam spread in various cases, see Fante (1975).

At a large distance, the beam no longer wanders as much, but it breaks up into a multitude of stringlike spots at random locations. The long exposure picture is a blurred version of the short exposure picture, but their radii are approximately the same. From (20-181), it is seen that $\langle \rho_c{}^2 \rangle$ should be much smaller than $\langle \rho_L{}^2 \rangle$ or $\langle \rho_s{}^2 \rangle$ in the strong fluctuation region.

(b) *Saturation of intensity fluctuation* The Rytov solution for the intensity fluctuation of a plane wave in the weak fluctuation region is given by

$$\sigma_{\ln I}^2 = \langle (\ln I - \langle \ln I \rangle)^2 \rangle = 4\sigma_\chi{}^2 = 1.23 C_n{}^2 k^{7/6} L^{11/6}. \tag{20-188}$$

This increases indefinitely as L or C_n increases. However, it is clear that the intensity fluctuation cannot increase indefinitely. We already indicated that the scintillation index m^2 should approach 1 as $\sigma_\chi{}^2 \to \infty$ for the thin screen case [see (20-164)]. This was known in astrophysics (Rumsey, 1975), but in optical propagation in the earth's atmosphere the saturation effect was first observed by Gracheva and Gurvich (1965). Since then, numerous experimental and theoretical investigations have been reported. (Deitz and Wright, 1969; de Wolf, 1968; Kerr, 1972b; Khmelevtsov, 1973; Khmelevtsov and Tsvyk, 1970).

General characteristics of the scintillation index $\sigma_I{}^2$ as a function of the Rytov solution $4\sigma_\chi{}^2$ are shown in Fig. 20-10. Here the variance $\sigma_I{}^2$ of the intensity I

$$\sigma_I{}^2 = \frac{\langle (I - \langle I \rangle^2 \rangle}{\langle I \rangle^2} \tag{20-189}$$

instead of ln I is plotted against the Rytov solution

$$4\sigma_\chi^2 = 1.23 C_n^2 k^{7/6} L^{11/6}. \tag{20-190}$$

Some aspects of this saturation have been discussed in Section 20-18.

Let us next consider the correlation of the intensities. The normalized covariance $b_I(x, \boldsymbol{\rho}_1, \boldsymbol{\rho}_2)$ is given by

$$b_I(x, \boldsymbol{\rho}_1, \boldsymbol{\rho}_2) = \frac{\langle I_1 I_2\rangle - \langle I_1\rangle\langle I_2\rangle}{\langle I_1\rangle\langle I_2\rangle} \tag{20-191}$$

where $I_1 = I(x, \boldsymbol{\rho}_1)$ and $I_2 = I(x, \boldsymbol{\rho}_2)$. A plot of $b_I(x, \rho)$, $\rho = |\boldsymbol{\rho}_1 - \boldsymbol{\rho}_2|$, for a plane wave case is shown in Fig. 20-11. In the weak fluctuation region, the correlation distance is given by ρ_0:

$$\rho_0 = [1.46 k^2 x C_n^2]^{-3/5}. \tag{20-192}$$

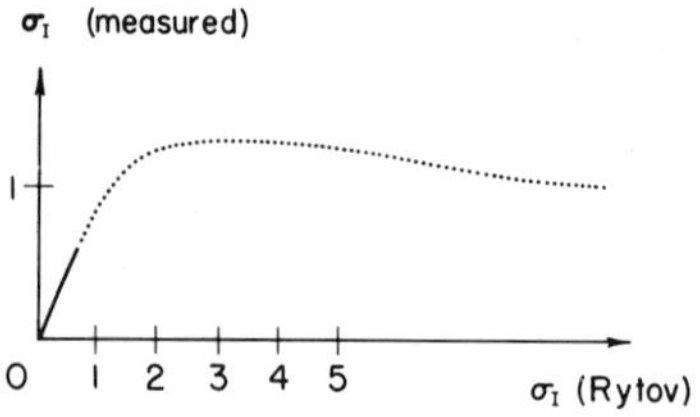

FIG. 20-10

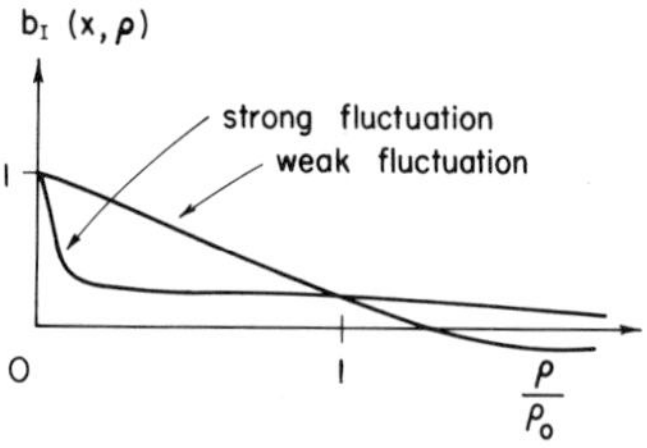

FIG. 20-11

FIG. 20-10 Measured values of the variance of the intensity fluctuation versus the Rytov solution.

FIG. 20-11 The covariance of the intensity. The correlation distance is ρ_0 in the weak fluctuation case, whereas in strong fluctuation, there is a short correlation distance and a long tail.

In the strong fluctuation region, the covariance is characterized by a short correlation distance and a long tail.

(c) *Aperture averaging* (Tatarski, 1971, p. 273; Lutomirski and Yura, 1969; Strohbehn, 1971) The intensity fluctuations σ_I^2 discussed in the preceding section are those observed by a point receiver. If the receiver has a finite size, the fluctuations are averaged over the receiving aperture and the fluctuations will be reduced. This phenomenon is called aperture averaging.

Consider the power P received by an aperture A. This is given by

$$P = \int_A I(x, \boldsymbol{\rho})\, d\boldsymbol{\rho}. \tag{20-193}$$

The mean square fluctuation $\sigma_p{}^2$ is then given by

$$\sigma_p{}^2 = \int_A d\mathbf{\rho}_1 \int_A d\mathbf{\rho}_2[\langle I(x, \mathbf{\rho}_1)I(x, \mathbf{\rho}_2)\rangle - \langle I(x, \mathbf{\rho}_1)\rangle\langle I(x, \mathbf{\rho}_2)\rangle]. \quad (20\text{-}194)$$

For a circular aperture of diameter D, (20-194) becomes (see Section 15-4)

$$\sigma_p{}^2(d) = \pi D^2 \int_0^D B_I(\rho)K(\rho)\rho\, d\rho \quad (20\text{-}195)$$

where

$$K(\rho) = \{\text{arc}\cos(\rho/D) - (\rho/D)[1 - (\rho^2/D^2)]^{1/2}\}$$

$$B_I(\rho) = \langle I(x, \mathbf{\rho}_1)I(x, \mathbf{\rho}_2)\rangle - \langle I(x, \mathbf{\rho}_1)\rangle\langle I(x, \mathbf{\rho}_2)\rangle, \qquad \rho = |\mathbf{\rho}_1 - \mathbf{\rho}_2|$$

For a point receiver, we let $B_I(\rho) = B_I(0)$ in (20-195) and obtain

$$\sigma_p{}^2(0) = \frac{(\pi D^2)^2}{16} B_I(0). \quad (20\text{-}196)$$

The ratio of $\sigma_p{}^2(D)$ to $\sigma_p{}^2(0)$ shows how the intensity fluctuations decrease as the aperture size increases.

$$G(D) = \frac{\sigma_p{}^2(D)}{\sigma_p{}^2(0)} = \frac{16}{\pi D^2}\int_0^D \frac{B_I(\rho)}{B_I(0)} K(\rho)\rho\, d\rho. \quad (20\text{-}197)$$

The ratio is plotted in Fig. 20-12. It shows that in the weak fluctuation region, the fluctuations of the power are significantly reduced if the aperture size D exceeds the correlation distance $(\lambda x)^{1/2}$ of the field. In the strong fluctuation region, the fluctuations are reduced when the aperture size D exceeds $0.36(\lambda x)^{1/2}[\sigma_{IR}^2]^{-3/5}$, where σ_{IR}^2 is the Rytov solution $4\sigma_\chi{}^2$ given in (20-190).

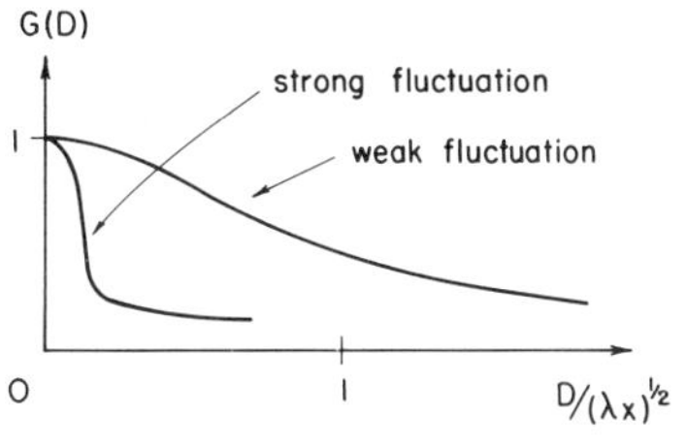

FIG. 20-12 Aperture averaging effects. The ratio of the power fluctuations *G(D)* of an aperture with size *D* to a point receiver.

(d) *Probability distribution of the intensity* (de Wolf, 1969; Strohbehn *et al.*, 1975; Furuhama, 1976) In the weak fluctuation region where the Rytov solution $\sigma_\chi{}^2 \ll 1$, the probability distribution has been found to be nearly log-normal. This implies that the log-amplitude χ is normally distributed. If a random variable z is normally distributed, the average of $\exp(z)$ is given by

$$\langle \exp(z) \rangle = \exp[\langle z \rangle + \tfrac{1}{2}\langle (z - \langle z \rangle)^2 \rangle]. \tag{20-198}$$

Therefore we get

$$\begin{aligned} \langle I^2 \rangle &= \langle \exp(4\chi) \rangle = \exp[4\langle \chi \rangle + 8\langle (\chi - \langle \chi \rangle)^2 \rangle] \\ \langle I \rangle &= \exp[4\langle \chi \rangle + 4\langle (\chi - \langle \chi \rangle)^2 \rangle]. \end{aligned} \tag{20-199}$$

From this, we obtain

$$\sigma_I{}^2 = \frac{\langle I^2 \rangle - \langle I \rangle^2}{\langle I \rangle^2} = \exp[4\langle (\chi - \langle \chi \rangle)^2 \rangle] - 1. \tag{20-200}$$

In the Rytov solution, we have

$$\langle (\chi - \langle \chi \rangle)^2 \rangle = 0.307 C_n{}^2 k^{7/6} L^{11/6}$$

and therefore we can calculate $\sigma_I{}^2$.

In the region where $0.3 < \sigma_{IR}^2 < 25$, where σ_{IR}^2 is the Rytov solution $4\sigma_\chi{}^2$, the probability distribution is neither log-normal nor Rayleigh. In the region $25 < \sigma_{IR}^2 < 100$, the probability distribution appears to be log-normal. And as $\sigma_{IR}^2 \to \infty$, the probability distribution should approach the Rayleigh distribution (Fante, 1975).

20-20 MODULATION TRANSFER FUNCTION OF A RANDOM MEDIUM†

In Section 15-4 we discussed the question of image resolution when a plane wave is propagated through a random distribution of scatterers and then received by an image forming lens. Using the point spread function, we showed the decrease of the coherent intensity and the increase of the incoherent intensity at the image plane, as the optical distance of the medium increases. In this section we present a more complete treatment of the imaging problem making use of the concept of modulation transfer function.

Let us consider an incoherent source with intensity $I_0(\boldsymbol{\rho}_0)$ in the object plane. The radiation from this object propagates through a random medium over a distance L and is incident on a lens of diameter D. The image intensity $I_i(\boldsymbol{\rho}_i)$ is observed at the image plane which is located at a distance d from the lens (see Fig. 20-13).

† For MTF, see Born and Wolf (1964, p. 526), Goodman (1968), Murata (1966), Levi (1968), and O'Neill (1963). For atmospheric effects on MTF, see Hufnagel and Stanley (1964), Fried (1966b), Lutomirski and Yura (1971a), Consortini *et al.* (1973), Shapiro (1976a,b, 1977), Tavis and Yura (1976), and Korff *et al.* (1975). For applications in astronomy, see Young (1974).

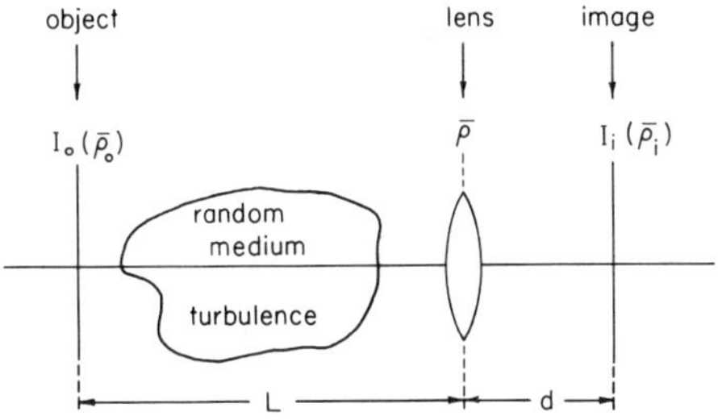

FIG. 20-13 Geometry showing object plane, lens, and image plane.

It is desired to relate the image intensity I_i and the object intensity I_o in the integral form

$$I_i(\boldsymbol{\rho}_i) = \int P(\boldsymbol{\rho}_i, \boldsymbol{\rho}_o) I_o(\boldsymbol{\rho}_o)\, d\boldsymbol{\rho}_o. \tag{20-201}$$

The function $P(\boldsymbol{\rho}_i, \boldsymbol{\rho}_o)$ represents the response of the total imaging system when the object intensity is a delta function $I_o(\boldsymbol{\rho}_o) = \delta(\boldsymbol{\rho}_o - \boldsymbol{\rho}_o')$ and is called the point spread function. In general $P(\boldsymbol{\rho}_i, \boldsymbol{\rho}_o)$ is a function of both $\boldsymbol{\rho}_i$ and $\boldsymbol{\rho}_o$, but as will be shown shortly, P becomes a function of $\boldsymbol{\rho}_i - \boldsymbol{\rho}_o'$ if we use the equivalent object vector $\boldsymbol{\rho}_o'$, taking into account the following magnification and inversion of the image through the lens:

$$\boldsymbol{\rho}_o' = -(d/L)\boldsymbol{\rho}_o \tag{20-202}$$

and if we choose the distance d so that the following lens law is satisfied:

$$1/L + 1/d = 1/f_0 \tag{20-203}$$

where f_0 is the focal distance of the lens. In this case the point spread function becomes a function of $\boldsymbol{\rho}_i - \boldsymbol{\rho}_o'$. The imaging system with $P = P(\boldsymbol{\rho}_i - \boldsymbol{\rho}_o')$ is called *isoplanatic.* We then write (20-201) in the form

$$I_i(\boldsymbol{\rho}_i) = \int P(\boldsymbol{\rho}_i - \boldsymbol{\rho}_o') I_o(\boldsymbol{\rho}_o')\, d\boldsymbol{\rho}_o'. \tag{20-204}$$

We now take a Fourier transform of this convolution integral and obtain

$$G_i(\mathbf{f}) = M(\mathbf{f}) G_o(\mathbf{f}). \tag{20-205}$$

where G_i and G_o are the normalized spectra:

$$G_i(\mathbf{f}) = \frac{\int I_i(\boldsymbol{\rho}_i) \exp(i2\pi\mathbf{f} \cdot \boldsymbol{\rho}_i)\, d\boldsymbol{\rho}_i}{\int I_i(\boldsymbol{\rho}_i)\, d\boldsymbol{\rho}_i}, \tag{20-206}$$

$$G_o(\mathbf{f}) = \frac{\int I_o(\boldsymbol{\rho}_o') \exp(i2\pi\mathbf{f} \cdot \boldsymbol{\rho}_o')\, d\boldsymbol{\rho}_o'}{\int I_o(\boldsymbol{\rho}_o)\, d\boldsymbol{\rho}_o'}, \tag{20-207}$$

and $M(\mathbf{f})$ is the normalized Fourier transform of P given by

$$M(\mathbf{f}) = \frac{\int P(\boldsymbol{\rho}) \exp(i2\pi \mathbf{f} \cdot \boldsymbol{\rho})\, d\boldsymbol{\rho}}{\int P(\boldsymbol{\rho})\, d\boldsymbol{\rho}}. \tag{20-208}$$

The function $M(\mathbf{f})$ is known as the optical transfer function and its modulus $|M(\mathbf{f})|$ is known as the modulation transfer function (MTF). The frequency $\mathbf{f}$ is measured in cycles per unit length (commonly cycles/mm). Alternatively $\mathbf{f}' = \mathbf{f}d$ can be used. In this case $\mathbf{f}'$ represents angular frequency and is measured in cycles per radian of field of view.

Let us return to the problem depicted in Fig. 20-13. The lens is assumed to be thin and to produce the phase transformation given by

$$\exp[-i(k/2f_0)\rho^2] \tag{20-209}$$

where $\boldsymbol{\rho}$ is the radial vector at the lens and f_0 the focal distance of the lens. The field behind the lens is equal to the field $U(\boldsymbol{\rho})$ incident on the lens with the additional phase given in (20-209): $U(\boldsymbol{\rho}) \exp[-i(k/2f_0)\rho^2]$. This field is then propagated through the distance d. Using a Kirchoff formula and the Fresnel approximation (see Section 15-4), we get the field $U_i(\boldsymbol{\rho}_i)$ at the image plane:

$$U_i(\boldsymbol{\rho}_i) = \frac{k \exp(ikd)}{2\pi i d} \int_s U(\boldsymbol{\rho}) \exp\left[\frac{ik|\boldsymbol{\rho}_i - \boldsymbol{\rho}|^2}{2d} - \frac{ik\rho^2}{2f_0}\right] d\boldsymbol{\rho} \tag{20-210}$$

where the integral is over the lens aperture s. The intensity $I_i(\boldsymbol{\rho}_i)$ at the image plane is therefore given by

$$I_i(\boldsymbol{\rho}_i) = \langle |U_i(\boldsymbol{\rho}_i)|^2 \rangle = \frac{k^2}{(2\pi d)^2} \int_s \int_s \Gamma(\boldsymbol{\rho}_1, \boldsymbol{\rho}_2) \exp(i\phi_1)\, d\boldsymbol{\rho}_1\, d\boldsymbol{\rho}_2 \tag{20-211}$$

where

$$\Gamma(\boldsymbol{\rho}_1, \boldsymbol{\rho}_2) = \langle U(\boldsymbol{\rho}_1) U^*(\boldsymbol{\rho}_2) \rangle, \qquad \phi_1 = -\frac{ik\boldsymbol{\rho}_i \cdot \boldsymbol{\rho}_d}{d} - \frac{ik}{2L}(\rho_1{}^2 - \rho_2{}^2).$$

We used (20-203) and $\boldsymbol{\rho}_d = \boldsymbol{\rho}_1 - \boldsymbol{\rho}_2$.

The mutual coherence function $\Gamma(\boldsymbol{\rho}_1, \boldsymbol{\rho}_2)$ is produced by $I_o(\boldsymbol{\rho}_o)$ and its expression has already been given in (20-75):

$$\Gamma(\boldsymbol{\rho}_1, \boldsymbol{\rho}_2) = \frac{1}{L^2} \int d\boldsymbol{\rho}_o I_o(\boldsymbol{\rho}_o) \exp(i\phi_2 - H)\, d\boldsymbol{\rho}_o$$

$$\phi_2 = -\frac{ik\boldsymbol{\rho}_o \cdot \boldsymbol{\rho}_d}{L} + \frac{ik}{2L}(\rho_1{}^2 - \rho_2{}^2). \tag{20-212}$$

The function $H = H(\rho_d)$ is given in (20-75) and is equal to $\frac{1}{2}D$ where D is the wave structure function for a spherical wave given in (20-64):

$$H(\rho_d) = \tfrac{1}{2}D(\rho_d) = 4\pi^2 k^2 \int_0^L dx' \int_0^\infty \left[1 - J_0\left(\kappa \rho_d \frac{x'}{L}\right)\right] \Phi_n(\kappa)\kappa \, d\kappa. \tag{20-213a}$$

For a turbulent medium, we have [see (20-81b)]

$$H(\rho_d) = 0.547 k^2 C_n{}^2 L \rho_d^{5/3}. \tag{20-213b}$$

Combining (20-211) and (20-212), we write

$$I_i(\boldsymbol{\rho}_i) = \int P(\boldsymbol{\rho}_i, \boldsymbol{\rho}_o) I_o(\boldsymbol{\rho}_o) \, d\boldsymbol{\rho}_o$$

$$P(\boldsymbol{\rho}_i, \boldsymbol{\rho}_o) = C \int_S \int_S \exp\left[-ik\boldsymbol{\rho}_d \cdot \left(\frac{\boldsymbol{\rho}_o}{L} + \frac{\boldsymbol{\rho}_i}{d}\right) - H(\rho_d)\right] d\boldsymbol{\rho}_1 \, d\boldsymbol{\rho}_2 \tag{20-214}$$

where C is constant $[k^2/(4\pi^2 d^2 L^2)]$. The function $P(\boldsymbol{\rho}_i, \boldsymbol{\rho}_o)$ in (20-214) does not satisfy the *isoplanatic* condition. However, we recognize that the image is magnified and inverted through the lens. Therefore we should use the object vector $\boldsymbol{\rho}_o'$ given in (20-202). We then satisfy the isoplanatic condition:

$$\begin{aligned} P &= P(\boldsymbol{\rho}_i - \boldsymbol{\rho}_o') \\ &= C \int_S \int_S \exp\left[-\frac{ik\boldsymbol{\rho}_d \cdot (\boldsymbol{\rho}_i - \boldsymbol{\rho}_o')}{d} - H(\rho_d)\right] d\boldsymbol{\rho}_1 \, d\boldsymbol{\rho}_2. \end{aligned} \tag{20-215}$$

If the lens has a circular aperture of diameter D, we can convert (20-215) using the circle function [see (20-195)]

$$\operatorname{circ}(\rho) = \begin{cases} 1 & \text{if } \rho \le 1 \\ 0 & \text{if } \rho > 1. \end{cases} \tag{20-216}$$

We then get

$$P(\boldsymbol{\rho}_i - \boldsymbol{\rho}_o') = \frac{C\pi D^2}{4} \int \exp\left[-\frac{ik\boldsymbol{\rho}_d \cdot (\boldsymbol{\rho}_i - \boldsymbol{\rho}_o')}{d} - H(\rho_d)\right] K\left(\frac{\rho_d}{D}\right) d\boldsymbol{\rho}_d \tag{20-217}$$

where

$$K\left(\frac{\rho_d}{D}\right) = \begin{cases} \dfrac{2}{\pi}\left\{\cos^{-1}\left(\dfrac{\rho_d}{D}\right) - \left(\dfrac{\rho_d}{D}\right)\left[1 - \left(\dfrac{\rho_d}{D}\right)^2\right]^{1/2}\right\} & \text{for } \rho_d \le D \\ 0 & \text{for } \rho_d > D. \end{cases}$$

We recognize that (20-217) is in the form of a Fourier transform and therefore noting (20-208), we get the following optical transfer function:

$$M(\mathbf{f}) = \exp[-H(\lambda\, d\mathbf{f})]K(\lambda\, d\mathbf{f}/D) \tag{20-218a}$$

where use is made of $2\pi\mathbf{f} = k\boldsymbol{\rho}_d/d$. Alternatively, we can use the angular frequency $\mathbf{f}' = d\mathbf{f}$ and write

$$M(\mathbf{f}') = \exp[-H(\lambda\mathbf{f}')]K(\lambda\mathbf{f}'/D). \tag{20-218b}$$

Let us examine $M(\mathbf{f})$. Note that if the random medium is absent, $H = 0$ and $M(\mathbf{f}) = K(\lambda\, d\mathbf{f}/D)$. Therefore $K(\lambda\, d\mathbf{f}/D)$ is the modulation transfer function of the imaging system with a circular aperture of diameter D, and is given by

$$K\left(\frac{f}{f_c}\right) = \begin{cases} \dfrac{2}{\pi}\left\{\cos^{-1}\left(\dfrac{f}{f_c}\right) - \left(\dfrac{f}{f_c}\right)\left[1 - \left(\dfrac{f}{f_c}\right)^2\right]^{1/2}\right\} & \text{for } f \le f_c, \\ 0 & \text{for } f > f_c, \end{cases} \tag{20-219}$$

where $f_c = D/\lambda d$ is the cutoff frequency of this aperture system. We can also express (20-219) using $f'/f_c' = f/f_c$. In this case the cutoff angular frequency f_c' is given by $f_c' = D/\lambda$ (see Fig. 20-14).

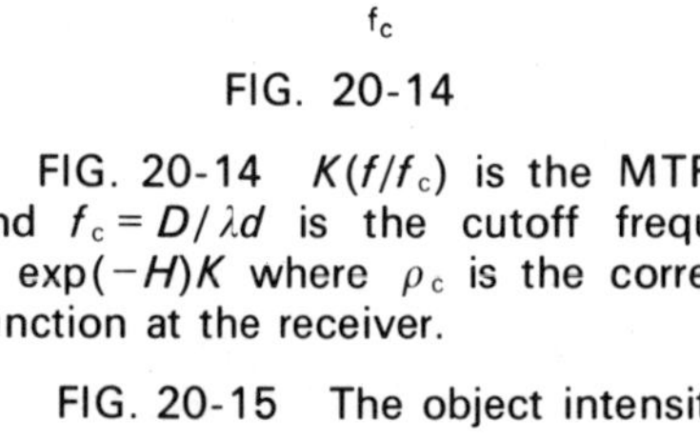

FIG. 20-14

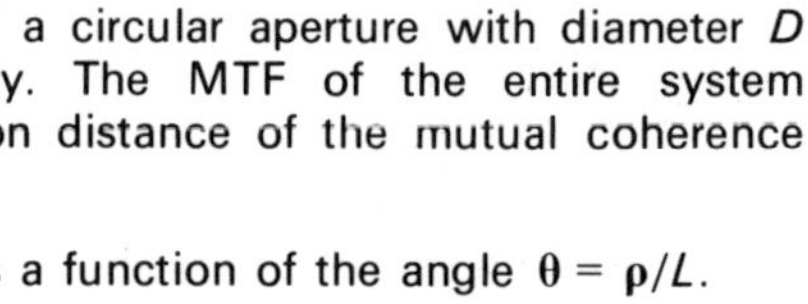

FIG. 20-15

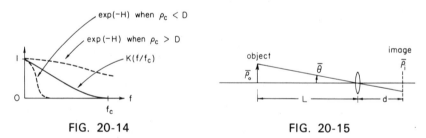

FIG. 20-14 $K(f/f_c)$ is the MTF for a circular aperture with diameter D and $f_c = D/\lambda d$ is the cutoff frequency. The MTF of the entire system is $\exp(-H)K$ where ρ_c is the correlation distance of the mutual coherence function at the receiver.

FIG. 20-15 The object intensity as a function of the angle $\boldsymbol{\theta} = \boldsymbol{\rho}/L$.

The effect of the random medium is given by $\exp[-H]$. Note that $L^{-2}\exp[-H]$ is the mutual coherence function of a spherical wave. For a turbulent medium, we have from (20-213b)

$$\exp[-H] = \exp[-0.547k^2C_n{}^2L(\lambda\, df)^{5/3}] = \exp\left[-\left(\frac{D}{\rho_c}\right)^{5/3}\left(\frac{f}{f_c}\right)^{5/3}\right] \tag{20-220}$$

where $\rho_c = (0.547k^2C_n{}^2L)^{-3/5}$. This is also pictured in Fig. 20-14. Note that if $\rho_c < D$, the atmospheric effect is dominant, whereas if $\rho_c > D$, the MTF is almost the same as the aperture MTF.

It is often more convenient to express the object intensity $I_o(\boldsymbol{\rho}_o)$ as a function of the angle $\boldsymbol{\theta} = \boldsymbol{\rho}_o/L$ observed from the receiver (see Fig. 20-15). Since $G_o(\mathbf{f})$ is the Fourier transform of $I_o(\boldsymbol{\rho}_o')$, we let $\boldsymbol{\rho}_o' = -d\boldsymbol{\rho}_o/L$ and obtain

$$G_o(\mathbf{f}') = \frac{\int I_o(\boldsymbol{\theta}) \exp(-i2\pi\mathbf{f}' \cdot \boldsymbol{\theta})\, d\boldsymbol{\theta}}{\int I_o(\boldsymbol{\theta})\, d\boldsymbol{\theta}}. \tag{20-221}$$

Also note that $G_i(\mathbf{f})$ is already given by (20-206).

In this section we have shown that the modulation transfer function of the random medium and the imaging system is a product of the mutual coherence function evaluated at $\rho_d = \lambda f'$ as shown in (20-220) and the modulation transfer function of the aperture given in (20-219).

The modulation transfer function gives the amount of modulation transferred through a system. For example, suppose that the object intensity is sinusoidally modulated in one dimension with unit magnitude. Taking the x axis in the propagation direction, we write

$$I_o(\boldsymbol{\theta}) = 1 + \cos(2\pi N\theta_2), \qquad \boldsymbol{\theta} = \theta_2\,\hat{\mathbf{y}} + \theta_3\,\hat{\mathbf{z}}. \tag{20-222}$$

The image intensity is then proportional to

$$I_i(\boldsymbol{\rho}_i) \sim 1 + M(N)\cos(2\pi N y_i/d) \tag{20-223}$$

where $M(N) = M(\mathbf{f}')$ evaluated at $\mathbf{f}' = N\hat{\mathbf{y}}$ and $\boldsymbol{\rho}_i = y_i\hat{\mathbf{y}} + z_i\hat{\mathbf{z}}$. It is therefore possible to determine the MTF by measuring the modulation of an image due to an object with a certain periodic pattern (Murata, 1966).

In this section we discussed the imaging of an incoherent object. If the object is coherent or partially coherent, it is possible to use a similar technique to relate the image mutual coherence function to the object mutual coherence function. In this case (Shapiro 1976a,b), however, the isoplanatism is satisfied only when the object is located within a small region in the object plane. The object is then said to be in an isoplanatic region.

The MTF just discussed is called the long-exposure MTF because we took the ensemble average to obtain the image intensity, and this corresponds to recording the image using a long exposure time. If a short exposure time is used, the wave front tilt causes a displacement of the image but it does not affect the sharpness of the image. Therefore the wave front tilt does not contribute to the short-exposure MTF. The short-exposure MTF is approximately given by (Fried, 1966b)

$$\begin{aligned} M(f') &= \exp\{-H(f')\,[1 - (f'/f_c')^{1/3}]\}K(f'/f_c') \quad \text{if} \quad D \gg (L\lambda)^{1/2} \\ &= \exp\{-H(f')[1 - \tfrac{1}{2}(f'/f_c')^{1/3}]\}K(f'/f_c') \quad \text{if} \quad D \ll (L\lambda)^{1/2}. \end{aligned} \tag{20-224}$$

From this, we note that the effect of the short exposure is to increase the magnitude of MTF, particularly for f' close to the cutoff frequency f_c', thus improving the sharpness of the image.

A measure of the performance of an imaging system is the quantity R,

$$R = \int d\mathbf{f} M(\mathbf{f}), \tag{20-225}$$

where $M(\mathbf{f})$ is shown in (20-218a). Physically this represents the spatial bandwidth of the random medium and the imaging system, and detailed study has been made on the effects of outer scales and short exposure (Fried, 1966b; Consortini *et al.*, 1973).

20-21 ADAPTIVE OPTICS

In this chapter we discussed in some detail the effect of the random medium or the turbulent atmosphere on wave propagation characteristics. As was discussed in Section 20-20, the image through the atmosphere is in general blurred due to the wave front distortion. In recent years vigorous attempts have been made to adapt the optical imaging system to compensate for the distorted wave front, thus achieving an almost ideal diffraction-limited performance even through a turbulent medium (for a review of this general problem, see Fried, 1977; J. Opt. Soc. Ann. special issue "Adaptive Optics," March 1977; Greenwood and Fried, 1976). In this section we give a short sketch of some typical techniques and a few references.

The core of the adaptive optics is to sense in real time the appropriate phase distortion and to introduce an electronically controlled phase shift, typically through the use of an array of piezoelectrically driven mirror elements. We will sketch a few typical techniques of adaptive optics.

(a) *Subaperture coherent optical heterodyne receiver* (Fried, 1976) The wave front distortion over a large receiving aperture results in a poor signal-to-noise ratio due to random interference between different parts of the aperture. If the aperture is subdivided into an array of small apertures, there is no significant wave front distortion across each subaperture. The phase differences between the outputs from different subapertures can be determined and they can be used to control phase shifters for each subaperture so that the outputs can be coherently summed (Fig. 20-16).

A more sophisticated system using a shear interferometer has been proposed by Wyant (1975).

(b) *Multidither receiver* (Fried, 1976) This also makes use of an array of subelements such as piezoelectrically driven mirrors. Each element ($i = 1, 2, 3, \ldots$) is mechanically "dithered" at a frequency ω_i ($i = 1, 2, 3, \ldots$) of that element. The phase-shifted optical signal is focused on a pinhole and the

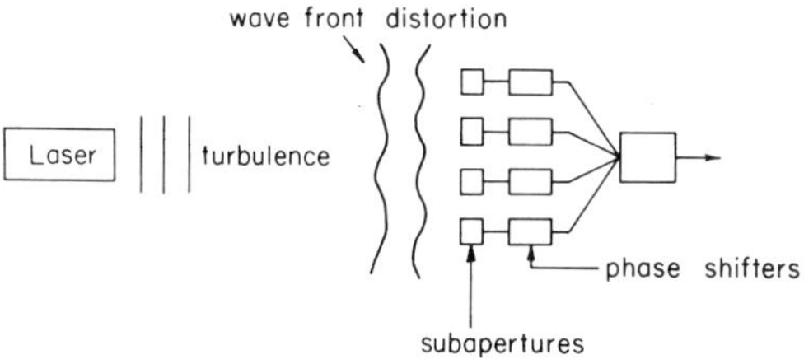

FIG. 20-16 The wave front distortion is small over each subaperture. The phase differences between different subapertures are determined to control phase shifters.

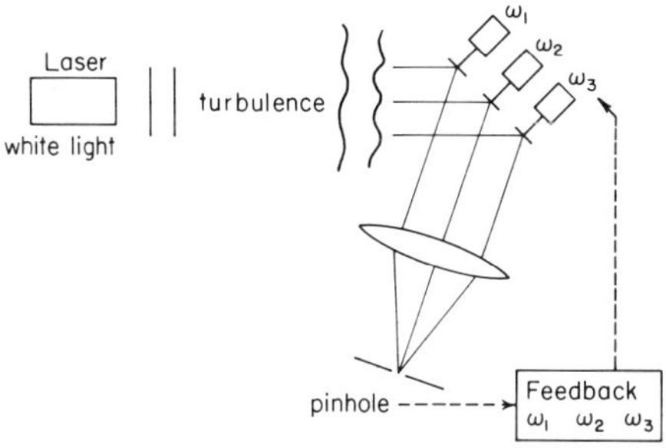

FIG. 20-17 Multidither receiver.

signal at ω_i is observed and used to compensate for the turbulence-induced phase for the element, thus achieving a nearly diffraction-limited image (see Fig. 20-17).

(c) *Image sharpening* (Dyson, 1975; Muller and Buffington, 1974) Instead of the dithering technique, it is possible to sense the image sharpness at the focal point and use this information to control the phase of each element. A typical configuration is similar to Fig. 20-17. As a measure of the sharpness, the integral of the intensity squared over the focal plane may be used. This technique has been considered for applications in astronomy.

(d) *COAT* (*Coherent optical adaptive techniques* (Bridges *et al.*, 1974; Pearson, 1976; Pearson *et al.*, 1976a,b) In Fig. 20-18, a sketch of a

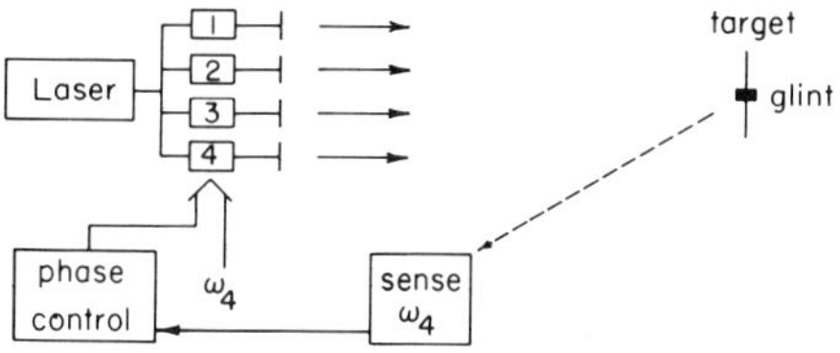

FIG. 20-18 Multidither COAT system.

multidither COAT system is shown. Each transmitting element is dithered at different frequencies ω_i ($i = 1, 2, 3, \ldots$) so that the phase of each element varies by a small amount ($\pm 10°$) with ω_i. Therefore the interference pattern at the target will also vary at frequency ω_i. If there is a bright point reflector, a "glint," at the target, the light intensity reflected from the glint is modulated at this frequency ω_i. This is sensed and used to control the phase shifter for this element. The transmitter wave front is then predistorted so that the intensity on the glint is maximized. In this way the wave arrives at the target as if the wave were propagated through free space. A similar technique using deformable mirrors has also been proposed (Primmerman and Fouche, 1976; Bradley and Herrmann, 1974).

APPENDIX 20A

1. *Functionals* (Gelfand and Formin, 1963, p. 1) A functional is a quantity that depends on a function. A typical example is the field quantity $u(\mathbf{r})$ which depends on the fluctuation of the dielectric constant $\varepsilon(\mathbf{r}')$. We may write this as

$$u(\varepsilon) = \int F(\mathbf{r}', \varepsilon)\, d\mathbf{r}'. \tag{20A-1}$$

2. *Functional derivative* (Gelfland and Formin, 1963, p. 27 and Chapter 3) (or variational derivative) Consider a one-dimensional case

$$u(\varepsilon) = \int_a^b F(x, \varepsilon)\, dx. \tag{20A-2}$$

We vary ε by $\delta\varepsilon$ in the neighborhood of x_0 and consider the corresponding variation of u (Fig. 20A-1). We next consider the ratio

$$\frac{u(\varepsilon + \delta\varepsilon) - u(\varepsilon)}{\Delta\sigma} \tag{20A-3}$$

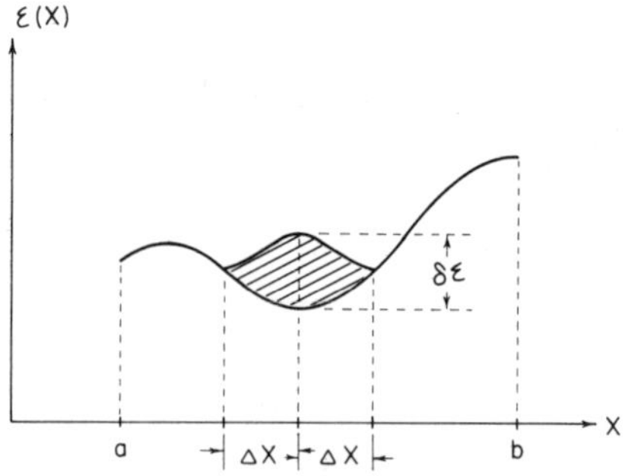

FIG. 20A-1 Functional derivative.

where $\Delta\sigma$ is the shaded area and is equal to $(\delta\varepsilon)\,\Delta x$.

In the limit as $\Delta\sigma \to 0$, we get the functional derivative

$$\frac{\delta u}{\delta \varepsilon} = \lim_{\Delta\sigma \to 0} \frac{u(\varepsilon + \delta\varepsilon) - u(\varepsilon)}{\Delta\sigma}. \tag{20A-4}$$

Noting that

$$u(\varepsilon + \delta\varepsilon) - u(\varepsilon) = \frac{\partial F}{\partial \varepsilon} \Delta x\, \delta\varepsilon \tag{20A-5}$$

we get

$$\frac{\delta u}{\delta \varepsilon} = \frac{\delta F}{\partial \varepsilon}. \tag{20A-6}$$

More generally, if

$$u(\varepsilon) = \int_a^b F(x, \varepsilon, \varepsilon')\, dx, \tag{20A-7}$$

then

$$\frac{\delta u}{\delta \varepsilon} = \frac{\partial F}{\partial \varepsilon} - \frac{d}{dx}\frac{\partial F}{\partial \varepsilon'}. \tag{20A-8}$$

APPENDIX 20B

To prove that (Novikov, 1965)

$$\langle \varepsilon(r) U(\mathbf{r}) \rangle = \int dV' \langle \varepsilon(\mathbf{r}) \varepsilon(\mathbf{r}') \rangle \left\langle \frac{\delta U(\mathbf{r})}{\delta \varepsilon(\mathbf{r}')} \right\rangle, \tag{20B-1}$$

where $\varepsilon(r)$ is a Gaussian random function with $\langle \varepsilon(r) \rangle = 0$, we represent the functional $U(\mathbf{r})$ in a functional Taylor's series:

$$U(\varepsilon) = U(0) + \sum_{n=1}^{\infty} \frac{1}{n!} \int \cdots \int dr_1 \cdots d\mathbf{r}_n \times \left| \frac{\delta^n U}{\delta \varepsilon_1(\mathbf{r}_1) \cdots \delta \varepsilon_n(\mathbf{r}_n)} \right|_{\varepsilon = 0} \varepsilon_1(\mathbf{r}_1) \cdots \varepsilon_n(\mathbf{r}_n). \tag{20B-2}$$

We multiply this by $\varepsilon(r)$ and take the average. We note that the average of the product of an even number of Gaussian random variables is equal to the sum of the products of the mean values of all possible combinations of pairs, and the average of the product of an odd number of such quantities is zero (Middletown, 1960, p. 343). Therefore, we write

$$\langle \varepsilon(\mathbf{r})\varepsilon_1(\mathbf{r}_1) \cdots \varepsilon_n(\mathbf{r}_n)\rangle = \sum_{\alpha=1}^{n} \langle \varepsilon(\mathbf{r})\varepsilon_\alpha(\mathbf{r}_\alpha)\rangle \langle \varepsilon_1(\mathbf{r}_1) \cdots \varepsilon_{\alpha-1}(\mathbf{r}_{\alpha-1})\varepsilon_\alpha(\mathbf{r}_\alpha) \cdots \varepsilon_n(\mathbf{r}_n)\rangle$$
$$= n\langle \varepsilon(\mathbf{r})\varepsilon_1(\mathbf{r}_1)\rangle\langle \varepsilon_2(\mathbf{r}_2) \cdots \varepsilon_n(\mathbf{r}_n)\rangle. \tag{20B-3}$$

Using this with (B-2), we get

$$\langle \varepsilon(\mathbf{r})U(\varepsilon)\rangle = \sum_{n=1}^{\infty} \frac{\langle \varepsilon(\mathbf{r})\varepsilon_1(\mathbf{r}_1)\rangle}{(n-1)!} \int \cdots \int d\mathbf{r}_1 \cdots d\mathbf{r}_n$$
$$\times \left. \frac{\delta^n U}{\delta\varepsilon_1(\mathbf{r}_1) \cdots \delta\varepsilon(\mathbf{r}_n)} \right|_{\varepsilon=0} \langle \varepsilon_2(\mathbf{r}_2) \cdots \varepsilon_n(\mathbf{r}_n)\rangle. \tag{20B-4}$$

On the other hand, taking the functional derivative of (20B-2), we get

$$\left\langle \frac{\delta U}{\delta \varepsilon} \right\rangle = \sum_{n=1}^{\infty} \frac{1}{(n-1)!} \int \cdots \int d\mathbf{r}_2 \cdots d\mathbf{r}_n$$
$$\times \left. \frac{\delta^n U}{\delta\varepsilon_1(\mathbf{r}_1) \cdots \delta\varepsilon_n(\mathbf{r}_n)} \right|_{\varepsilon=0} \langle \varepsilon(\mathbf{r}_2) \cdots \varepsilon_n(\mathbf{r}_n)\rangle, \tag{20B-5}$$

where we used the symmetry of U in all $\varepsilon_1 \cdots \varepsilon_n$. Substituting this into the right-hand side of (20B-1), we see that (20B-1) is identical to (20B-4).

APPENDIX 20C

To calculate $\delta U/\delta\varepsilon$, we first obtain $U(\mathbf{r})$ by integrating (20-7) with respect to x from $x = 0$ to x:

$$2ikU(x, \boldsymbol{\rho}) - 2ikU(0, \boldsymbol{\rho}) + \nabla_t^2 \int_0^x U(\xi, \boldsymbol{\rho})\, d\xi + k \int_0^x \varepsilon_1(\xi, \boldsymbol{\rho})U(\xi, \boldsymbol{\rho})\, d\xi = 0. \tag{20C-1}$$

Now, we take the functional derivative of (20C-1):

$$2ik \frac{\delta U(x, \boldsymbol{\rho})}{\delta\varepsilon_1(x', \boldsymbol{\rho}')} + \nabla_t^2 \frac{\delta}{\delta\varepsilon_1(x', \boldsymbol{\rho}')} \int_0^x U(\xi, \boldsymbol{\rho})\, d\xi$$
$$+ k^2 \frac{\delta}{\delta\varepsilon_1(x'\boldsymbol{\rho}')} \int_0^x \varepsilon_1(\xi, \boldsymbol{\rho})U(\xi, \boldsymbol{\rho})\, d\xi = 0 \tag{20C-2}$$

and let $x' \to x$.

We next show that the second term of (20C-2) becomes zero. As is evident from (20C-1), the field at $(x, \boldsymbol{\rho})$ depends on the inhomogeneity

$\varepsilon_1(x', \boldsymbol{\rho}')$ in the range $x' < x$. Physically, this means that the backscattering from $\varepsilon_1(x', \boldsymbol{\rho}')$ in the region $x' > x$ is neglected. Under this assumption, we have

$$\frac{\delta U(x, \boldsymbol{\rho})}{\delta \varepsilon(x', \boldsymbol{\rho}')} = 0 \qquad \text{if} \quad x' > x. \tag{20C-3}$$

Therefore

$$\frac{\delta}{\delta \varepsilon(x', \boldsymbol{\rho}')} \int_0^x U(\xi, \boldsymbol{\rho})\, d\xi = \int_{x'}^x \frac{\delta U(\xi, \boldsymbol{\rho})}{\delta \varepsilon_1(x', \boldsymbol{\rho}')}\, d\xi \tag{20C-4}$$

which becomes zero as $x' \to x$.

The last term in (20C-2) needs some additional considerations because of $\varepsilon_1(\xi, \boldsymbol{\rho})$ in the integrand. We write this term using the unit step function $H(\xi)$:

$$H(\xi) = \begin{cases} 0 & \text{for} \quad \xi < 0 \\ \frac{1}{2} & \text{for} \quad \xi = 0 \\ 1 & \text{for} \quad \xi > 0 \end{cases} \tag{20C-5}$$

and the delta function $\delta(\boldsymbol{\rho} - \boldsymbol{\rho}'')$:

$$k^2 \frac{\delta}{\delta \varepsilon_1(x', \boldsymbol{\rho}')} \int_0^\infty d\xi \iint_{-\infty}^{\infty} dy''\, dz'' [H(x, \xi)\, \delta(\boldsymbol{\rho} - \boldsymbol{\rho}'') \varepsilon_1(\xi, \boldsymbol{\rho}'')] U(\xi, \boldsymbol{\rho}''). \tag{20C-6}$$

As is shown in Appendix 20A, the functional derivative of a functional of the form

$$I(x) = \int_a^b F(x, \xi, \varepsilon_1(\xi))\, d\xi \tag{20C-7}$$

is given by

$$\frac{\delta I(x)}{\delta \varepsilon_1(x')} = \frac{F(x, x', \varepsilon_1(x'))}{\partial \varepsilon_1(x')}. \tag{20C-8}$$

Using this, (20C-6) becomes

$$k^2 H(x - x')\, \delta(\boldsymbol{\rho} - \boldsymbol{\rho}') U(x', \boldsymbol{\rho}')$$

$$+ k^2 \int_0^\infty d\xi \iint_{-\infty}^{\infty} dy''\, dz'' [H(x - \xi)\, \delta(\boldsymbol{\rho} - \boldsymbol{\rho}'') \varepsilon_1(\xi, \boldsymbol{\rho}'')] \frac{\delta U(\xi, \boldsymbol{\rho}'')}{\delta \varepsilon_1(x', \boldsymbol{\rho}')}. \tag{20C-9}$$

As $x' \to x$, the second term of (20C-9) vanishes because of the assumption (20C-3), and the first term becomes

$$\frac{k^2}{2}\delta(\boldsymbol{\rho} - \boldsymbol{\rho}')U(x', \boldsymbol{\rho}'). \tag{20C-10}$$

Therefore, substituting (20C-10) into (20C-2), we obtain

$$\frac{\delta U(x, \boldsymbol{\rho})}{\delta \varepsilon_1(x', \boldsymbol{\rho}')} = \frac{ik}{4}\delta(\boldsymbol{\rho} - \boldsymbol{\rho}')U(x, \boldsymbol{\rho}'). \tag{20C-11}$$

PART V □ ROUGH SURFACE SCATTERING AND REMOTE SENSING

CHAPTER 21 ROUGH SURFACE SCATTERING

Many natural and biological surfaces are rough in varying degrees, and this roughness affects the propagation and scattering characteristics of a wave. For example, the propagation characteristics of a wave over such a surface are different from the characteristics over a smooth surface. A wave incident on a rough surface is not only reflected in a specular direction but is also scattered in all directions. When the rough surface is in motion, the scattered wave contains Doppler-shifted frequency components.

There are a variety of engineering and scientific problems which require thorough understanding of the rough surface scattering characteristics of a wave. For example, radio communication over the ocean is affected by the surface roughness. In radio oceanography, radar returns from the sea surface are used to probe the ocean wave characteristics (Swift, 1977). In radar astronomy, waves scattered from planetary surfaces are analyzed to deduce the surface characteristics. In biological media, rough interfaces between different organs and tissues affect propagation and scattering characteristics of a wave. Another example is the effects of surface tolerance on the performance of reflector antennas (Shifrin, 1971).

In rough surface scattering, it is important to recognize that the roughness of a surface depends on the wavelength and the direction of wave propagation and scattering. Consider a wave incident on a rough surface as shown in Fig. 21-1. If the surface is completely smooth, two rays are specularly reflected, the reflected rays are in phase, and the reflection angle is equal to the incident angle θ_i. If the surface becomes rough, two rays are no longer in phase, and the phase difference $\Delta\phi$ is given by

$$\Delta\phi = 2kh \cos \theta_i. \tag{21-1}$$

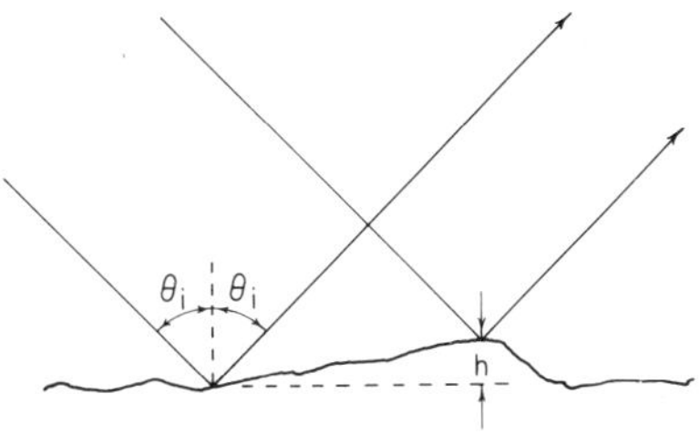

FIG. 21-1 Rayleigh criterion of surface roughness.

If this phase difference is negligible compared with 2π, then the surface may be regarded as "smooth." If the phase difference is significant, then the specular reflection is reduced due to the interference and a part of the wave is scattered in all directions. Rayleigh used a criterion that a surface may be considered rough or smooth depending on whether the phase difference (21-1) is greater or smaller than $\pi/2$. In terms of h, the Rayleigh criterion is

$$h \gtrless \lambda/(8 \cos \theta_i). \tag{21-2}$$

Statistically, the height h used here should be the standard deviation of the height variation of the rough surface.

Before we go into detailed mathematical analysis, it may be instructive to consider general characteristics of rough surface scattering. Suppose that a wave is radiated from a transmitter and is incident on a surface (see Fig. 21-2). If the surface is smooth, the reflected wave is identical to the

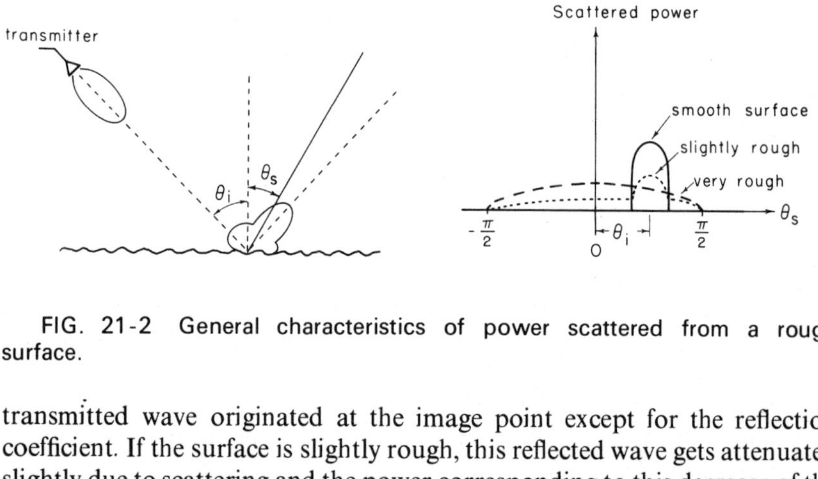

FIG. 21-2 General characteristics of power scattered from a rough surface.

transmitted wave originated at the image point except for the reflection coefficient. If the surface is slightly rough, this reflected wave gets attenuated slightly due to scattering and the power corresponding to this decrease of the reflected power is scattered in all directions. This reflected power is sometimes called the *specular* component, and the scattered power is called the *diffuse* component. The specular reflection from a rough surface corresponds to the coherent power in a random medium and is equal to the square of the coherent (average) field. The diffuse component corresponds to the incoherent field in a random medium. If the surface becomes very rough, the specular (coherent) component almost disappears and the diffuse (incoherent) field dominates (see Fig. 21-2).

At present, there are two general approaches to rough surface scattering problems: the "perturbation technique" and the "Kirchhoff approximation."

The perturbation technique† applies to a surface which is slightly rough and whose surface slope is generally smaller than unity. The Kirchhoff approximation technique‡ is applicable to a surface whose radius of curvature is much greater than a wavelength. We present an introduction to these two techniques in this chapter.

21-1 RECEIVED POWER AND SCATTERING CROSS SECTION PER UNIT AREA OF ROUGH SURFACE

Consider a transmitter radiating a wave which is incident on a rough surface (see Figs. 21-3a and 21-3b). The coherent received power P_{rcoh} is equal to that which is specularly reflected from a smooth surface and is

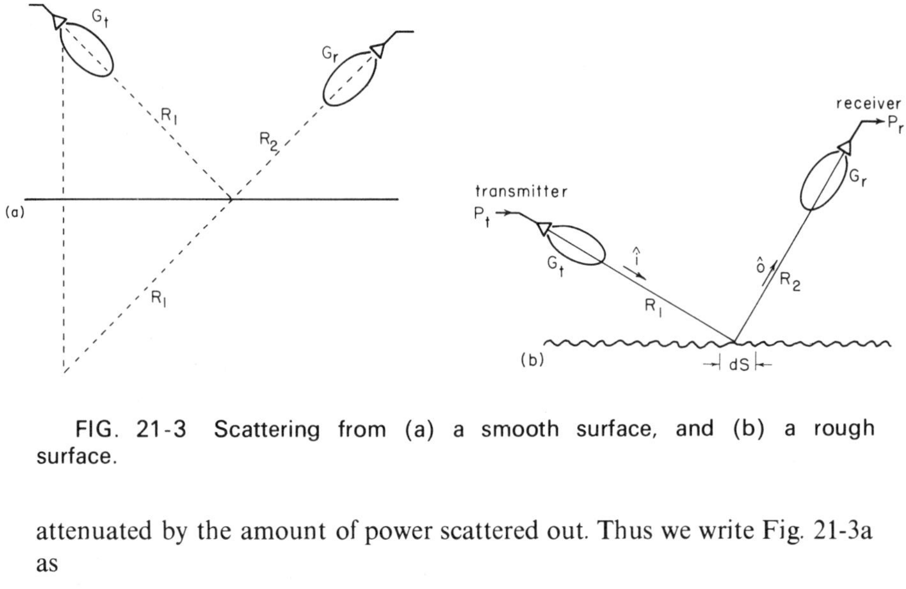

FIG. 21-3 Scattering from (a) a smooth surface, and (b) a rough surface.

attenuated by the amount of power scattered out. Thus we write Fig. 21-3a as

$$\frac{P_{rcoh}}{P_t} = \frac{\lambda^2}{(4\pi)^2} \frac{G_t G_r}{(R_1 + R_2)^2} R_f{}^2 |\chi|^2, \qquad |\chi|^2 \le 1 \tag{21-3}$$

† The perturbation technique was used by Rayleigh and extended by Rice (1951). See contributions by Barrick and Peake (1968), Barrick (1971, 1972), Guinard and Daley (1970), Valenzuela (1968a,b, 1970), Valenzuela and Laing (1971, 1972), Wright (1968,), Crombie (1955), a survey paper by Fortuin (1970), and Krishen (1970, 1971, 1973).

‡See Beckmann and Spizzichino (1963). For earlier works on rough surface scattering, see Ament (1953), Twersky (1957), and Beard (1961, 1967). For related periodic surface scattering, see Wait (1971), DeSanto (1971), and Zipfel and DeSanto (1972). See DeSanto (1973) for the diagram approach.

where R_f^2 is the power reflection coefficient of the smooth surface and $|\chi|^2$ is the effect of the roughness.

Next, we consider the diffuse (incoherent) component. This can be expressed most conveniently by the scattering cross section per unit area of the rough surface. We note an obvious analogy between this cross section and the scattering cross section per unit volume of a random medium discussed in the preceding chapters.

Let us consider an incident plane wave with unit amplitude. It is incident on a patch of surface ΔS in the direction $\hat{\mathbf{i}}$. If $\mathbf{E}_s$ is the scattered wave in the direction $\hat{\mathbf{0}}$ at a large distance R, then the bistatic cross section of this patch ΔS is given by $4\pi R^2 |E_s|^2$ and therefore the scattering cross section per unit area of rough surface is given by

$$\sigma = \sigma(\hat{\mathbf{0}}, \hat{\mathbf{i}}) = \frac{4\pi R^2 \langle |\mathbf{E}_s|^2 \rangle}{\Delta S}. \tag{21-4}$$

If a transmitter and a receiver are located in a far zone of each other, then the incoherent scattered power P_r, when the transmitted power is P_t, is given by the radar equation (see Fig. 21-3b)

$$\frac{P_r}{P_t} = \frac{\lambda^2}{(4\pi)^3} \int \frac{G_t(\hat{\mathbf{i}})G_r(\hat{\mathbf{0}})}{R_1^2 R_2^2} \sigma(\hat{\mathbf{0}}, \hat{\mathbf{i}})\, ds \tag{21-5}$$

where G_t and G_r are the gain functions of the transmitter and the receiver, respectively.

We note that in this formulation, the patch ΔS in (21-4) must be small enough so that the incident wave is almost planar within ΔS, but it should be large enough to include all the statistical characteristics of the rough surface. Specifically, the size of ΔS must be much greater than the correlation distance of the height variation.

We note also that if the surface varies slowly in time, then (21-5) can be modified to

$$\frac{B_s(\tau)}{P_t} = \frac{\lambda^2}{(4\pi)^3} \int \frac{G_t(\hat{\mathbf{i}})G_r(\hat{\mathbf{0}})}{R_1^2 R_2^2} \sigma(\hat{\mathbf{0}}, \hat{\mathbf{i}}, \tau)\, ds \tag{21-6a}$$

where $B_s(\tau)$ is the temporal correlation function of the output voltage of the receiver and is proportional to the correlation function of the field $\mathbf{E}_r$ at the receiver:

$$B_s(\tau) = \frac{\langle E_r(t+\tau)E_r^*(t)\rangle}{\langle |E_r|^2 \rangle} P_r. \tag{21-6b}$$

The cross section $\sigma(\hat{\mathbf{0}}, \hat{\mathbf{i}}, \tau)$ is given by

$$\sigma(\hat{\mathbf{0}}, \hat{\mathbf{i}}, \tau) = \frac{4\pi R^2 \langle E_s(t+\tau)E_s^*(t)\rangle}{\Delta S} \tag{21-7}$$

where E_s is the scattered wave for an incident wave with unit amplitude.

By taking a Fourier transform with respect to τ, we obtain the temporal frequency spectrum of the received wave:

$$\frac{W_s(\omega)}{P_t} = \frac{\lambda^2}{(4\pi)^3} \int \frac{P_t(\hat{\mathbf{i}})P_r(\hat{\mathbf{0}})}{R_1{}^2 R_2{}^2} \sigma(\hat{\mathbf{0}}, \hat{\mathbf{i}}, \omega)\, ds \tag{21-8}$$

where

$$W_s(\omega) = 2\int_{-\infty}^{\infty} B_s(\tau) e^{-i\omega\tau}\, d\tau, \qquad \sigma(\hat{\mathbf{0}}, \hat{\mathbf{i}}, \omega) = 2\int_{-\infty}^{\infty} \sigma(\hat{\mathbf{0}}, \hat{\mathbf{i}}, \tau) e^{-i\omega\tau}\, d\tau.$$

Note that ω in (21-8) is the frequency deviation from the carrier frequency ω_0 and therefore (21-8) represents the spectrum broadening. Also note that in the preceding, we assumed that the frequency deviation ω is much smaller than the carrier frequency ω_0 and therefore the gains of the antennas are those at the carrier frequency ω_0. This assumption is valid for almost all practical situations.

21-2 FIRST ORDER PERTURBATION SOLUTION FOR HORIZONTALLY POLARIZED INCIDENT WAVE

It was noted in Section 21-1 that a central problem in rough surface scattering is to find the scattering cross section per unit area of a rough surface. In this section, we present a derivation of the scattering cross section using the perturbation method. This method is applicable to a slightly rough surface defined as follows:

Let the height of a rough surface be given by

$$z = \zeta(x, y). \tag{21-9a}$$

We choose $z = 0$ so that (21-9a) represents the deviation from the average height:

$$\langle \zeta(x, y) \rangle = 0. \tag{21-9b}$$

Then the perturbation method is applicable when the phase difference due to the height variation is much smaller than 2π, and the slope is much smaller than unity. Mathematically, we write

$$|k\zeta \cos \theta_i| \ll 1 \tag{21-10a}$$

$$\left|\frac{\partial \zeta}{\partial x}\right| \ll 1, \qquad \left|\frac{\partial \zeta}{\partial y}\right| \ll 1. \tag{21-10b}$$

We first consider a patch of rough surface ΔS. We take ΔS to be a square area with the side L (see Fig. 21-4). We assume that the incident wave is

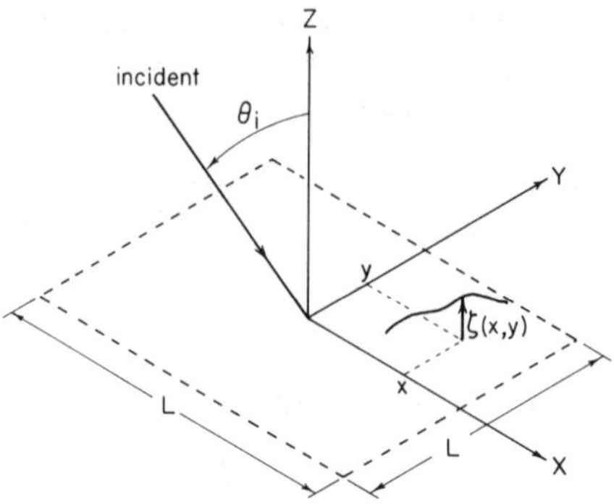

FIG. 21-4 Rough surface ΔS with the side L. The surface height is given by $\zeta(x, y)$.

horizontally polarized and that the plane of incidence is in the xz plane. We also assume that the rough surface is perfectly conducting.

The incident wave has only one component and is given by

$$E_{yi}(x, z) = \exp(i\beta x - i\gamma z) \tag{21-11}$$

where $\beta = k \sin \theta_i$ and $\gamma = k \cos \theta_i$. If the surface is smooth, then the reflected wave is simply given by

$$E_{yr}(x, z) = -\exp(i\beta x + i\gamma z). \tag{21-12}$$

For a rough surface, the total field is the sum of (21-11) and (21-12) plus the wave scattered by the rough surface. Since we have a patch of area L^2, any scattered field can be expressed in terms of two-dimensional Fourier series in the x and y directions with the period L:

$$E_y(x, z) = E_{yi}(x, z) + E_{yr}(x, z) + E_{ys}(x, z) \tag{21-13a}$$

$$E_{ys}(x, z) = \sum_m \sum_n B_{mn} E(v + m, n; z) \tag{21-13b}$$

$$E(v + m, n; z) = \exp[i(2\pi/L)(v + m)x + i(2\pi/L)ny + ib(m + v, n)z]$$
$$\beta = k \sin \theta_i = (2\pi/L)v. \tag{21-13c}$$

Since E_y satisfies a wave equation, $b(m + v, n)$ should satisfy

$$[(2\pi/L)(v + m)]^2 + (2\pi n/L)^2 + [b(m + v, n)]^2 = k^2. \tag{21-14}$$

This Fourier series, Eqs. (21-13b) and (21-13c), is called the space harmonic representation.

We note that (21-13c) contains $\exp(ibz)$, which is the wave propagating in the positive z direction, but it does not include $\exp(-ibz)$, which is the wave propagating in the negative z direction. Strictly speaking, (21-13c) is not complete because the wave between the peak and the bottom of the rough surface should consist of the waves traveling in the $+z$ and $-z$ directions. An exact solution should therefore require an additional term with $\exp(-ibz)$. Various numerical investigations show (Fortuin, 1970), however, that (21-13c) gives a good approximation as long as the slope of the surface is no greater than about 0.4. This condition is usually met in many practical problems. This assumption that the scattered wave can be represented by traveling waves in the $+z$ direction as given by (21-13c) is called the Rayleigh hypothesis.

The other field components are given by

$$E_x(x, z) = \sum_m \sum_n A_{mn} E(v + m, n; z) \tag{21-15}$$

$$E_z(x, z) = \sum_m \sum_n C_{mn} E(v + m, n; z). \tag{21-16}$$

The coefficients A_{mn}, B_{mn}, and C_{mn} are not independent. If two of them are known, the third can be found by using the divergence equation $\nabla \cdot \mathbf{E} = 0$:

$$(2\pi/L)(v + m)A_{mn} + (2\pi n/L)B_{mn} + b(v + m, n)C_{mn} = 0. \tag{21-17}$$

Let us consider the boundary conditions. Let $\hat{\mathbf{N}}$ be the unit vector normal to the surface. Then the tangential component of $\mathbf{E}$ is given by

$$\mathbf{E}_t = \mathbf{E} - \hat{\mathbf{N}}(\mathbf{E} \cdot \hat{\mathbf{N}}). \tag{21-18}$$

The boundary condition that $\mathbf{E}_t = 0$ on the surface is therefore given by taking the x and y components of (21-18) and equating them to zero:

$$E_x - N_x(\mathbf{E} \cdot \hat{\mathbf{N}}) = 0 \tag{21-19a}$$

$$E_y - N_y(\mathbf{E} \cdot \hat{\mathbf{N}}) = 0. \tag{21-19b}$$

The z component of (21-18) is not independent of (20-19a) and (20-19b) since it is obtained from $\mathbf{E}_t \cdot \hat{\mathbf{N}} = 0$.

The normal unit vector $\hat{\mathbf{N}}$ is related to the surface height $\zeta(x, y)$ through

$$\frac{\partial \zeta}{\partial x} = -\frac{N_x}{N_z} \quad \text{and} \quad \frac{\partial \zeta}{\partial y} = -\frac{N_y}{N_z}. \tag{21-20}$$

Noting that $\hat{\mathbf{N}} \cdot \hat{\mathbf{N}} = 1$, we obtain

$$\hat{\mathbf{N}} = \left[1 + \left(\frac{\partial \zeta}{\partial x}\right)^2 + \left(\frac{\partial \zeta}{\partial y}\right)^2\right]^{-1/2} \left[-\frac{\partial \zeta}{\partial x}\hat{\mathbf{x}} - \frac{\partial \zeta}{\partial y}\hat{\mathbf{y}} + \hat{\mathbf{z}}\right]. \tag{21-21}$$

We now have a completely general expression for the fields (21-13), (21-15), and (21-16) with unknown coefficients A_{mn}, B_{mn}, and C_{mn} and the boundary conditions (21-19a) and (21-19b). In general, it is impossible to apply these boundary conditions and to obtain an analytical solution. However, it is possible to obtain a simple expression for the first order perturbation solution.

The perturbation solution is obtained by seeking the terms of the first power of a small quantity ε. This small quantity ε may be the height of the surface in wavelength or the slope of the surface:

$$\varepsilon \sim |k\zeta| \sim \left|\frac{\partial\zeta}{\partial x}\right| \sim \left|\frac{\partial\zeta}{\partial y}\right|. \tag{21-22}$$

We expand $\hat{\mathbf{N}}$ in (21-21) in powers of ε and obtain

$$N_x = -\frac{\partial\zeta}{\partial x} + O(\varepsilon^3), \qquad N_y = -\frac{\partial\zeta}{\partial y} + O(\varepsilon^3), \qquad N_z = 1 + O(\varepsilon^2). \tag{21-23}$$

Using (21-23), we expand the boundary conditions (21-19a) and (21-19b) in powers of ε and keep the lowest order:

$$E_x + \frac{\partial\zeta}{\partial x} E_z = 0, \tag{21-24a}$$

$$E_y + \frac{\partial\zeta}{\partial y} E_z = 0. \tag{21-24b}$$

We note from (21-16) that for horizontal polarization, E_z vanishes for a smooth surface ($k\zeta \to 0$) and therefore E_z is at least on the order of ε. From (21-24), we then have

$$E_z \sim \varepsilon, \qquad E_x \sim \varepsilon^2, \qquad E_y \sim \varepsilon^2. \tag{21-25}$$

For vertical polarization, $E_z \to$ const for a smooth surface ($k\zeta \to 0$) and therefore using (21-24), we get

$$E_z \sim \text{const}, \qquad E_x \sim \varepsilon, \qquad E_y \sim \varepsilon. \tag{21-26}$$

Let us next consider the expressions for E_x, E_y, and E_z given in (21-13), (21-15), and (21-16). On the surface $z = \zeta(x, y)$, we can expand $E(v + m, n; z)$ in powers of ε:

$$E(v + m, n; \zeta) = E(v + m, n; 0)(1 + ib\zeta + \cdots). \tag{21-27}$$

We also expand the coefficients A_{mn}, B_{mn}, and C_{mn} in a series of powers of ε:

$$A_{mn} = A_{mn}^{(1)} + A_{mn}^{(2)} + \cdots, \tag{21-28}$$

where $A_{mn}^{(1)}$ is on the order of ε, $A_{mn}^{(2)}$ is on the order of ε^2, and so on. Similarly we expand B_{mn} and C_{mn} in powers of ε. We note that since all the space harmonics disappear as the surface becomes smooth ($\zeta \to 0$), all the coefficients A_{mn}, B_{mn}, and C_{mn} are at least on the order of ε or higher.

Using these expansions, we write a general expression for E_x, E_y, and E_z on the surface $z = \zeta(x, y)$ in powers of ε:

$$E_x = \sum_m \sum_n [A_{mn}^{(1)} + A_{mn}^{(2)} + \cdots]E(v + m, n; 0)(1 + ib\zeta + \cdots) \quad (21\text{-}29)$$

$$E_y = 2i(\gamma\zeta + \cdots)e^{i\beta x} + \sum_m \sum_n [B_{mn}^{(1)} + B_{mn}^{(2)} + \cdots]$$

$$\times E(v + m, n; 0)(1 + ib\zeta + \cdots) \quad (21\text{-}30)$$

$$E_z = \sum_m \sum_n [C_{mn}^{(1)} + C_{mn}^{(2)} + \cdots]E(v + m, n; 0)(1 + ib\zeta + \cdots). \quad (21\text{-}31)$$

Substituting (21-29) and (21-31) into (21-24a) and equating the first power of ε to zero, we get

$$\sum_m \sum_n A_{mn}^{(1)} E(v + m, n; 0) = 0. \quad (21\text{-}32)$$

Since this is a Fourier series in x and y, all the coefficients must be zero:

$$A_{mn}^{(1)} = 0. \quad (21\text{-}33)$$

Similarly, we can equate the second power of ε to zero, and obtain the relationship for the second order coefficients (Valenzuela, 1968a,b, 1970; Barrick, 1971, 1972). However, we are interested only in the first order solution and we will not pursue the detailed analysis for the second and higher order solutions.

Next, we substitute (21-30) and (21-31) into (21-24b), and obtain the following relationship between the height ζ and $B_{mn}^{(1)}$:

$$2i\gamma\zeta e^{i\beta x} + \sum_m \sum_n B_{mn}^{(1)} E(v + m, n; 0) = 0. \quad (21\text{-}34)$$

Now, we express the surface height $\zeta = \zeta(x, y)$ in a two-dimensional Fourier series:

$$\zeta(x, y) = \sum_m \sum_n P(m, n) \exp\left(i\frac{2\pi m}{L}x + i\frac{2\pi n}{L}y\right). \quad (21\text{-}35)$$

Then the first term of (21-34) becomes

$$2i\gamma\zeta e^{i\beta x} = 2i\gamma \sum_m \sum_n P(m, n)E(v + m, n; 0). \quad (21\text{-}36)$$

Substituting (21-36) into (21-34), we obtain

$$B_{mn}^{(1)} = -2i\gamma P(m, n). \quad (21\text{-}37)$$

Using (21-17), we obtain

$$C^{(1)}_{mn} = \frac{(2i\gamma)(2\pi n/L)}{b} P(m, n). \tag{21-38}$$

The first order solution is therefore given by

$$\mathbf{E} = \mathbf{E}_i + \mathbf{E}_r + \mathbf{E}_s, \qquad \mathbf{H} = \mathbf{H}_i + \mathbf{H}_r + \mathbf{H}_s, \tag{21-39}$$

where $\mathbf{E}_i$ and $\mathbf{H}_i$ are the incident wave, $\mathbf{E}_r$ and $\mathbf{H}_r$ are the wave reflected from a smooth surface, and $\mathbf{E}_s$ and $\mathbf{H}_s$ are the effect of the rough surface:

$$E_{iy} + E_{ry} = 2i \sin \gamma z e^{i\beta x}, \qquad E_{sx} = 0 \tag{21-40a}$$

$$E_{sy} = \sum_m \sum_n B^{(1)}_{mn} E(\nu + m, n; z), \qquad E_{sz} = \sum_m \sum_n C^{(1)}_{mn} E(\nu + m, n; z).$$

The magnetic field is obtained from (21-40a) by using one of Maxwell's equations $\nabla \times \mathbf{E} = i\omega\mu_0 \mathbf{H}$:

$$\sqrt{\frac{\mu_0}{\varepsilon_0}}(H_{ix} + H_{rx}) = -\frac{2\gamma \cos \gamma z}{k} e^{i\beta x},$$

$$\sqrt{\frac{\mu_0}{\varepsilon_0}}(H_{iz} + H_{rz}) = \frac{2i\beta \sin \gamma z}{k} e^{i\beta x}$$

$$\sqrt{\frac{\mu_0}{\varepsilon_0}} H_{sx} = \sum_m \sum_n D^{(1)}_{mn} E(\nu + m, n; z), \tag{21-40b}$$

$$\sqrt{\frac{\mu_0}{\varepsilon_0}} H_{sy} = \sum_m \sum_n E^{(1)}_{mn} E(\nu + m, n; z)$$

$$\sqrt{\frac{\mu_0}{\varepsilon_0}} H_{sz} = \sum_m \sum_n F^{(1)}_{mn} E(\nu + m, n; z),$$

where

$$D^{(1)}_{mn} = \left(\frac{2i\gamma}{bk}\right)\left\{\left(\frac{2\pi n}{L}\right)^2 + b^2\right\} P(m, n)$$

$$E^{(1)}_{mn} = \left(\frac{-2i\gamma}{bk}\right)\left(\frac{2\pi n}{L}\right)\left[\frac{2\pi}{L}(\nu + m)\right] P(m, n)$$

$$F^{(1)}_{mn} = -\left(\frac{2i\gamma}{k}\right)\left[\frac{2\pi}{L}(\nu + m)\right] P(m, n).$$

Equations (21-39) and (21-40) constitute the complete first order solution for the field when a horizontally polarized wave is incident on a rough surface. The rough surface height is characterized by its two-dimensional Fourier expansion (21-35) with coefficients $P(m, n)$.

When a vertically polarized wave is incident on a surface, using the same procedure, we get

$$E_{ix} + E_{rx} = -2i \cos\theta_i \sin\gamma z e^{i\beta x}$$

$$E_{iz} + E_{rz} = 2 \sin\theta_i \cos\gamma z e^{i\beta x}$$

$$E_{sx} = \sum_m \sum_n A_{mn}^{(1)} E(\nu + m, n; z), \qquad E_{sy} = \sum_m \sum_n B_{mn}^{(1)} E(\nu + m, n; z)$$

$$E_{sz} = \sum_m \sum_n C_{mn}^{(1)} E(\nu + m, n; z) \tag{21-41}$$

$$\sqrt{\frac{\mu_0}{\varepsilon_0}}(H_{iy} + H_{ry}) = -2 \cos\gamma z, \qquad \sqrt{\frac{\mu_0}{\varepsilon_0}} H_{sx} = \sum_m \sum_n D_{mn}^{(1)} E(\nu + m, n; z)$$

$$\sqrt{\frac{\mu_0}{\varepsilon_0}} H_{sy} = \sum_m \sum_n E_{mn}^{(1)} E(\nu + m, n; z)$$

$$\sqrt{\frac{\mu_0}{\varepsilon_0}} H_{sz} = \sum_m \sum_n F_{mn}^{(1)} E(\nu + m, n; z)$$

where

$$A_{mn}^{(1)} = 2i\left[k \cos^2\theta_i - \left(\frac{2\pi m}{L}\right)\sin\theta_i\right] P(m, n)$$

$$B_{mn}^{(1)} = -2i\left(\frac{2\pi n}{L}\right)\sin\theta_i P(m, n)$$

$$C_{mn}^{(1)} = -\frac{1}{b}\left[\frac{2\pi\cdot(\nu + m)}{L} A_{mn}^{(1)} + \frac{2\pi n}{L} B_{mn}^{(1)}\right]$$

$$D_{mn}^{(1)} = \frac{1}{k}\left[\left(\frac{2\pi n}{L}\right) C_{mn}^{(1)} - b B_{mn}^{(1)}\right], \qquad E_{mn}^{(1)} = \frac{1}{k}\left[b A_{mn}^{(1)} - \frac{2\pi m}{L} C_{mn}^{(1)}\right]$$

$$F_{mn}^{(1)} = \frac{1}{k}\left[\left(\frac{2\pi m}{L}\right) B_{mn}^{(1)} - \left(\frac{2\pi n}{L}\right) A_{mn}^{(1)}\right].$$

We note that this solution is applicable not only to a rough surface but to any deterministic surface with a period L.

21-3 DERIVATION OF THE FIRST ORDER SCATTERING CROSS SECTION PER UNIT AREA

In the preceding section we derived the first order solution for the field over the patch $\Delta S = L^2$ of the rough surface. This field gives rise to the scattered field in all directions. Let us consider the field at a distance R from

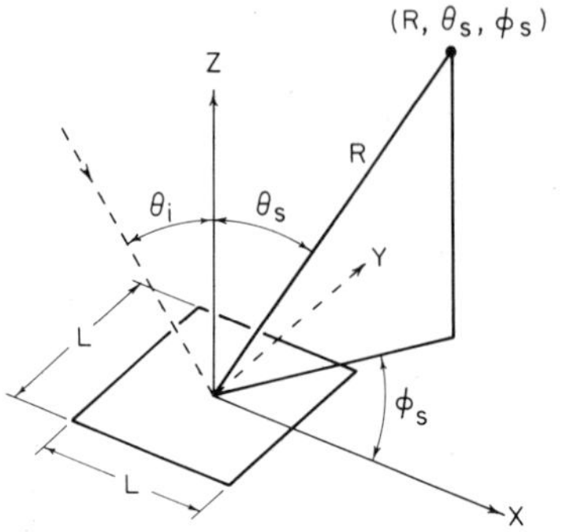

FIG. 21-5 A wave is incident in the direction $(\theta_i, 0)$ and the scattered wave is observed in the direction (θ_s, ϕ_s) at a large distance R from thè area L^2.

the patch in the direction (θ_s, ϕ_s) (see Fig. 21-5). We take R to be in the far zone of the patch:

$$R \gg L^2/\lambda \qquad \text{and} \qquad L \gg \text{correlation distance.}$$

Let us consider the field at (R, θ_s, ϕ_s). It should in general consist of the coherent field and the incoherent field. The coherent field in the first order approximation is simply the field which exists if the surface is completely smooth, and therefore the received coherent power P_{rcoh} given in (21-3) becomes

$$\frac{P_{\text{rcoh}}}{P_t} = \frac{\lambda^2}{(4\pi)^2} \frac{G_t G_r}{(R_1 + R_2)^2} |R_f|^2 \tag{21-42}$$

where $|R_f|^2$ is the power reflection coefficient of the surface. Note that in this first order solution, χ in (21-3) is simply unity.

The incoherent field (diffuse field) is produced by the scattered portion of the field on the rough surface which is given by the coefficients $A_{mn}^{(1)}, B_{mn}^{(1)}, \ldots$. In order to obtain the incoherent field at (R, θ_s, ϕ_s) we need to relate the field on the surface ΔS and the far field. In general, the field at $\mathbf{r}$ due to the field at $\mathbf{r}'$on a surface s is given by (Tai, 1972)

$$\mathbf{E}(\mathbf{r}) = \nabla \times \int_s [\hat{\mathbf{N}} \times \mathbf{E}(\mathbf{r}')] G_0(\mathbf{r}, \mathbf{r}')\, ds' + \frac{i}{\omega\varepsilon_0} \nabla \times \nabla \times \int_s \hat{\mathbf{N}} \times \mathbf{H}(\mathbf{r}')] G_0(\mathbf{r}, \mathbf{r}')\, ds' \tag{21-43}$$

$$\mathbf{H}(\mathbf{r}) = \nabla \times \int_s [\hat{\mathbf{N}} \times \mathbf{H}(r')] G_0(\mathbf{r}, \mathbf{r}')\, ds' - \frac{i}{\omega\mu_0} \nabla \times \nabla \times \int_s [\hat{\mathbf{N}} \times \mathbf{E}(\mathbf{r}')] G_0(\mathbf{r}, \mathbf{r}')\, ds'$$

where

$$G_0(\mathbf{r}, \mathbf{r}') = \frac{\exp(ik|\mathbf{r} - \mathbf{r}'|)}{4\pi|\mathbf{r} - \mathbf{r}'|}.$$

We take the observation point $\mathbf{r}$ to be in the far zone of the area s and let the area s be the patch $\Delta S = L^2$ at $z = 0$. Then, the far field at (R, θ_s, ϕ_s) is given by the fields E_x, E_y, H_x, and H_y on the surface Δs:

$$E_\theta = \frac{ik}{4\pi R} e^{ikR} I_\theta, \qquad E_\phi = \frac{ik}{4\pi R} e^{ikR} I_\phi$$

$$I_\theta = \int_{\Delta S} [-(E_x \cos\phi_s + E_y \sin\phi_s) + Z_0(H_x \sin\phi_s - H_y \cos\phi_s) \cos\theta_s] \exp(-ik\mathbf{r}' \cdot \hat{\mathbf{r}}_s)\, dx'\, dy' \tag{21-44}$$

$$I_\phi = \int_{\Delta S} [E_x \sin\phi_s - E_y \cos\phi_s) \cos\theta_s + Z_0(H_x \cos\phi_s + H_y \sin\phi_s)] \exp(-ik\mathbf{r}' \cdot \hat{\mathbf{r}}_s)\, dx'\, dy'$$

where $\mathbf{r}' \cdot \hat{\mathbf{r}}_s = x' \sin\theta_s \cos\phi_s + y' \sin\theta_s \sin\phi_s$, and $Z_0 = (\mu_0/\varepsilon_0)^{1/2}$.

In terms of (21-44) we can define the cross section per unit area of the rough surface in the following manner. Consider the solution (21-40) for the horizontally polarized incident wave. Then the scattering cross section per unit area when the receiver receives the horizontally polarized wave is given by

$$\sigma_{hh} = \frac{4\pi |E_\phi|^2 R^2}{|E_{in}|^2 L^2} = \frac{k^2 I_\phi I_\phi^*}{4\pi L^2}. \tag{21-45a}$$

The first subscript h means the horizontal polarization received by the receiver, and the second subscript h means the horizontal polarization of the incident wave. Similarly, for a receiver that receives vertical polarization E_θ, we have

$$\sigma_{vh} = k^2 I_\theta I_\theta^* / 4\pi L^2. \tag{21-45b}$$

In these two expressions, we used (21-40) to evaluate I_ϕ and I_θ.

When the incident wave is vertically polarized, we use (21-41) and obtain

$$\sigma_{hv} = k^2 I_\phi I_\phi^* / 4\pi L^2 \tag{21-45c}$$

$$\sigma_{vv} = k^2 I_\theta I_\theta^* / 4\pi L^2. \tag{21-45d}$$

For a rough surface, I_ϕ and I_θ are random functions and therefore we should take an ensemble average. For example,

$$\sigma_{hh} = (k^2 \langle I_\phi I_\phi^* \rangle)/4\pi L^2, \qquad \sigma_{vh} = (k^2 \langle I_\phi I_\phi^* \rangle)/4\pi L^2. \tag{21-46}$$

In order to calculate the ensemble average in (21-46) we need to examine the statistical description of the rough surface. This is presented in the following section.

21-4 STATISTICAL DESCRIPTION OF A ROUGH SURFACE

Up to this point, we have not introduced any statistical consideration and thus the description is valid both for deterministic and rough surfaces. Now we introduce the statistical nature of the surface. The surface height has been given in a two-dimensional Fourier series:

$$\zeta(x, y) = \sum_m \sum_n P(m, n) \exp\left(i\frac{2\pi m}{L}x + i\frac{2\pi n}{L}y\right). \tag{21-47}$$

Since $\zeta(x, y)$ is a random function, $P(m, n)$ is a random variable. We assume that the coefficient $P(m, n)$ satisfies the following conditions:

(1) $P(m, n)$ is a random variable with zero mean, and
(2) $P(m, n)$ are uncorrelated for different spatial frequencies.

Mathematically, we write

$$\begin{aligned} \langle P(m, n)\rangle &= 0 \\ \langle P(m, n)P^*(m'n')\rangle &= 0 \qquad \text{if} \quad m \neq m' \quad \text{and} \quad n \neq n' \\ &= \left(\frac{2\pi}{L}\right)^2 \frac{1}{4} W\left(\frac{2\pi m}{L}, \frac{2\pi n}{L}\right) \qquad \text{if} \quad m = m' \quad \text{and} \quad n = n'. \end{aligned} \tag{21-48a}$$

In addition, since the height ζ is a real function, $P(m, n)$ satisfies

$$P(m, n) = P^*(-m, -n). \tag{21-48b}$$

The function $W(p, q)$ with $p = 2\pi m/L$ and $q = 2\pi n/L$ is called the spectral density of the rough surface height function and is a Fourier transform of the surface height correlation function. To show this, consider the correlation function

$$\begin{aligned} \langle \zeta(x_1, y_1)\zeta(x_2, y_2)\rangle &= \sum_m \sum_n \sum_{m'} \sum_{n'} \langle P(m, n)P^*(m', n')\rangle \\ &\quad \times \exp\left[i\frac{2\pi}{L}(mx_1 - m'x_2) + i\frac{2\pi}{L}(ny_1 - n'y_2)\right] \\ &= \sum_m \sum_n \left(\frac{2\pi}{L}\right)^2 \frac{1}{4} W(p, q) \\ &\quad \times \exp[ip(x_1 - x_2) + iq(y_1 - y_2)] \end{aligned} \tag{21-49}$$

where $p = 2\pi m/L$ and $q = 2\pi n/L$. Now, by letting $L \to \infty$ we convert this series to the integral†

$$\langle\zeta(x_1, y_1)\zeta(x_2, y_2)\rangle$$
$$= \tfrac{1}{4}\int_{-\infty}^{\infty} dp \int_{-\infty}^{\infty} dq\, W(p, q) \exp[ip(x_1 - x_2) + iq(y_1 - y_2)]. \quad (21\text{-}50)$$

This conversion to an integral is justified because L is much greater than the correlation distance. The spectral density $W(p, q)$ is a real positive function of p and q and since ζ is real, $W(p, q)$ is an even function of p and q. The inverse of (21-50) is given by

$$W(p, q) = \frac{1}{\pi^2}\int_{-\infty}^{\infty} dx_d \int_{-\infty}^{\infty} dy_d \langle\zeta(x_1, y_1)\zeta(x_2, y_2)\rangle \exp(-ipx_d - iqy_d) \quad (21\text{-}51)$$

where $x_d = x_1 - x_2$ and $y_d = y_1 - y_2$.

As can be seen from (21-50) the correlation function is a function of the differences x_d and y_d only and therefore the rough surface under the assumption (21-48a) is statistically homogeneous. We note that (21-48a) is equivalent to the spectral representation of a homogeneous random function (see Appendix A). The physical meaning of the spectral density $W(p, q)$ is that $W(p, q)\,dp\,dq$ is the amount of the component of the rough surface having the spatial wave number between p and $p + dp$ in the x direction and between q and $q + dq$ in the y direction.

21-5 BISTATIC CROSS SECTION OF A ROUGH SURFACE

We now evaluate the cross section given in (21-46) making use of the statistical description of a rough surface developed in the preceding section for a horizontally polarized incident wave. We substitute E_s and H_s in (21-40) into (21-44) and express I_θ and I_ϕ in terms of the Fourier series containing $P(m, n)$. Noting that E_{sx}, E_{sy}, H_{sx}, and H_{sy} always contain $P(m, n)E(v + m, n; z)$, we can write

$$\begin{Bmatrix} I_\theta \\ I_\phi \end{Bmatrix} = \int_{\Delta S} dx'\, dy' \sum_m \sum_n \begin{Bmatrix} f_\theta(m, n) \\ f_\phi(m, n) \end{Bmatrix} P(m, n) \exp(ip'x' + iq'y') \quad (21\text{-}52)$$

† Note that $f(x) = \sum_{-\infty}^{\infty} c(n)e^{i(2\pi/L)nx}$ and $c(n) = L^{-1}\int_0^L f(x)e^{-i(2\pi/L)nx}\,dx$ can be converted to a Fourier transform by letting $(2\pi/L)n \to \omega$, $2\pi/L \to d\omega$, $L \to \infty$. We get $f(x) = \int_{-\infty}^{\infty} [(L/2\pi)c(n)]e^{i\omega x}\,d\omega$, $(L/2\pi)c(n) = (2\pi)^{-1}\int_{-\infty}^{\infty} f(x)e^{-i\omega x}\,dx$.

where

$$p' = \frac{2\pi}{L}(v+m) - k \sin\theta_s \cos\phi_s, \qquad q' = \frac{2\pi}{L} n - k \sin\theta_s \sin\phi_s$$

$$f_\theta = -[e_x \cos\phi_s + e_y \sin\phi_s] + (h_x \sin\phi_s - h_y \cos\phi_s)\cos\theta_s$$

$$f_\phi = (e_x \sin\phi_s - e_y \cos\phi_s)\cos\theta_s + (h_x \cos\phi_s + h_y \sin\phi_s)$$

$$e_x = \frac{A_{mn}^{(1)}}{P(m,n)}, \qquad e_y = \frac{B_{mn}^{(1)}}{P(m,n)}, \qquad h_x = \frac{D_{mn}^{(1)}}{P(m,n)}, \qquad h_y = \frac{E_{mn}^{(1)}}{P(m,n)}.$$

Now we consider $\langle I_\phi I_\phi^* \rangle$. Using (21-48a), we obtain

$$\langle I_\phi I_\phi^* \rangle = \int dx' \int dy' \int dx'' \int dy'' \sum_m \sum_n |f_\phi(m,n)|^2 \times \left(\frac{2\pi}{L}\right)^2 \frac{1}{4} W\left(\frac{2\pi m}{L}, \frac{2\pi n}{L}\right) \exp(ip'x_d + iq'y_d) \quad (21\text{-}53)$$

where $x_d = x' - x''$ and $y_d = y' - y''$. We note that

$$\int dx' \int dx'' = \int dx_d \int dx_c$$

where $x_c = \frac{1}{2}(x' + x'')$ and $y_c = \frac{1}{2}(y' + y'')$. Also we note that

$$\int \exp(iKx_d)\, dx_d \to 2\pi\, \delta(K) \qquad \text{as} \quad L \to \infty.$$

Converting the summation in (21-53) to integrals, we obtain

$$\langle I_\phi I_\phi^* \rangle = L^2 \int dp \int dq\, |f_\phi(p,q)|^2 \tfrac{1}{4} W(p,q)(2\pi)^2\, \delta(p')\, \delta(q') \quad (21\text{-}54)$$

where

$$p = \frac{2\pi m}{L}, \qquad q = \frac{2\pi n}{L}, \qquad \beta = k \sin\theta_i = \frac{2\pi}{L} v$$

$$p' = p + \beta - k \sin\theta_s \cos\phi_s, \qquad q' = q - k \sin\theta_s \sin\phi_s.$$

Evaluating the integrals in (21-54) and substituting $\langle I_\phi I_\phi^* \rangle$ into (21-46), we get

$$\sigma_{hh} = \frac{\pi k^2}{4} |f_\phi(p,q)|^2 W(p,q) \quad (21\text{-}55)$$

where p and q are given by

$$p = k \sin\theta_s \cos\phi_s - k \sin\theta_i, \qquad q = k \sin\theta_s \sin\phi_s. \quad (21\text{-}56)$$

Similarly, we have

$$\sigma_{vh} = \frac{\pi k^2}{4} \,|\, f_\theta(p, q) \,|^2 W(p, q). \tag{21-57}$$

For a vertical incident wave, using (21-41) in (21-44) we get

$$\sigma_{hv} = \frac{\pi k^2}{4} \,|\, f_\phi(p, q) \,|^2 W(p, q) \tag{21-58}$$

$$\sigma_{vv} = \frac{\pi k^2}{4} \,|\, f_\theta(p, q) \,|^2 W(p, q). \tag{21-59}$$

We can easily obtain f_θ and f_ϕ in this expression from (21-52). For example, f_ϕ for the horizontally polarized incidence is given by

$$f_\phi = (e_x \sin \phi_s - e_y \cos \phi_s) \cos \theta_s + (h_x \cos \phi_s + h_y \sin \phi_s). \tag{21-60}$$

From (21-33), (21-37), and (21-40b), we get

$$e_x = 0, \qquad e_y = -2i\gamma, \qquad h_x = \left(\frac{2i\gamma}{bk}\right)(q^2 + b^2), \qquad h_y = \left(-\frac{2i\gamma}{bk}\right) q(p + \beta) \tag{21-61}$$

where $\gamma = k \cos \theta_i$, $\beta = k \sin \theta_i$, $p + \beta = k \sin \theta_s \cos \phi_s$, $q = k \sin \theta_s \sin \phi_s$, and $b = k \cos \theta_s$. Substituting (21-61) into (21-60), we get

$$f_\phi = 4ik \cos \theta_i \cos \theta_s \cos \phi_s. \tag{21-62}$$

Therefore, from (21-55), we get

$$\sigma_{hh} = 4\pi k^4 \cos^2 \theta_i \cos^2 \theta_s \cos^2 \phi_s W(p, q). \tag{21-63}$$

Similarly, we obtain

$$\begin{aligned} \sigma_{vh} &= 4\pi k^4 \cos^2 \theta_i \sin^2 \phi_s W(p, q) \\ \sigma_{hv} &= 4\pi k^4 \cos^2 \theta_s \sin^2 \phi_s W(p, q) \\ \sigma_{vv} &= 4\pi k^4 (\sin \theta_i \sin \theta_s - \cos \phi_s)^2 W(p, q) \end{aligned} \tag{21-64}$$

where $W(p, q)$ are evaluated at $p = k \sin \theta_s \cos \phi_s - k \sin \theta_i$ and $q = k \sin \theta_s \sin \phi_s$.

For backscattering ($\theta_s = \theta_i$ and $\phi_s = \pi$), we have

$$\begin{aligned} &\sigma_{hh} = 4\pi k^4 \cos^4 \theta_i W(-2k \sin \theta_i, 0), \qquad \sigma_{vh} = 0, \qquad \sigma_{hv} = 0 \\ &\sigma_{vv} = 4\pi k^4 (1 + \sin^2 \theta_i)^2 W(-2k \sin \theta_i, 0). \end{aligned} \tag{21-65}$$

The physical meaning of (21-65) is that the backscattering is proportional to that portion of the rough surface spectrum where $p = -2k \sin \theta_1$. p is the spatial wave number of the surface. Therefore, the backscattering comes from the spectrum with the spatial wave number with a period d in the x direction such that

$$p = 2\pi/d = -2k \sin \theta_i. \tag{21-66}$$

This also means that two rays hitting the surface separated by the spacing d produce the phase shift equal to 2π (see Fig. 21-6). Similarly, for a general case,

$$p = k \sin \theta_s \cos \phi_s - k \sin \theta_i \quad \text{and} \quad q = k \sin \theta_s \sin \phi_s$$

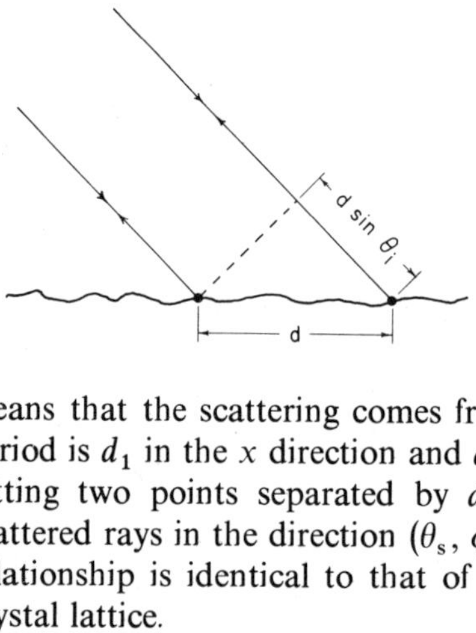

FIG. 21-6 Backscattering comes from the portion of the spectrum with a period d such that $2d \sin \theta_i$ is equal to a wavelength.

means that the scattering comes from the portion of the spectrum whose period is d_1 in the x direction and d_2 in the y direction such that two rays hitting two points separated by d_1 and d_2 on the surface produce the scattered rays in the direction (θ_s, ϕ_s) with the phase difference of 2π. This relationship is identical to that of the Bragg diffraction of x-rays from a crystal lattice.

This analysis can be extended to include the scattering from a rough interface. Letting ε_r be the complex relative dielectric constant of the medium below the interface, we get (Ruck *et al.*, 1970, Vol. 2, p. 706; Valenzuela, 1967)

$$\sigma_{hh} = 4\pi k^4 \cos^2 \theta_i \cos^2 \theta_s |\alpha_{hh}|^2 W(p, q) \tag{21-67}$$

where

$$\alpha_{hh} = -\frac{(\varepsilon_r - 1)\cos \phi_s}{[\cos \theta_i + (\varepsilon_r - \sin^2 \theta_i)^{1/2}][\cos \theta_s + (\varepsilon_r - \sin^2 \theta_s)^{1/2}]}.$$

Similarly, for σ_{vh}, σ_{hv}, and σ_{vv}, we use (21-67) with α_{hh} replaced by α_{vh}, α_{hv}, and α_{vv}, respectively:

$$\alpha_{vh} = -\frac{(\varepsilon_r - 1)(\varepsilon_r - \sin^2\theta_s)^{1/2}\sin\phi_s}{[\cos\theta_i + (\varepsilon_r - \sin^2\theta_i)^{1/2}][\varepsilon_r\cos\theta_s + (\varepsilon_r - \sin^2\theta_s)^{1/2}]}$$

$$\alpha_{hv} = \frac{\sin\phi_s(\varepsilon_r - 1)(\varepsilon_r - \sin^2\theta_i)^{1/2}}{[\varepsilon_r\cos\theta_i + (\varepsilon_r - \sin^2\theta_i)^{1/2}][\cos\theta_s + (\varepsilon_r - \sin^2\theta_s)^{1/2}]}$$

$$\alpha_{vv} = \frac{(\varepsilon_r - 1)[\varepsilon_r\sin\theta_i\sin\theta_s - \cos\phi_s(\varepsilon_r - \sin^2\theta_i)^{1/2}(\varepsilon_r - \sin^2\theta_s)^{1/2}]}{[\varepsilon_r\cos\theta_i + (\varepsilon_r - \sin^2\theta_i)^{1/2}][\varepsilon_r\cos\theta_s + (\varepsilon_r - \sin^2\theta_s)^{1/2}]}. \tag{21-68}$$

21-6 EFFECT OF TEMPORAL VARIATION OF A ROUGH SURFACE

In Section 21-4 we discussed the statistical representation of a time invariant rough surface. In this section, we extend the results to include a time-varying rough surface. We note that each space harmonic component given in (21-47) may move with a definite phase velocity. For example, deep-water ocean waves of wavelength l travel with a phase velocity $v_p = \sqrt{gl/2\pi}$ where $g = 9.81$ m/sec^2 is the gravitational acceleration. The angular frequency ω_r corresponding to this phase velocity is

$$\omega_r = (2\pi/l)v_p = [g(2\pi/l)]^{1/2}. \tag{21-69}$$

Consider the space harmonic representation of the rough surface height given in (21-47). Since each component has a wave number $2\pi m/L$ and $2\pi n/L$ in the x and y directions, respectively, the angular frequency ω_{mn} for this component for deep-water waves is given by

$$\omega_{mn} = \{g[(2\pi m/L)^2 + (2\pi n/L)^2]^{1/2}\}^{1/2}. \tag{21-70}$$

The rough surface height as a function of x, y, and t is therefore given by

$$\zeta(x, y, t) = \sum_m \sum_n P(m, n)\exp\left(i\frac{2\pi m}{L}x + i\frac{2\pi n}{L}y - i\omega_{mn}t\right). \tag{21-71}$$

Since $\zeta(x, y, t)$ is real, we require, in addition to (21-48), that ω_{mn} is an odd function of m and n:

$$\omega_{-m,-n} = -\omega_{m,n}. \tag{21-72}$$

The correlation function (21-50) then becomes

$$\langle\zeta(x_1, y_1, t_1)\zeta(x_2, y_2, t_2)\rangle$$
$$= \tfrac{1}{4}\int_{-\infty}^{\infty} dp \int_{-\infty}^{\infty} dq W(p, q) \exp(ipx_d + iqy_d - i\omega_r\tau) \quad (21\text{-}73)$$

where $\tau = t_1 - t_2$. The spectral density $W(p, q)$ is an even function of p and q. We note that (21-71) represents the surface moving in the $+x$ direction if ω_{mn} is positive and in the $-x$ direction if ω_{mn} is negative. Therefore, when an ocean wave is advancing in the $+x$ direction, we have

$$\omega_r = [g(p^2 + q^2)^{1/2}]^{1/2}, \quad (21\text{-}74a)$$

and when an ocean wave is advancing in the $-x$ direction, we have

$$\omega_r = -[g(p^2 + q^2)^{1/2}]^{1/2}. \quad (21\text{-}74b)$$

We note that the only difference between (21-73) and the time invariant correlation function (21-50) is that $W(p, q)$ for the latter is replaced by $W(p, q) \exp(-i\omega_r\tau)$ for the former. Therefore, the time-varying scattering cross section σ is given by

$$\sigma = (\pi k^2/4)|f(p, q)|^2 W(p, q) \exp(-i\omega_r\tau). \quad (21\text{-}75)$$

The temporal frequency spectrum of the cross section in (21-8) is given by

$$\sigma(\omega) = (\pi k^2/4)|f(p, q)|^2 W(p, q) 4\pi\, \delta(\omega + \omega_r) \quad (21\text{-}76)$$

where ω is the deviation from the carrier frequency ω_0, ω_r is given in (21-74), and p and q are given by (21-56). For example, the backscattering cross section given in (21-65) has the frequency spectrum

$$\sigma_{hh}(\omega) = 4\pi k^4 \cos^4 \theta_i W(-2k \sin\theta_i, 0) 4\pi\, \delta(\omega + \omega_r)$$

where ω_r is given in (21-74).

Extensive theoretical and experimental studies have been conducted on the scattering from ocean surface based on this analysis (Barrick, 1972; Valenzuela, 1968).

21-7 OCEAN WAVE SPECTRA

The roughness of an ocean surface is caused by winds. We may consider the following two general cases of ocean waves: (a) fully developed waves and (b) swell (Barrick, 1972; Guinard and Daley, 1970; Valenzuela *et al.*, 1972). In the first case, winds have been blowing several hours prior to the observation time so that the ocean wave is fully developed. In general, for wind velocities of 20–30 knots, it takes several hours to fully develop ocean waves of wavelengths 20–200 m. In the second case, ocean waves have been produced at other times and locations and have propagated to the observation point. This is called the swell.

For fully developed ocean waves due to wind speed U (m/sec), Phillips and Munk proposed an empirical spectrum (Phillips, 1969)

$$W(p, q) = \begin{cases} \dfrac{2 \times 10^{-2}}{\pi(p^2 + q^2)^2} & \text{for} \quad (p^2 + q^2)^{1/2} > \dfrac{g}{U^2} \\ 0 & \text{for} \quad (p^2 + q^2)^{1/2} < \dfrac{g}{U^2}. \end{cases} \tag{21-77}$$

Other spectra, including the power law, and their relationships with wind velocity have been studied (Valenzuela *et al.*, 1970; Valenzuela and Laing, 1972; Laing, 1971).

21-8 OTHER RELATED PROBLEMS

We have already discussed the derivation of the scattering cross sections per unit area of the rough surface for different polarizations. In addition to scattering, there are two other important problems: (a) propagation along a rough surface and (b) transmission of a wave through a rough interface. Both problems can be solved by applying the perturbation technique.

In this chapter, we have discussed only electromagnetic scattering. However, the perturbation technique is equally applicable to acoustic scattering problems. If the rough surface is an interface between two fluids such as air and water, then we can use acoustic potential or acoustic pressure. On the other hand, if the surface is an interface between liquid and solid, we need to consider scalar and vector potentials.

If the surface consists of a superposition of two rough surfaces, this is called the composite rough surface. An example is ocean waves with long wavelengths carrying capillary waves with short wavelengths. Since these two rough surfaces are statistically almost independent, the scattering cross section of a composite rough surface may be well approximated by the sum of cross sections for each surface. The scattering from composite rough surfaces has been studied extensively (Peake *et al.*, 1970; Fung and Chan, 1969).

We also mention here earlier studies on rough surface scattering by Ament (1953), Beard (1961, 1967), and Twersky (1957), and more recent works by Krishen (1970) and Fung (1968, 1970). The relationship between the Kirchhoff approach discussed in the following section and the perturbation analysis has also been discussed (Valenzuela and Laing, 1972; Leader, 1971). We also add extensive literature on Rayleigh hypothesis (Bates, 1969; Burrows, 1969a,b; Millar, 1969, 1971; Millar *et al.*, 1969).

21-9 KIRCHHOFF APPROXIMATION–SCATTERING OF SOUND WAVES FROM A ROUGH SURFACE

Up to this point, we have considered the first order perturbation solution to the rough surface scattering problem. In the first order solution, the coherent power is identical to that of a smooth surface and the incoherent power is given in terms of the scattering cross section per unit area of the rough surface. When the surface height is not negligible compared with a wavelength, the coherent power tends to decrease and the incoherent (diffuse) power increases. At present, no general theory exists which is applicable to a very rough surface. However, if the surface is slowly varying so that the radius of curvature is much greater than a wavelength, the Kirchhoff approximation may be used to obtain a reasonably simple solution.

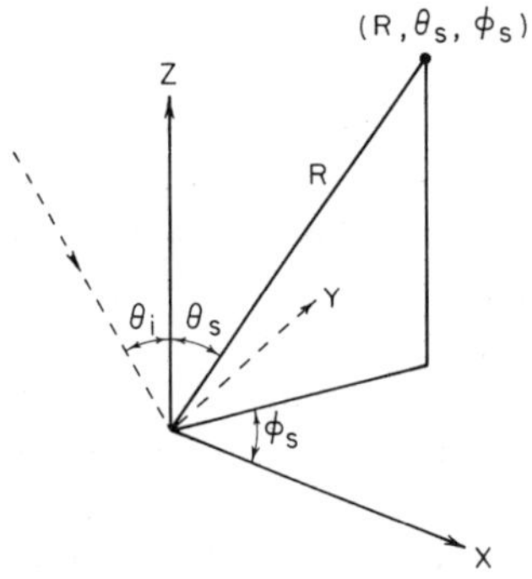

FIG. 21-7 Scattering in the direction (θ_s, ϕ_s) from a rough surface $\zeta = \zeta(x, y)$ illuminated by a plane wave in the direction $(\theta_i, 0)$.

Let us consider an acoustic wave incident on a rough surface (see Fig. 21-7). The surface separates two fluid media such as air and water. We choose the xz plane to be the plane of incidence. The incident pressure p_i is then given by

$$p_i(\mathbf{r}) = \exp(i\beta x - i\gamma z) = \exp(i\mathbf{k}_i \cdot \mathbf{r}) \tag{21-78}$$

where $\mathbf{k}_i = k \sin\theta_i \hat{\mathbf{x}} - k\cos\theta_i \hat{\mathbf{z}} = \beta\hat{\mathbf{x}} - \gamma\hat{\mathbf{z}}$ and $\mathbf{r} = x\hat{\mathbf{x}} + y\hat{\mathbf{y}} + z\hat{\mathbf{z}}$. The surface height $\zeta(x, y)$ is measured from the average height, and therefore

$$\langle \zeta(x, y) \rangle = 0. \tag{21-79}$$

Let us consider the field $p(\mathbf{r})$ at $\mathbf{r}$ produced by the field $p(\mathbf{r}')$ on the rough surface S. This has been shown for electromagnetic waves in (21-43). For a scalar wave, we have the following equivalent formula based on Green's theorem:

$$p(\mathbf{r}) = \int_s \left[p(\mathbf{r}') \frac{\partial G_0(\mathbf{r}, \mathbf{r}')}{\partial n'} - G_0(\mathbf{r}, \mathbf{r}') \frac{\partial p(\mathbf{r}')}{\partial n'} \right] ds' \tag{21-80}$$

where $G_0(\mathbf{r}, \mathbf{r}') = [\exp(ik|\mathbf{r} - \mathbf{r}'|)]/(4\pi|\mathbf{r} - \mathbf{r}'|)$, $\partial/\partial n'$ is the normal derivative, and the positive n' is in the direction of the space containing $\mathbf{r}$.

We now take the observation point $\mathbf{r}(R, \theta_s, \phi_s)$ to be in the far zone of the surface S. Then we can approximate Green's function in (21-80) by

$$G_0 \approx \frac{\exp(ikR - i\mathbf{k}_s \cdot \mathbf{r}')}{4\pi R}, \qquad \frac{\partial G_0}{\partial n'} \approx G_0 \frac{\partial(-i\mathbf{k}_s \cdot \mathbf{r}')}{\partial n'} = -i\mathbf{k}_s \cdot \hat{\mathbf{N}} G_0 \tag{21-81}$$

where $\hat{\mathbf{N}}$ is the unit vector normal to the surface and

$$\mathbf{k}_s = (k \sin\theta_s \cos\phi_s)\hat{\mathbf{x}} + (k \sin\theta_s \sin\phi_s)\hat{\mathbf{y}} + (k \cos\theta_s)\hat{\mathbf{z}}.$$

The field $p(\mathbf{r}')$ and its normal derivative $\partial p(\mathbf{r}')/\partial n'$ on the surface in (21-80) are unknown, and their exact determination requires solution of the boundary value problem. This was done in the preceding sections using the perturbation technique. In this section, however, we use the Kirchhoff approximation to obtain the field on the surface. In the Kirchhoff approximation, we assume that the surface is locally plane and that the field on the surface is approximated by the field which would exist if the surface were a plane tangent to the actual surface at that point. We then have

$$p(\mathbf{r}') = p_i(\mathbf{r}')(1 + R_f), \qquad \frac{\partial p(\mathbf{r}')}{\partial n'} = i\mathbf{k}_i \cdot \hat{\mathbf{N}} p_i(\mathbf{r}') - i(\mathbf{k}_i \cdot \hat{\mathbf{N}}) R_f p_i(\mathbf{r}') \tag{21-82}$$

where R_f is the reflection coefficient at $\mathbf{r}'$ and in general a function of the incident angle θ_i and $\hat{\mathbf{N}}$. If the wave is incident from the medium with density ρ_1 and acoustic velocity c_1 upon the medium with density ρ_2 and acoustic velocity c_2, then the reflection coefficient R_f is given by

$$R_f = \frac{m \cos\theta_1 - n \cos\theta_2}{m \cos\theta_1 + n \cos\theta_2} \tag{21-83}$$

where $m = \rho_2/\rho_1$, $n = c_1/c_2$, and θ_1 is the incident angle at the point $\mathbf{r}'$ and is defined by the direction of $\mathbf{k}_i$ and $\hat{\mathbf{N}}$. The angle θ_2 is the transmission angle given by

$$(\sin\theta_1)/(\sin\theta_2) = c_1/c_2. \tag{21-84}$$

Substituting (21-81) and (21-82) into (21-80), we get

$$p(\mathbf{r}) = \frac{i \exp(ikR)}{4\pi R} \int_S (\mathbf{v} \cdot \hat{\mathbf{N}} R_f - \mathbf{w} \cdot \hat{\mathbf{N}}) \exp(i\mathbf{v} \cdot \mathbf{r}')\, ds' \tag{21-85}$$

where

$$\mathbf{v} = \mathbf{k}_i - \mathbf{k}_s = v_x \hat{\mathbf{x}} + v_y \hat{\mathbf{y}} + v_z \hat{\mathbf{z}}, \qquad \mathbf{w} = \mathbf{k}_i + \mathbf{k}_s = w_x \hat{\mathbf{x}} + w_y \hat{\mathbf{y}} + w_z \hat{\mathbf{z}}$$

$$v_x = -k(\sin\theta_s \cos\phi_s - \sin\theta_i), \qquad v_y = -k \sin\theta_s \sin\phi_s$$

$$v_z = -k(\cos\theta_i + \cos\theta_s)$$

$$w_x = k(\sin\theta_s \cos\phi_s + \sin\theta_i), \qquad w_y = k \sin\theta_s \sin\phi_s$$

$$w_z = k(\cos\theta_s - \cos\theta_i), \qquad \mathbf{v} \cdot \mathbf{r}' = v_x x' + v_y y' + v_z \zeta(x, y).$$

Equation (21-85) is the Kirchhoff formula for the scattered field $p(\mathbf{r})$ when the wave $p_i(\mathbf{r})$ is incident on the surface S.

The reflection coefficient R_f is given by (21-83) and is, in general, a function of x and y. We can write R_f as a sum of the reflection coefficient R_{f0} for a smooth surface and the effect of the roughness R_{fr}. If the surface is slowly varying, R_{fr} should be small compared with R_{f0} and therefore we can approximate R_f by R_{f0}:

$$R_f \approx R_{f0} = \frac{m \cos\theta_i - n \cos\theta_t}{m \cos\theta_i + n \cos\theta_t} \tag{21-86}$$

where $(\sin\theta_i)/(\sin\theta_t) = c_1/c_2$. We also note that since $ds' = (dx'\,dy')/N_z$,

$$\begin{aligned} \mathbf{v} \cdot \hat{\mathbf{N}}\, ds' &= (v_x N_x + v_y N_y + v_z N_z) \frac{dx'\,dy'}{N_z} \\ &= \left(-v_x \frac{\partial \zeta}{\partial x} - v_y \frac{\partial \zeta}{\partial y} + v_z\right) dx'\,dy' \\ \mathbf{w} \cdot \hat{\mathbf{N}}\, ds' &= \left(-w_x \frac{\partial \zeta}{\partial x} - w_y \frac{\partial \zeta}{\partial y} + w_z\right) dx'\,dy'. \end{aligned} \tag{21-87}$$

Substituting (21-86) and (21-87) into (21-85), we observe that (21-85) contains the integral of the form

$$\int dx' \int dy' \frac{\partial \zeta}{\partial x} \exp(i\mathbf{v} \cdot \mathbf{r}').$$

Integrating by parts with respect to x', we obtain

$$\int dy' \left[\frac{\exp(i\mathbf{v} \cdot \mathbf{r}')}{i v_z}\right]_{x_1}^{x_2} - \int dx'\,dy' \left(\frac{v_x}{v_z}\right) \exp(i\mathbf{v} \cdot \mathbf{r}').$$

The first term is the difference of the quantity in the square bracket evaluated at the limits $x' = x_2$ and $x' = x_1$, but the second term is the integral with respect to x' and thus as the area S becomes large, the first term should become negligible compared with the second. Therefore, we have approximately

$$\int_S dx'\,dy' \frac{\partial \zeta}{\partial x} \exp(i\mathbf{v}\cdot\mathbf{r}') \approx -\frac{v_x}{v_z}\int_S dx'\,dy' \exp(i\mathbf{v}\cdot\mathbf{r}'). \tag{21-88a}$$

Similarly we have

$$\int_S dx'\,dy' \frac{\partial \zeta}{\partial y} \exp(i\mathbf{v}\cdot\mathbf{r}') \approx -\frac{v_y}{v_z}\int_S dx'\,dy' \exp(i\mathbf{v}\cdot\mathbf{r}'). \tag{21-88b}$$

Using (21-86)–(21-88), we write (21-85) in the form

$$p(\mathbf{r}) = \frac{i\exp(ikR)}{4\pi R} F \int_S \exp(i\mathbf{v}\cdot\mathbf{r}')\,dx'\,dy' \tag{21-89}$$

where

$$F = -\frac{2k[1+\cos\theta_i\cos\theta_s - \sin\theta_i\sin\theta_s\cos\phi_s]}{(\cos\theta_i + \cos\theta_s)} R_{f0}.$$

If the surface is flat, then $\mathbf{v}\cdot\mathbf{r}' = v_x x' + v_y y'$ and the field in the specular direction ($\theta_s = \theta_i$ and $\phi_s = 0$) is given by

$$p_0(\mathbf{r}) = \frac{i\exp(ikR)}{4\pi R}(-2k\cos\theta_i)R_{f0}\int_S dx'\,dy'. \tag{21-90}$$

Using the specularly reflected field $p_0(\mathbf{r}) = p_0(R, \theta_i, 0)$ in (21-90), the field $p(\mathbf{r}) = p(R, \theta_s, \phi_s)$ is given by

$$p(R, \theta_s, \phi_s) = p_0(R, \theta_i, 0) f(\theta_i, \theta_s, \phi_s)\frac{1}{S}\int_S \exp(i\mathbf{v}\cdot\mathbf{r}')\,dx'\,dy' \tag{21-91}$$

where

$$f(\theta_i, \theta_s, \phi_s) = \frac{1+\cos\theta_i\cos\theta_s - \sin\theta_i\sin\theta_s\cos\phi_s}{\cos\theta_i(\cos\theta_i + \cos\theta_s)}.$$

This is a general expression of the field in the Kirchhoff approximation.

21-10 COHERENT FIELD IN THE KIRCHHOFF APPROXIMATION

Let us consider a transmitter illuminating a rough surface. Consider the coherent field in the specular direction ($\theta_s = \theta_i$, $\phi_s = 0$). Noting that $f = 1$ in (21-91) and $v_x = v_y = 0$, we have

$$\langle p(R, \theta_i, 0)\rangle = p_0(R, \theta_i, 0)\chi(v_z) \tag{21-92}$$

where $p_0(R, \theta_i, 0)$ is the field reflected from a flat surface. $\chi(v_z)$ is the characteristic function of the random function $\zeta(x, y)$ and is given by

$$\chi(v_z) = \langle \exp(iv_z\zeta) \rangle, \qquad v_z = -2k \cos \theta_i. \tag{21-93}$$

In terms of the probability density function $W_0(\zeta)$, we have

$$\chi(v_z) = \int_{-\infty}^{\infty} \exp(iv_z\zeta) W_0(\zeta)\, d\zeta. \tag{21-94}$$

If the height ζ is normally distributed with the variance σ_0^2, then we have

$$W_0(\zeta) = \frac{1}{(2\pi)^{1/2}\sigma_0} \exp\left(-\frac{\zeta^2}{2\sigma_0^2}\right) \tag{21-95}$$

$$\chi(v_z) = \exp\left(-\frac{\sigma_0^2 v_z^2}{2}\right) = \exp(-2\sigma_0^2 k^2 \cos^2 \theta_i). \tag{21-96}$$

Referring to Section 21-1, the coherent power is given by (21-3):

$$\frac{P_{\text{rcoh}}}{P_t} = \frac{\lambda^2}{(4\pi)^2} \frac{G_t G_r}{(R_1 + R_2)^2} R_f \chi(v_z) \tag{21-97}$$

with $\chi(v_z)$ given by (21-93).

Let us next consider the coherent field in the direction (θ_s, ϕ_s). Taking the average of (21-91), we obtain

$$\langle p \rangle = \left[p_0 f \frac{1}{S} \int_S \exp(iv_x x' + iv_y y')\, dx'\, dy' \right] \chi(v_z) \tag{21-98}$$

where v_x, v_y, and v_z are given in (21-85). Note that the quantity in the brackets in (21-98) is the field scattered from a flat surface, and therefore the average field is equal to the field for a flat surface multiplied by $\chi(v_z)$.

21-11 SCATTERING CROSS SECTION PER UNIT AREA OF ROUGH SURFACE

We now consider the incoherent (diffuse) field. The scattering cross section per unit area is given by

$$\sigma = \frac{4\pi R^2}{S} \langle |p - \langle p \rangle|^2 \rangle. \tag{21-99}$$

Using (21-90), (21-91), and (21-98), we get

$$\sigma = \frac{k^2 \cos^2 \theta_i}{\pi} R_{f0}^2 f^2 I \tag{21-100}$$

$$I = \frac{1}{S}\int_S dx'\,dy' \int_S dx''\,dy'' \exp[iv_x(x' - x'') + iv_y(y' - y'')]$$
$$\times [\chi_2(v_z, -v_z) - |\chi(v_z)|^2]$$
$$= \int dx_d\,dy_d \exp(iv_x x_d + iv_y y_d)[\chi_2(v_z, -v_z) - |\chi(v_z)|^2] \quad (21\text{-}101)$$

where $x_d = x' - x''$ and $y_d = y' - y''$. $\chi_2(v_1, v_2)$ is a joint characteristic function of the height function ζ:

$$\chi(v_1, v_2) = \langle \exp(iv_1\zeta_1 + iv_2\zeta_2)\rangle. \quad (21\text{-}102)$$

Under the assumption that the surface is statistically homogeneous and isotropic, $\chi(v_1, v_2)$ is a function of $(x_d^2 + y_d^2)^{1/2}$ only.

If the surface is normally distributed, we use the joint probability density function:

$$W_0(\zeta_1, \zeta_2) = \frac{1}{2\pi\sigma_0^2(1 - C^2)^{1/2}} \exp\left[-\frac{\zeta_1^2 - 2C\zeta_1\zeta_2 + \zeta_2^2}{2\sigma_0^2(1 - C^2)}\right] \quad (21\text{-}103)$$

where σ_0^2 is the variance and $C = C(\rho)$ is the correlation coefficient with $\rho = (x_d^2 + y_d^2)^{1/2}$:

$$\langle \zeta(x_1, y_1)\zeta(x_2, y_2)\rangle = \sigma_0^2 C(\rho). \quad (21\text{-}104)$$

For this case, the joint characteristic function (21-102) becomes

$$\chi(v_1, v_2) = \int d\zeta_1 \int d\zeta_2\, W_0(\zeta_1, \zeta_2) \exp(iv_1\zeta_1 + iv_2\zeta_2)$$

$$\chi(v_z, -v_z) = \exp\{-v_z^2\sigma_0^2[1 - C(\rho)]\}. \quad (21\text{-}105)$$

Assuming normal distribution given in (21-95) and (21-103), we have

$$\chi_2(v_z, -v_z) - |\chi(v_z)|^2 = \exp\{-v_z^2\sigma_0^2[1 - C(\rho)]\} - \exp(-v_z^2\sigma_0^2). \quad (21\text{-}106)$$

If the surface is slightly rough, then (21-106) becomes $[\exp(-v_z^2\sigma_0^2)]v_z^2\sigma_0^2 C(\rho)$. Substituting this into (21-100) and (21-101) and using the definition of $W(p, q)$ given in (21-51),† we get

$$I = \pi^2 W(p, q) \quad (21\text{-}107)$$
$$\sigma = \pi k^4 R_{f0}^2[1 + \cos\theta_i \cos\theta_s - \sin\theta_i \sin\theta_s \cos\phi_s]^2 W(p, q)$$

† $W(p, q)$ is the spectral density of the rough surface height function.

where $p = -v_x = k \sin\theta_s \cos\phi_s - k \sin\theta_i$ and $q = -v_y = k \sin\theta_s \sin\phi_s$ are the same quantities used in (21-56). For backscattering ($\theta_s = \theta_i$ and $\phi_s = \pi$), we have

$$\sigma = 4\pi k^4 R_{f0}^2 W(-2k \sin\theta_i, 0). \tag{21-108}$$

We note that (21-108) is identical to the first order solution (21-65) except for the reflection coefficient R_{f0}^2 and the $\cos^4\theta_i$ dependence on horizontal polarization. The $\cos^4\theta_i$ dependence comes from the fact that the tangential magnetic current on the surface is proportional to $\cos\theta_i$ and the scattering pattern has an additional $\cos\theta_i$ pattern. Therefore, the scattered field pattern is proportional to $\cos^2\theta_i$ and the power pattern is proportional to $\cos^4\theta_i$.

For a slightly rough surface, the Kirchhoff approximation yields almost the same result as the first order perturbation solution. This is, of course, to be expected. However, it should be noted that the perturbation method yields better polarization effects because it is based on the correct boundary condition on the surface. The Kirchhoff approximation makes use of the reflection coefficient for a flat surface and therefore it does not include the complete polarization effect (Valenzuela and Laing, 1972).

If the surface is very rough, the coherent power $|\langle p \rangle|^2$ disappears almost completely and we have

$$\chi_2(v_z, -v_z) - |\chi(v_z)|^2 \approx \chi_2(v_z, -v_z)$$

$$I = \int dx_d\, dy_d \exp(iv_x x_d + iv_y y_d) \exp\{-v_z^2 \sigma_0^2 [1 - C(\rho)]\}$$

$$= 2\pi \int_0^\infty \rho\, d\rho J_0(v\rho) \exp\{-v_z^2 \sigma_0^2 [1 - C(\rho)]\} \tag{21-109}$$

where $v = (v_x^2 + v_y^2)^{1/2}$ and $\rho = (x_d^2 + y_d^2)^{1/2}$. For a very rough surface, $v_z^2 \sigma_0^2 \gg 1$ and therefore the major contribution to the integral in (21-109) comes from the neighborhood of $\rho = 0$. Thus we expand $1 - C(\rho)$ in a series of powers of ρ^2 and keep the first term:

$$1 - C(\rho) = \rho^2/l^2 + \cdots, \tag{21-110}$$

where l is the correlation distance of the surface height function. Substituting (21-110) into (21-109), we obtain

$$\sigma = \frac{k^2 \cos^2\theta_i}{\pi} R_{f0}^2 f^2 I, \qquad I = \frac{\pi l^2}{v_z^2 \sigma_0^2} \exp\left(-\frac{v^2 l^2}{4 v_z^2 \sigma_0^2}\right) \tag{21-111}$$

where f is given in (21-91) and

$$v_z = -k(\cos\theta_i + \cos\theta_s), \qquad v^2 = p^2 + q^2$$

$$p = k(\sin\theta_s \cos\phi_s - \sin\theta_i), \qquad q = k \sin\theta_s \sin\phi_s.$$

The backscattering cross section ($\theta_s = \theta_i$ and $\phi_s = \pi$) is given by

$$\sigma = \left(\frac{R_{f0}^2}{4\cos^4\theta_i}\right)\frac{l^2}{\sigma_0{}^2}\exp\left(-\frac{l^2}{\sigma_0{}^2}\frac{\tan^2\theta_i}{4}\right). \tag{21-112}$$

This can also be written as

$$\sigma = \frac{R_{f0}^2}{\cos^4\theta_i}\cot^2\beta_0\exp\left(-\frac{\tan^2\theta_i}{\tan^2\beta_0}\right) \tag{21-113}$$

where $\tan\beta_0 = 2\sigma_0/l$ is interpreted as a mean value of the slope of the rough surface.

The incoherent power P_s is then given by (21-5). For example, assume that the transmitter has a narrow transmitting pattern given by

$$G_t(\hat{\mathbf{i}}) = G_t(0)\exp(-\theta^2/\theta_0{}^2) \tag{21-114}$$

where θ is measured from the direction of the peak of the pattern (see Fig. 21-8). Noting that $R_1 = R_2 \approx R$, we get

$$\frac{P_s}{P_t} = \frac{\lambda^2\pi\theta_0{}^2\sigma G_t(0)^2}{(4\pi)^3 2R^2\cos\theta_i} \tag{21-115}$$

where σ is given in (21-112).

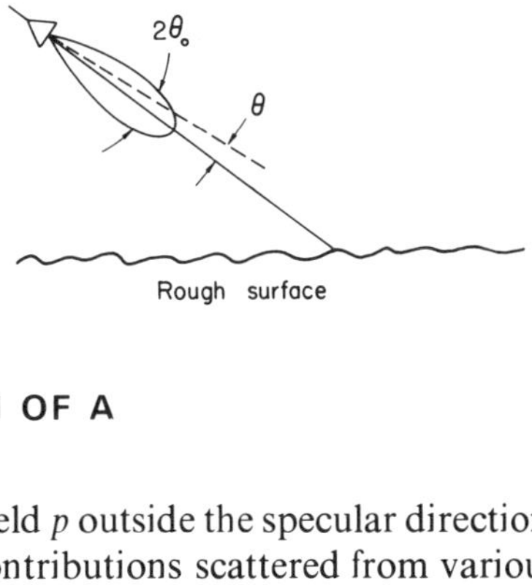

FIG. 21-8 Backscattering from a rough surface illuminated by a narrow beam transmitter.

21-12 PROBABILITY DISTRIBUTION OF A SCATTERED FIELD

Let us consider the diffuse scattered field p outside the specular direction. In this case $\langle p\rangle = 0$ and p consists of contributions scattered from various parts of the rough surface. If we let $p = X + iY$, then X and Y are sums of many independent contributions X_i and Y_i:

$$X = \sum_i^N X_i \qquad \text{and} \qquad Y = \sum_i^N Y_i. \tag{21-116}$$

According to the "central limit theorem," the distribution of a random variable which is a sum of N independent random variables approaches normal as $N \to \infty$ regardless of the distribution of each random variable.

This situation is identical to that discussed in Section 4-9. Therefore the probability distribution of the amplitude A of the field p is the "Rayleigh" distribution.

In the direction in which the field consists of the coherent field $\langle p \rangle$ and the diffuse field, we need to reexamine (21-116). If we let $\langle p \rangle = A_0 \exp(i\phi_0)$ and choose the phase reference so that $\phi_0 = 0$, then the incoherent field $p_s = p - \langle p \rangle$ can be written as a sum of many contributions

$$p_s = (X - A_0) + iY, \qquad X - A_0 = \sum_i^N X_i, \qquad Y = \sum_i^N Y_i. \quad (21\text{-}117)$$

Using the central limit theorem, we write the probability density function $W_s(X, Y)$ of the field p:

$$W_s(X, Y) = \frac{1}{2\pi\sigma_s^2} \exp\left[-\frac{(X - A_0)^2 + Y^2}{2\sigma_s^2}\right] \quad (21\text{-}118)$$

where σ_s^2 is the variance for $X - A_0$ and Y.

The probability density for the amplitude A is then given by [see Eqs. (4-80) and (4-81); and Hoffman, 1960]

$$W_s(A) = \int_0^{2\pi} W_s(X, Y)A\, d\phi = \frac{A}{\sigma_s^2} \exp\left(-\frac{A^2 + A_0^2}{2\sigma_s^2}\right) I_0\left(\frac{A_0 A}{\sigma_s^2}\right) \quad (21\text{-}119)$$

where I_0 is the modified Bessel function. This distribution is called the Rice–Nakagami distribution, first derived by Nakagami in 1940 and later by Rice in 1944. Note that when $A_0 = 0$, it reduces to the Rayleigh distribution.

Another distribution in which the variance for $X - A_0$ is different from the variance for Y has also been investigated and is called the Hoyt distribution (Beckmann and Spizzichino, 1963, Chapter 7). When a surface is illuminated by a coherent light, the surface seems to look granular. This is called the speckle pattern and is fundamentally the same as the rough surface scattering. For an up-to-date discussion on this subject, see the special issue on "Speckle in Optics," *Journal of the Optical Society of America*, November 1976 (for an excellent review, see Goodman, 1976).

CHAPTER 22 □ REMOTE SENSING AND INVERSION TECHNIQUES

Remote sensing of the earth's environment by wave propagation and scattering techniques is important because it provides a tool in the study of the structure of the environment and its effects on communication through it, and the detection and identification of objects in various environments. For example, remote sensing of the troposphere is useful in weather forecasting, weather modification, pollution studies, storm warning, and air traffic safety. A principal reason for the importance and usefulness of remote probing is the need for the collection of environmental data over a wide range of space within a limited time. This is either prohibitively expensive or virtually impossible by conventional means (Yates, 1970).

In this chapter we first discuss some aspects of the remote sensing of the troposphere. We give some accounts of the probing of the atmospheric turbulence and wind velocity. In remote sensing, measurement error is often greatly amplified in the process of extracting the desired information. This is called the ill-posed problem and various inversion techniques are needed to reduce the error in the desired information within reasonable bounds. We present three representative inversion techniques: the smoothing (regularization) method, the statistical inversion method, and the Backus–Gilbert inversion technique.

We also briefly discuss remote sensing of geophysical data in the earth's atmosphere, ocean and land environment, and subsurface environment. Problems associated with remote sensing of planetary atmosphere and solar wind are touched upon. Remote sensing of particle size distribution is important in atmospheric and ocean physics and will be discussed in connection with inversion techniques. Remote sensing of biological media is an important area of study, but is not discussed here (Wells, 1969; Baker, 1970; Erickson *et al.*, 1974).

22-1 REMOTE SENSING OF THE TROPOSPHERE

Remote sensing of the troposphere (Derr, 1972; Yates, 1970; Derr and Little, 1970; Lawrence, 1972b; Wait, 1972) may be divided into two categories: passive and active. A passive sensing system is one in which the observing instrument simply receives the natural radiation from the environment, such as the radiation from gas and aerosol constituents of the atmosphere or the radiation from the sun, the moon, or the planets. An example is radiometric techniques. An active sensing system is one in which a signal is sent out from the transmitter, interacts with the environment or the target, and after interaction, is observed and measured. Examples include radar, lidar, acoustic sounding, line-of-sight propagation, imaging radars, and holographic techniques.

In the probing of the atmosphere (Derr and Little, 1970), we may distinguish the following two types of interactions: "spectroscopic" and "refractive index fluctuation." In the spectroscopic interaction, the molecules in the atmosphere react with the wave and cause absorption and scattering of the wave. The absorption and scattering cross sections are dependent on the properties of the specific molecule, the frequency, and the environment (pressure, temperature). The scattering and absorption characteristics of dielectric spheres such as haze, fog, and rain droplets can be calculated exactly by the Mie theory. If the particle size is much smaller than a wavelength, the Rayleigh scattering formulas can be used (see Chapter 2). The size distribution and the refractive index of particles can be inferred from the measurement of the scattering characteristics. However, this requires the use of inversion techniques. We will discuss this further in Sections 22-4–22-8.

The Mie scattering and the Rayleigh scattering can take place at any wavelength. "Fluorescent" scattering can occur only when the frequency of the incident wave is in an absorption band of the particle. This causes transition to a higher energy state and emission at frequencies equal to or lower than the incident frequency. This is called Stokes fluorescence. When the emission is at higher frequencies, it is called anti-Stokes fluorescence. "Raman" scattering is weaker than any of the scattering processes just mentioned, but it is useful because it occurs regardless of the incident frequency. The Raman lines occur at a series of frequencies $\nu \pm \nu_1, \nu \pm \nu_2, \ldots,$ where ν is the incident frequency. The lower frequencies are called Stokes lines and the higher ones anti-Stokes lines (for details, see Szymanski, 1967).

Pulsed Doppler radar can be used to study the atmospheric motion (Derr, 1972, Chapter 13; for ultrasonic blood flow sensor, see Baker, 1970). Since a pulsed Doppler radar can measure the phase, it allows estimation of the radial component of the velocity spectrum of the particle and the turbulence in the scattering volume. It also gives information on cross sections, reflectivity, and density. The Doppler radar has been used in cloud physics, in the estimation of storm systems, wind field, and turbulence field. A network of two or more Doppler radars is used to triangulate on a storm and to resolve its velocity components.

High power radar (Derr, 1972, Chapters 13 and 14) and acoustic (Derr, 1972, Chapters 19 and 20) echo sounder have been used in the study of clear air turbulence. Lidar (laser radar) has been used in the measurement of tropospheric aerosols (Derr, 1972, Chapter 23). Temperature profiles in the atmosphere have been obtained from passive microwave radiometric measurement and vertical water vapor distribution has been determined by satellite infrared spectrometer measurement (Derr, 1972, Chapter 15).

Atmospheric turbulence causes fluctuations of the refractive index of air, which in turn produces fluctuations in amplitude and phase of an optical beam propagating through it. Therefore, it should be possible to deduce the characteristics of turbulence from observations of fluctuations of the wave. We will use this problem as an example of remote probing and discuss several known remote probing techniques in the following section.

The turbulence characteristics to be probed include the strength of turbulence represented by the structure constant C_n and the wind velocity. Remote sensing of these quantities may be divided into two areas: One is to find the average quantities over the total path length and the other is to obtain the profile of these quantities as functions of position along the path. At present, remote sensing studies have been almost entirely based on weak fluctuation theory, and therefore at optical frequencies it is applicable only within a distance of a few kilometers. Beyond this distance, the strong fluctuation theory must be used. No serious study of remote sensing in the strong fluctuation region seems to have been reported in the literature.

22-2 REMOTE SENSING OF THE AVERAGE STRUCTURE CONSTANT C_n OVER THE PATH

In a well-developed turbulence, velocity fluctuations are known to have the Kolmogorov spectrum. Certain quantities such as potential temperature and mass of water vapor move with the velocity field in the turbulence without appreciable change and therefore their fluctuation characteristics also obey the Kolmogorov spectrum. At optical frequencies, humidity effects are negligible and therefore temperature field is known to have the Kolmogorov spectrum. The index of refraction field n is related to the temperature field by

$$N = (n - 1)10^6 = 79P/T \tag{22-1}$$

where P is the atmospheric pressure in millibars and T is the temperature in degrees Kelvin. The relationship between the pressure and the temperature depends on the process of heat transfer in turbulence. Considering the short lifetime of turbulence, we may regard the process as adiabatic, and under this assumption we have

$$\frac{\delta P}{P} = \frac{\gamma}{\gamma - 1} \frac{\delta T}{T} \tag{22-2}$$

where γ is the ratio of specific heats ($= C_p/C_v = 1.4$ for air). Therefore, the structure function of the index of refraction $D_n(r)$ is related to the temperature structure function $D_T(r)$ through

$$D_n(r) = C_n{}^2 r^{2/3}, \qquad D_T(r) = C_T{}^2 r^{2/3}, \qquad C_n = \left[\frac{79}{\gamma - 1}\frac{P}{T^2} 10^{-6}\right] C_T. \tag{22-3}$$

It is important to note that remote sensing by an optical beam detects refractive index fluctuations in turbulence, and not velocity field fluctuation itself. These two are closely related to each other as just noted. However, in some cases, such as neutral atmosphere, strong velocity fluctuations may exist with little optical effect (Lawrence, 1972b).

The average structure constant C_n of a turbulence can be obtained by measuring the variance of log-intensity fluctuation of an optical wave propagating through it, and using the following formulas, obtained in Chapter 18:

$$\begin{aligned} \sigma_{\ln I}^2 &= 1.228 k^{7/6} L^{11/6} C_n{}^2 \qquad \text{plane wave} \\ &= 0.496 k^{7/6} L^{11/6} C_n{}^2 \qquad \text{spherical wave} \end{aligned} \tag{22-4}$$

where $k = 2\pi/\lambda$ and L is the propagation distance. For a beam wave, we use

$$\sigma_{\ln I}^2 = 4\sigma_\chi{}^2(L, \rho) \tag{22-5}$$

where $\sigma_\chi{}^2(L, \rho)$ is shown in Chapter 18.

The measurement of the log-intensity fluctuations must be made with an aperture small compared with the correlation distance of the wave, which is approximately equal to $\sqrt{\lambda L}$ for plane and spherical waves. For a collimated beam wave, the correlation distance is also approximately $\sqrt{\lambda L}$, but for a focused beam, it can be considerably smaller than $\sqrt{\lambda L}$.

22-3 REMOTE SENSING OF THE AVERAGE WIND VELOCITY OVER THE PATH

Remote sensing of the average wind across a wave propagation path has been studied in recent years (Ishimaru, 1972; Lawrence *et al.*, 1972c; Shen, 1970). In this section, we discuss the following three methods of remotely sensing the wind velocity: (a) the temporal frequency spectrum method, (b) the time delay method, and (c) the correlation slope method.

(a) *Temporal frequency spectrum method* The scintillation pattern at the receiver drifts with the transverse wind, and the higher the wind velocity, the faster the drift of the scintillation. Therefore we expect that as the cross wind velocity increases, the frequency spectra of the amplitude and phase fluctuations contain higher frequency components. Consequently, it should be possible to obtain the average wind velocity by measuring the spectrum of the fluctuations. We will outline the following three ways of obtaining the wind velocity: (1) spectrum shape, (2) ratio of spectra at two frequencies, and (3) coherence between two frequencies.

The theoretical spectrum shape of the log-amplitude fluctuation is that for low frequency the spectrum is relatively constant, and for high frequency the spectrum tends to decrease as $f^{-8/3}$. For plane and spherical waves, the spectra are obtained in Section 19-2 and their asymptotic forms are

for plane wave

$$W_{\chi\mathrm{pl}} \to 0.8506(C_n{}^2/V)k^{2/3}L^{7/3} \qquad \text{as } f\to 0 \tag{22-6a}$$

$$\to 2.192(C_n{}^2/V)k^{2/3}L^{7/3}(f/f_0)^{-8/3} \qquad \text{as } f\to \infty \tag{22-6b}$$

for spherical wave

$$W_{\chi\mathrm{sp}} \to 0.1905(C_n{}^2/V)k^{2/3}L^{7/3} \qquad \text{as } f\to 0 \tag{22-6c}$$

$$\to 2.192(C_n{}^2/V)k^{2/3}L^{7/3}(f/f_0)^{-8/3} \qquad \text{as } f\to \infty \tag{22-6d}$$

where $f_0 = (V/2\pi)(k/L)^{1/2} = 0.4V/\sqrt{\lambda L}$ and V is the cross wind velocity.

The shapes of the spectra obtained experimentally generally follow (22-6a)–(22-6d) and therefore the shape and particularly the frequency f_c at which these two asymptotics meet give a good indicator with which to obtain the wind velocity. Note that this frequency f_c is $1.43f_0$ for a plane wave and $2.60f_0$ for a spherical wave.

This technique requires the measurement of fluctuations at a single operating frequency. In practice, however, the breakpoint frequency f_c is not easy to determine experimentally. If we can transmit two waves with different operating frequencies, then we have additional data with which to extract more information.

We can compare spectra at two different operating frequencies k_1 and k_2 and take their ratio. From (22-6a)–(22-6d), we note that

$$\frac{W_\chi(k_2)}{W_\chi(k_1)} = (k_2/k_1)^{2/3} \qquad \text{as } f\to 0 \tag{22-7a}$$

$$= (k_2/k_1)^2 \qquad \text{as } f\to \infty. \tag{22-7b}$$

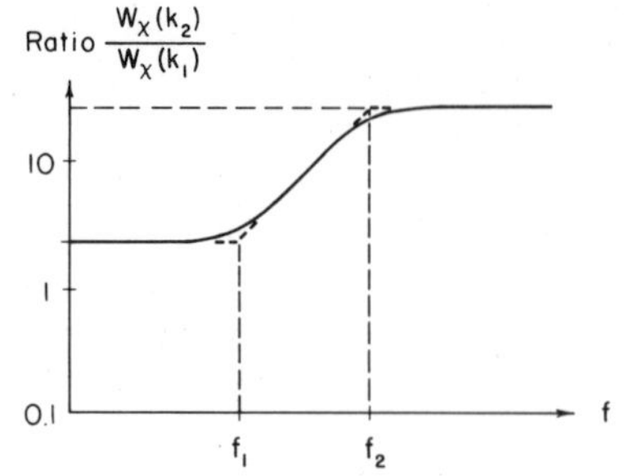

FIG. 22-1 A typical shape of the ratio of spectra at two different frequencies and two breakpoint frequencies f_1 and f_2.

The general shape of this ratio is shown in Fig. 22-1. Two breakpoint frequencies f_1 and f_2 are

$$f_i = 1.43(V/2\pi)(k_i/L)^{1/2}, \qquad i = 1, 2, \tag{22-8a}$$

for a plane wave and

$$f_i = 2.60(V/2\pi)(k_i/L)^{1/2}, \qquad i = 1, 2, \tag{22-8b}$$

for a spherical wave.

Therefore, it may be possible to obtain the wind velocity V by measuring the spectra at k_1 and k_2 and obtaining the ratio and the breakpoint frequencies f_i. This has been done for microwave frequencies (Ishimaru, 1972).

It is also possible to make use of correlation characteristics of fluctuations at two operating frequencies in determining the wind velocity. The square of the normalized cross spectrum

$$[W_\chi(\omega, k_1, k_2)]^2/[W_\chi(\omega, k_1)W_\chi(\omega, k_2)]$$

is called the coherence in the analysis of random data (Bendat and Piersol, 1971). The coherence approaches the following constant value as $\omega \to 0$:

$$\frac{[W_\chi(\omega, k_1, k_2)]^2}{W_\chi(\omega, k_1)W_\chi(\omega, k_2)} \to \left(\frac{k_2}{k_1}\right)^{4/3}\left[\left(\frac{1 + k_1/k_2}{2}\right)^{4/3} - \left(\frac{1 - k_1/k_2}{2}\right)^{4/3}\right]^2 \tag{22-9}$$

where $k_1 < k_2$.

The coherence is almost constant at the value given by (22-9) in a frequency band up to a certain frequency determined by the wind velocity and then it drops to a negligible value beyond this point (Mandics *et al.*, 1974; Ishimaru, 1972). A typical behavior is shown in Fig. 19-3. Therefore, it should be possible to obtain the wind velocity from the measurement of the cross spectrum.

(b) *Time delay method* If two detectors are placed perpendicular to the propagation path, parallel to the direction of wind velocity (Fig. 22-2), the fluctuation of a wave drifts with the wind. The fluctuation at the receiver

A at t is drifted to the fluctuation at the receiver B at a delayed time $t + \tau$. Therefore, we expect that at a certain delay time τ, the fluctuations at A and B are strongly correlated. This is pictured in Fig. 22-2. The delay time τ of the peak of the correlation is related to the wind velocity and is approximately equal to the correlation distance of the field divided by the wind velocity. The correlation distance is approximately equal to $\sqrt{\lambda L}$.

The general expression for the correlation function of the log-amplitude fluctuation χ is given in Chapter 18. For a spherical wave it reduces to

$$B_\chi(\rho, \tau) = 8\pi^2 k^2 \int_0^L d\eta \int_0^\infty \kappa \, d\kappa J_0\left(\kappa \left| \frac{\rho\eta}{L} - V\tau \right| \right) \times \sin^2\left[\frac{\eta(L-\eta)}{2kL}\kappa^2\right]\Phi_n(\kappa, \eta). \tag{22-10}$$

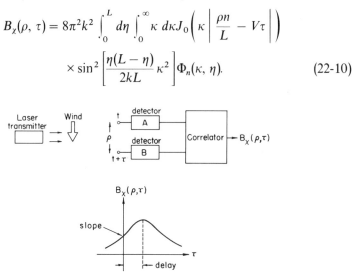

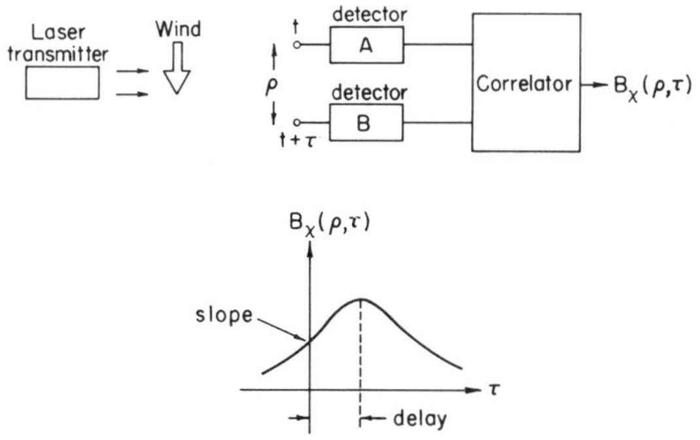

FIG. 22-2 Time delay method and correlation slope method.

Using the Kolmogorov spectrum

$$\Phi_n(\kappa, \eta) = 0.033 C_n^2(\eta)\kappa^{-11/3}, \tag{22-11}$$

we can calculate the shape of the correlation function (22-10) and compare with the experimental data. By noting the time delay of the peak, we can calculate the wind velocity. We note that Eq. (22-11) is based on the assumption that $l_0 \ll \sqrt{\lambda L} \ll L$ where l_0 and L_0 are the inner and outer scales of turbulence, and therefore the spectrum is that of the inertial subrange, and that the intensity of the turbulence is represented by the structure constant C_n. Also we note that the correlation function B_χ is in fact the covariance function because the average $\langle\chi\rangle$ is assumed to be zero in this analysis.

It has been found (Lawrence *et al.*, 1972) that the variations in $C_n(\eta)$ and the wind velocity $V(\eta)$ cause severe distortions in the shape of the correlation function and thus contribute to a considerable error. However, the slope of the correlation function at $\tau = 0$ seems to be relatively insensitive to these variations and may be preferable.

(c) *Correlation slope method* The slope of the correlation function at $\tau = 0$ is given by

$$\left.\frac{\partial B_\chi}{\partial \tau}\right|_{\tau=0} = 8\pi^2 k^2 \int_0^L d\eta \int_0^\infty \kappa^2 \, d\kappa V J_1\left(\kappa \frac{\rho\eta}{L}\right) \sin^2\left[\frac{\eta(L-\eta)}{2KL}\kappa^2\right] \Phi_n(\kappa, \eta). \tag{22-12}$$

It has been found that it is preferable to use $\rho = 0.33\sqrt{\lambda L}$ because it gives the most uniform weighting of C_n and V along the propagation path to the slope given in (22-12).

Optically measured wind velocity is compared with the average of several anemometer readings of the wind velocities along the path and good agreements have been obtained.

22-4 REMOTE SENSING OF THE PROFILE OF THE STRUCTURE CONSTANT AND THE ILL-POSED PROBLEM

Up to this point, we considered remote sensing of the average structure constant and the average wind velocity over the propagation path. Suppose that we wish to find the profile of the structure constant as a function of position along the path by means of the measurement of wave fluctuations. It should be obvious that to find the profile, rather than the average, requires more data taking. Typically, the number of detectors should be increased. To illustrate this point, let us consider the correlation function given in (22-10). Using (22-11), we write

$$B_\chi(\rho, \tau) = \int_0^L d\eta \, K(\rho, \tau, \eta) C_n^2(\eta) \tag{22-13}$$

where

$$K(\rho, \tau, \eta) = 8\pi^2 k^2 (0.033) \int_0^\infty \kappa^{-8/3} J_0\left(\kappa \left|\frac{\rho\eta}{L} - V\tau\right|\right) \sin^2\left[\frac{\eta(L-\eta)}{2kL}\kappa^2\right] d\kappa.$$

We now make a series of measurements and obtain M different values of the correlation function $B_\chi(\rho_i, \tau_i)$ where $i = 1, 2, \ldots, M$. We also approximate the integral in (22-13) by a series and obtain

$$B_\chi(\rho_i, \tau_i) = \sum_{j=1}^{N} a(\rho_i, \tau_i, \eta_j) C_n^2(\eta_j) \tag{22-14}$$

where $a(\rho_i, \tau_i, \eta_j) = W_j K(\rho_i, \tau_i, \eta_j)$, and W_j is an appropriate weighting function depending on a type of quadrature formula used to obtain the series (22-14) from the integral (22-13).

Let us write (22-14) in the matrix form

$$g = Af \tag{22-15}$$

where $g = (g_i)$ is an $M \times 1$ column matrix, $A = (a_{ij})$ is an $M \times N$ rectangular matrix, $f = (f_j)$ is an $N \times 1$ column matrix, and

$$g_i = B_\chi(\rho_i, \tau_i), \qquad a_{ij} = W_j K(\rho_i, \tau_i, \eta_j), \qquad f_j = C_n{}^2(\eta_j).$$

In (22-15), g_i $(i = 1, \ldots, M)$ are the data obtained from M different measurements, (a_{ij}) are the known matrix, and f_j $(j = 1, \ldots, N)$ are the unknown quantities.

In general, the number of measurements M must be at least equal to the number of unknowns N, and in general $M > N$. The simplest method with which to obtain the best possible unknown f from the measured data g is the least-squares procedure.

The least-squares procedure consists of finding f such that the square of the difference $g - Af$ is minimized:

$$|g - Af|^2 = \text{minimum}. \tag{22-16}$$

A solution to (22-16) is well known (Franklin, 1968, p. 50):

$$f = (A^+A)^{-1}A^+g \tag{22-17}$$

where A^+ is the complex conjugate of the transpose of A. If $M = N$, (22-17) is reduced to the usual inversion

$$f = A^{-1}g. \tag{22-18}$$

For many problems of remote sensing, it will soon be discovered that (22-17) or (22-18) does not yield a useful solution because a small error in the measured data g causes an extremely large error in the unknown. To illustrate this point, consider (22-17) and assume a certain error δg in g, and the resulting error δf in f. We then have

$$\delta f = (A^+A)^{-1}A^+\,\delta g. \tag{22-19}$$

As a percentage error Δ_f in f, we take

$$\Delta_f = \frac{\|\delta f\|}{\|f\|} \tag{22-20}$$

where $\|f\|$ is the norm of f defined by

$$\|f\| = \max|f_j|. \tag{22-21}$$

Using (22-19), we obtain

$$\Delta_f = \frac{\|\delta f\|}{\|f\|} < \|(A^+A)^{-1}\| \, \|A^+\| \frac{\|\delta g\|}{\|g\|} \tag{22-22}$$

where $\|A\|$ is the norm of a matrix A defined by

$$\|A\| = \max_i \sum_j |a_{ij}|, \tag{22-23}$$

and therefore the maximum percentage error $\Delta_{f\,\max}$ is given by

$$\Delta_{f\,\max} = \|(A^+A)^{-1}\| \, \|A^+\| \, \Delta g \tag{22-24}$$

where Δ_g is the percentage error in $g = \|\delta g\|/\|g\|$. In many remote-sensing problems, the elements a_{ij} are not very different from each other and therefore the norm $\|(A^+A)^{-1}\|$ can be extremely high and often on the order of magnitude of tens of powers of 10. This situation is called unstable and the problem (Franklin, 1970) is "ill-posed." Mathematically this is a result of an extremely small determinant $|A^+A|$. This is also equivalent to the large ratio of the maximum $\mu_{\max}$ to the minimum $\mu_{\min}$ eigenvalues of A^+A. In general the instability increases with the increase of $\mu_{\max}/\mu_{\min}$.

As an example, consider the simple problem

$$A = \begin{pmatrix} 1 & 1+\varepsilon \\ 1+\varepsilon & 1 \end{pmatrix}, \qquad g = \begin{pmatrix} 1 \\ 1 \end{pmatrix} \tag{22-25}$$

where $|\varepsilon| \ll 1$. The exact solution to $g = Af$ is

$$f = \begin{pmatrix} \dfrac{1}{2+\varepsilon} \\ \dfrac{1}{2+\varepsilon} \end{pmatrix} \tag{22-26}$$

Now we introduce some error δg in the measured data and estimate the error δf in the solution f. We note that $\|A\| \sim 2$. Also, since

$$A^{-1} = -\frac{1}{\varepsilon(\varepsilon+2)} \begin{pmatrix} 1 & -1-\varepsilon \\ -1-\varepsilon & 1 \end{pmatrix}, \tag{22-27}$$

we get $\|A^{-1}\| \sim |\varepsilon|^{-1}$. Therefore from (22-24) we obtain

$$\Delta_{f\,\max} = 2|\varepsilon|^{-1} \Delta g. \tag{22-28}$$

This means that if $|\varepsilon| = 10^{-3}$, a 1% error in the measured data g is magnified to 2000% error in the unknown f if we simply use $f = A^{-1}g_d$. We can also see this by letting

$$g_d = \begin{pmatrix} 1 \\ 1 \end{pmatrix} + \begin{pmatrix} \Delta g_1 \\ \Delta g_2 \end{pmatrix},$$

where Δg_1 and Δg_2 are on the order of $\pm 10^{-2}$. We then get

$$f = A^{-1}g_d \approx \begin{pmatrix} \frac{1}{2} \\ \frac{1}{2} \end{pmatrix} + \begin{pmatrix} \Delta f_1 \\ \Delta f_2 \end{pmatrix},$$

where $|\Delta f_1| \sim |\Delta f_2| \sim 10$.

This instability problem has been encountered by Shen (1970) who used only a limited number of unknowns to avoid this problem. In a more general case, however, it is necessary to devise a method by which a stable inversion of (22-15) can be accomplished. This is discussed in the following sections.

We will add here a few other techniques of remote sensing the atmospheric characteristics. Clifford *et al.* (1975) reported on the extension of the technique used by Lawrence *et al.* (1972). They discussed a technique which makes use of the naturally occurring ambient illumination of a scene such as a mountainside and clouds. Thus this technique does not require an active light source such as a laser or headlight.

Two promising methods of remote sensing the profile of the structure constant and the wind velocity have been recently proposed. They are (a) the crossed beam method and (b) the spatially filtered aperture technique.

(a) *Crossed beam method* (Fisher and Krause, 1967) Wang *et al.* (1974) proposed the use of two transmitters and two receivers so that two beams cross at a certain location in the path. By measuring the cross correlation between the outputs of two receivers, it is possible to obtain the turbulence characteristics at this particular beam crossing point.

(b) *Spatially filtered aperture technique* Lee (1974) proposed that if the transmitting and receiving apertures are appropriately weighted (or spatially filtered), then the output can be made sensitive to one particular spatial wave number which in turn selects one particular location in the path. He conducted simple experiments using a Fresnel lens of aperture 0.5 × 0.6 m across which were taped 11-mm-wide vertical strips of black paper separated by 11 mm. This forms a spatial filter. Using a similar lens for a receiving aperture, he obtained the wind velocity which agreed well with anemometer measurements (Lee and Harp, 1969; Ochs *et al.*, 1976).

22-5 INVERSE PROBLEM

In the preceding section we pointed out that the problem of remote sensing the structure constant profile can constitute an ill-posed problem. Many other remote-sensing problems are also ill-posed. An example is the problem of extracting geophysical information about the internal structure of the earth from a finite set of measured data (Backus and Gilbert, 1970). Another example is the problem of determining the temperature profile of the atmosphere from radiometric measurement (Derr, 1972, Chapter 16).

The determination of particle size distribution from scattering measurement is also an example of the ill-posed problem (see Section 22-8). The scattered intensity g may be measured as a function of wavelength λ and scattering angle θ. The scattered intensity is related to the particle size distribution $n(D)$ and the differential scattering cross section σ which is a function of the scattering angle θ, the wavelength λ, and the size D. We can then write

$$g(\lambda, \theta) = \int_0^\infty K(\lambda, \theta, D) f(D)\, dD \tag{22-29}$$

where $K(\lambda, \theta, D)$ is proportional to the differential cross section $\sigma(\lambda, \theta, D)$ and $f(D) = n(D)$.

We make a series of measurements at different θ_i and λ_i $(i = 1, 2, \ldots, M)$, and following the procedure in Section 22-4, we can write (22-29) in the form

$$g = Af. \tag{22-30}$$

These problems are often unstable and in the following sections we outline three commonly used inversion techniques to obtain stable solutions.

22-6 SMOOTHING (REGULARIZATION) METHOD

The inversion of an ill-posed problem (22-15) has been studied using several methods. In this section we briefly outline the method developed by Phillips (1961), Twomey (1963), and Tihonov (1963). For an excellent summary article, see Turchin *et al.* (1971), as well as Deschamps and Cabayan (1972) and Mittra (1973, p. 386).

Let us consider an ill-posed problem

$$g = Af. \tag{22-31}$$

In practice, the true g is never known because it always contains a certain experimental error n. The measured data g_d is therefore

$$g_d = g + n \tag{22-32}$$

where the $M \times 1$ matrix g_d is known and the $M \times 1$ matrix n is the experimental error.

We consider minimizing the following positive quantity U:

$$U = (g_d - Af)^+(g_d - Af) + \gamma_1(f - f_0)^+(f - f_0) + \gamma_2(Bf)^+(Bf) \quad (22\text{-}33)$$

where + denotes the complex conjugate of the transpose, γ_1 and γ_2 are positive constants, f_0 a "preconceived-notion function," and B an $M \times M$ matrix describing some smoothing of f. The first term of (22-33) is a measure of the accuracy of f as given by (22-31). Note that if there is no experimental error $g_d = g$, $\gamma_1 = \gamma_2 = 0$, $U = 0$ is the minimum value of U, and the solution is $f = A^{-1}g$. The second term of (22-33) indicates the departure of f from the preconceived-notion function f_0, and the third term indicates the departure of f from ideal smoothness $(Bf) = 0$. (Bf) commonly describes first or second derivatives (Derr, 1972, Chapter 16).

Consider the case $\gamma_2 = 0$. We differentiate U with respect to f and obtain the solution

$$f = (A^+A + \gamma_1 I)^{-1}(A^+g_d + \gamma_1 f_0) \quad (22\text{-}34)$$

where I is an $N \times N$ square matrix.

The choice of γ_1 should be made to give a reasonable compromise between the first and the second term of (22-33). If γ_1 is small, the first term dominates and the solution is too oscillatory. If γ_1 is too large, the second term dominates and the solution is oversmoothed. For further details, see Deschamps and Cabayan (1972).

For the case $\gamma_1 = 0$, the solution is given by

$$f = (A^+A + \gamma_2 B^+B)^{-1}A^+g_d. \quad (22\text{-}35)$$

The matrix B can be chosen to be

$$B = \begin{pmatrix} -2 & 1 & 0 & \cdots & \cdots & 0 \\ 1 & -2 & 1 & 0 & & \\ 0 & 1 & -2 & 1 & & \\ \vdots & & & & & \\ & 0 & 1 & -2 & 1 & 0 \\ & & 0 & 1 & -2 & 1 \\ 0 & \cdots & \cdots & 0 & 1 & -2 \end{pmatrix}$$

Bf is the discrete analog of the second derivative of f. For more details on this technique, see Chow and Tien (1976) and Derr (1972, Chapter 16).

22-7 STATISTICAL INVERSION TECHNIQUE

In many remote-sensing problems, errors encountered are statistical and it is more natural to consider the inversion problem by taking into account the statistical nature of the experimental errors and other statistical information. In this section, we discuss an element of the statistical inverse technique (Franklin, 1970; Edenhofer *et al.*, 1973; Strand and Westwater, 1968; Heneghan and Ishimaru, 1974).

Let us consider an ill-posed problem

$$g = Af. \tag{22-36}$$

In practice, the true g is never known because it always contains a certain experimental error n. The measured data g_d is therefore

$$g_d = g + n \tag{22-37}$$

where the $M \times 1$ matrix g_d is the known measured value and the $M \times 1$ matrix n is the experimental error. We may know some statistical characteristics of n.

Substituting (22-37) into (22-36), we write

$$g_d = Af + n. \tag{22-38a}$$

We regard (22-38a) as a stochastic equation and g_d, f, and n as random variables with zero mean:

$$\langle g_d \rangle = 0, \qquad \langle f \rangle = 0, \qquad \langle n \rangle = 0. \tag{22-38b}$$

The actual values of g_d, f, and n for a particular experiment are then considered as one outcome (member) of the ensemble.

The inversion problem may be stated as follows: Find an $N \times M$ matrix B such that Bg_d is as close to the desired unknown f as possible. This closeness may be stated mathematically as minimizing the average of the scalar product of $f - Bg_d$ and an $N \times 1$ vector a_1 for any arbitrary vector a_1:

$$\langle |(f - Bg_d)^+ a_1|^2 \rangle = \text{minimum}. \tag{22-39}$$

To find the solution, we let $(Bg_d)^+ a_1 = g_d{}^+ B^+ a_1$ and $B^+ a_1 = a_2$. We then write the left-hand side of (22-39) as

$$\langle |f^+ a_1 - g_d{}^+ a_2|^2 \rangle = a_1{}^+ R_{ff} a_1 - a_2{}^+ R_{gf} a_1 - a_1{}^+ R_{fg} a_2 + a_2{}^+ R_{gg} a_2 \tag{22-40}$$

where $R_{ff} = \langle ff^+ \rangle$ is an $N \times N$ matrix representing the covariance of f and is called the covariance matrix. Similarly, $R_{gf} = \langle g_d f^+ \rangle$, $R_{fg} = \langle f g_d{}^+ \rangle = R_{gf}^+$, and $R_{gg} = \langle g_d g_d{}^+ \rangle$.

We rewrite the right side of (22-40) in the form

$$(a_2 - R_{gg}^{-1} R_{gf} a_1)^+ R_{gg} (a_2 - R_{gg}^{-1} R_{gf} a_1) + a_1{}^+ (R_{ff} - R_{gf}^+ R_{gg}^{-1} R_{gf}) a_1.$$

If R_{gg} is positive definite, the first term is always nonnegative. Also, the matrix between $a_1{}^+$ and a_1 in the second term is Hermitian and therefore the second term is always positive. Therefore, we can minimize this quantity by choosing a_2 in such a manner that the first term becomes zero. This choice gives

$$a_2 = B^+ a_1 = R_{gg}^{-1} R_{gf} a_1. \tag{22-41}$$

From this we obtain the desired B:

$$B = R_{gf}^+ (R_{gg}^+)^{-1}. \tag{22-42}$$

Now we note that

$$R_{gf}^+ = \langle f g_d{}^+ \rangle = \langle f(Af + n)^+ \rangle = R_{ff} A^+ + R_{fn} \tag{22-43}$$

$$R_{gg} = \langle (Af + n)(Af + n)^+ \rangle = AR_{ff} A^+ + R_{nf} A + AR_{fn} + R_{nn} \tag{22-44}$$

where $R_{fn} = \langle fn^+ \rangle$ is an $N \times M$ covariance matrix and $R_{nn} = \langle nn^+ \rangle$ is an $M \times M$ covariance matrix.

Substituting (22-43) and (22-44) into (22-42), we obtain the final solution:

$$f = Bg_d, \qquad B = (R_{ff} A^+ + R_{fn})(AR_{ff} A^+ + R_{fn}^+ A^+ + AR_{fn} + R_{nn})^{-1}. \tag{22-45}$$

In many practical problems, the measurement error n and the unknown quantity f are independent of each other and in this case we have $R_{fn} = 0$ and (22-45) is reduced to

$$B = (R_{ff} A^+)(AR_{ff} A^+ + R_{nn})^{-1}. \tag{22-46}$$

The maximum percentage error $\Delta_{f\,\max}$ in the unknown f is related to the percentage error Δ_n in the measurement:

$$\Delta_{f\,\max} = \|A\| \, \|B\| \, \Delta_n \tag{22-47}$$

where $\Delta_f = \|\delta f\|/\|f\|$ and $\Delta_n = \|n\|/\|g_d\|$. The norm of B as given in (22-45) or (22-46) is usually quite small compared with the norm of A^{-1} and therefore the error $\Delta_{f\,\max}$ is comparable in magnitude to the experimental error Δ_n, and thus this procedure yields a stable solution.

As can be seen from (22-46), the effectiveness of this procedure depends on the choice of the covariance matrices R_{ff} and R_{nn}. It is obvious that if one knows something about the statistical properties of the unknown f and the experimental error n, we can make use of this information to construct the covariance matrices R_{ff} and R_{nn}. Clearly, the more we know about f and n, the better choice of R_{ff} and R_{nn} can be made, which should result in a better solution.

As an example consider the problem given in (22-25) and (22-26). Suppose we know that the magnitude of the elements of f is approximately unity [the exact value is $1/(2+\varepsilon)$]. We assume the following form for R_{ff}:

$$R_{ff} = \begin{pmatrix} \sigma_f^2 & 0 \\ 0 & \sigma_f^2 \end{pmatrix} \tag{22-48}$$

where σ_f^2 is on the order of unity and the correlation between the elements of f (off-diagonal terms) is assumed to be negligibly small. Also the experimental error is assumed to be 1%. Therefore we assume

$$R_{nn} = \begin{pmatrix} \sigma_n^2 & 0 \\ 0 & \sigma_n^2 \end{pmatrix} \tag{22-49}$$

where $\sigma_n = 10^{-2}$ and we also assume that the correlation between the noises (off-diagonal terms) is negligible. Substituting (22-48) and (22-49) into (22-46), we obtain

$$B \approx \tfrac{1}{4}\begin{pmatrix} 1 & 1 \\ 1 & 1 \end{pmatrix}. \tag{22-50}$$

The error in the unknown Δf as calculated from Bg_{d} is of the same order as the error in the measured data Δg. Thus Bg_{d} gives a stable solution.

22-8 BACKUS–GILBERT INVERSION TECHNIQUE

The regularization (smoothing) method in Section 22-6 requires a judicious choice of a parameter γ_1 or γ_2 to obtain a stable solution. The statistical technique in Section 22-7 requires statistical knowledge of the measurement errors and the unknowns. These requirements are not serious disadvantages. In fact in many practical problems these requirements may be met reasonably well. In 1970, Backus and Gilbert proposed an inversion technique which does not require an initial guess for the unknown. In addition it determines the resolution (spread) and the accuracy (variance) as functions of the measurement noise and thus it can control the trade-off between the spread and the variance. We outline this technique in this section (Backus and Gilbert, 1970; Westwater and Cohen, 1973; Post, 1976; Chow and Tien, 1976).

Let us consider the inversion problem (22-29) and write it in the form

$$g_i = \int_a^b K_i(r) f(r)\, dr \tag{22-51}$$

where $f(r)$ is the unknown to be determined, $K_i(r)$ a known function, and g_i quantity to be observed. The observed data from N measurements are g_{di}, $i = 1, 2, \ldots, N$. Since the true g_i is never known and g_{di} always contains some experimental errors (noise), we write

$$g_{di} = g_i + n_i \tag{22-52}$$

where n_i, $i = 1, 2, \ldots, N$, are the experimental errors. We assume that the average $\langle n_i \rangle$ is zero and the error covariance S_n is known:

$$S_n = \langle nn^+ \rangle \dagger \tag{22-53}$$

where $n = (n_i)$ is an $N \times 1$ column matrix and n^+ is the transpose of n.

Let us try to find a solution $f(r)$ from the N data g_{di}, $i = 1, 2, \ldots, N$. It is obvious that we cannot hope to calculate $f(r)$ from the finite numbers of g_{di}. However, we can hope to calculate a weighted average of $f(r)$ at different r with heavy weighting close to a point r_0 where we wish to estimate $f(r_0)$. We use the following averaging to calculate the local average of $f(r)$ in the neighborhood of r_0:

$$\langle f(r_0) \rangle = \int_a^b A(r_0, r) f(r)\, dr \tag{22-54}$$

where $A(r_0, r)$ is called the averaging kernel. $A(r_0, r)$ is small except near $r = r_0$ and satisfies

$$\int_a^b A(r_0, r)\, dr = 1. \tag{22-55}$$

Ideally, $A(r_0, r)$ should be a delta function, $A(r_0, r) = \delta(r_0 - r)$. Then $\langle f(r_0) \rangle$ reproduces $f(r)$ exactly. Of course, this is impossible in actual problems.

The Backus–Gilbert technique makes use of the following averaging kernel:

$$A(r_0, r) = \sum_{i=1}^{N} a_i(r_0) K_i(r) \tag{22-56}$$

where $K_i(r)$ is the known function in (22-51) and $a_i(r_0)$ is a constant to be chosen appropriately so that $A(r_0, r)$ resembles a delta function. Substituting (22-56) into (22-54), we obtain

$$\langle f(r_0) \rangle = \sum_{i=1}^{N} a_i(r_0) g_i. \tag{22-57}$$

† S_n is the same as R_{nn} used in the preceding section. Here we use S to conform with the commonly used notation.

We note that if $A(r_0, r)$ were a delta function, (22-57) would be exactly equal to $f(r)$. In reality $A(r_0, r)$ is never a delta function and g_i is never known. Instead we only know the measured data g_{di}. If we write

$$\langle f(r_0)\rangle_d = \sum_{i=1}^{N} a_i(r_0) g_{di}, \tag{22-58}$$

this is obviously not exact. However, if $a_i(r_0)$ is chosen appropriately, (22-58) should give a reasonable solution in some sense. As a measure of how reasonable this solution may be, we use the following two quantities: the spread and the variance.

The spread is defined by

$$s(r_0) = 12 \int_a^b (r - r_0)^2 A^2(r_0, r)\, dr. \tag{22-59}$$

The constant is chosen so that if $A(r_0, r)$ has a rectangular shape with height l^{-1} and width l centered at r_0, then $s(r_0) = l$. Substituting (22-56) into (22-59), we obtain

$$s(r_0) = a^+ S a \tag{22-60}$$

where $a = (a_i(r_0))$ is an $N \times 1$ column matrix and S is an $N \times N$ matrix with elements

$$S_{ij}(r_0) = 12 \int_a^b (r - r_0)^2 K_i(r) K_j(r)\, dr. \tag{22-61}$$

Equation (22-60) gives the spread for a given $a = (a_i)$.

Let us next consider the variance σ^2. From (22-57) and (22-58), the error in the unknown $\Delta_f = \langle f(r_0)\rangle_d - \langle f(r_0)\rangle$ is given by the measurement error (noise) n_i:

$$\Delta_f = \sum_{i=1}^{N} a_i(r_0) n_i. \tag{22-62}$$

The variance $\sigma^2(r_0)$ is therefore given by

$$\sigma^2(r_0) = \langle {\Delta_f}^2 \rangle = a^+ S_n a \tag{22-63}$$

where S_n is an $N \times N$ covariance matrix in (22-53).

Now we may expect that for a given noise S_n, if the spread $s(r_0)$ is small, the variance $\sigma^2(r_0)$ may become large. This was shown to be true by Backus and Gilbert. Stated differently, improvement in accuracy (reduction in variance) is achieved only by degrading the resolution (increasing the spread). This suggests that we need to choose $a_i(r_0)$ in such a manner that for a given problem both the spread and the variance may be reduced, and at the same time some compromise between accuracy and resolution can be achieved.

The solution obtained by Backus and Gilbert can be stated in the form

$$a = \frac{W^{-1}U}{U^{+}W^{-1}U} \tag{22-64}$$

where $U = (U_i)$ is an $N \times 1$ column matrix and W is an $N \times N$ matrix given by

$$U_i = \int_a^b K_i(r)\, dr, \qquad W = S \cos\theta + wS_n \sin\theta$$

where w is a constant. The parameter θ where $0 \le \theta \le \pi/2$ controls the trade-off between the spread and the variance. For example, $\theta = 0$ yields the minimum spread and the maximum variance and $\theta = \pi/2$ yields the maximum spread and the minimum variance. We can express this in a typical trade-off curve shown in Fig. 22-3. The constant w should be chosen so that both s and $w\sigma^2$ have approximately the same order of magnitude. With a_i given by (22-64), the final solution is given by (22-58).

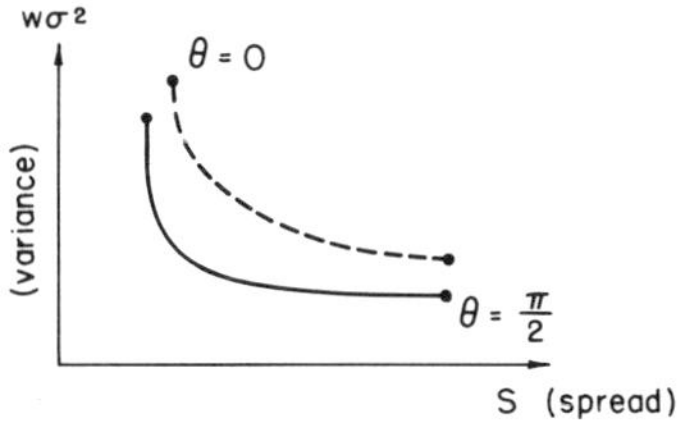

FIG. 22-3 Trade-off curve for different measurement errors. The solid curve indicates less error than the dashed curve.

The Backus–Gilbert technique has been compared with other techniques (Chow and Tien, 1976). If the unknown $f(r)$ is known *a priori* to be a smooth function, the smoothing method (Phillips–Twomey) appears to be more stable to errors in g. However, the Backus–Gilbert procedure does not require knowledge of the unknown $f(r)$ and it provides a measure of how good the solution is in terms of the spread and the variance. It should be noted that since the averaging kernel (22-56) should resemble a delta function, it requires a certain number of terms N to cancel out the value of $A(r_0, r)$ at r except near r_0. This number N may be somewhat larger than the number of oscillations exhibited by $K(r)$. The effect on resolution when the number of measurements is increased, and the effect of measurement error on resolution have been discussed by Westwater and Cohen (1973).

22-9 REMOTE SENSING OF OBSERVABLES IN GEOPHYSICS

In Sections 22-1–22-3, we discussed some aspects of the remote-sensing problem in the troposphere. In this section we briefly outline other remote-sensing problems.

Remote sensing of the sea surface has been actively studied in recent years. Barrick (Wait, 1971, 1972) summarized the problem as shown in Table 22-1. Detection of oil spills and discrimination of ice types are other examples of the remote sensing of the ocean environment (Krishen, 1971, 1973; Valenzuela and Laing, 1972).

TABLE 22-1

Frequency	Mechanism	Ocean Parameters
MF and HF ground wave	Signal spectrum Doppler shift	Wave height and direction, currents, tides
HF sky wave	Scattered signal spectrum	Same as preceding, but via second order wave height effects
UHF	Two-frequency correlation and phase detection	Short gravity waves that are superimposed on and coupled to longer waves
Microwave	Altimeter (backscatter)	Nonsymmetrical and non-Gaussian wave height spectra, mean sea level, tides, etc.
	Scatterometer (near specular scatter)	Capillary waves and possible strong dependence on wind speed
	Scatterometer (quasi-specular scatter)	Nonsymmetrical and non-Gaussian wave slopes
	Radiometer	Surface temperature, foam distribution, and sea state

Thermal emissions and radiometric measurements can be used to probe the internal structure of the earth as well as rain attenuation (Tsang and Kong, 1976; Stogryn, 1970; England, 1975; Gurvich *et al.*, 1973; Basharinov *et al.*, 1970; for a study of multiple scattering effects on radiometric measurements of rain attenuation, see Zavody, 1974). Extensive studies on radar backscatter from vegetation have been conducted (Ulaby and Batilivala, 1976; Dickey *et al.*, 1974). We also add electromagnetic geophysical prospecting (Wait, 1971; Bahar, 1971). In addition, the radio occultation technique (Woo and Ishimaru, 1973, 1974; Woo *et al.*, 1974; Woo, 1975, 1976) has been used in the remote sensing of planetary turbulence characteristics and temporal frequency spectra have been used to infer the solar wind characteristics.

APPENDIX A □ SPECTRAL REPRESENTATIONS OF A RANDOM FUNCTION

A-1 STATIONARY COMPLEX RANDOM FUNCTION

Let us consider a stationary complex random function $f(t)$ which satisfies

$$\langle f(t)\rangle = 0 \tag{A-1a}$$

$$\langle f(t_1)f^*(t_2)\rangle = B_f(t_1 - t_2). \tag{A-1b}$$

Here we have assumed that the average is zero and, therefore, B_f should be called the covariance. If the average is not zero, the fluctuation $f(t) - \langle f(t)\rangle$ satisfies the condition (A-1).

We now attempt to develop a spectral representation of a random function satisfying (A-1) (Yaglom, 1962; Tatarski, 1961, 1971). We may be tempted to write a Fourier transform representation

$$f(t) = \int_{-\infty}^{\infty} F(\omega)e^{-i\omega t}\, d\omega. \tag{A-2}$$

But mathematically this contradicts the assumption of stationarity, because in order to have a valid Fourier transform representation, it is required that Dirichlet's condition

$$\int_{-\infty}^{\infty} |f(t)|\, dt = \text{finite} \tag{A-3}$$

be satisfied. But (A-3) becomes divergent for a stationary random function.

To avoid this difficulty, a random function is represented by a stochastic Fourier Stieltjes integral

$$f(t) = \int_{-\infty}^{\infty} e^{-i\omega t}\, dv(\omega), \tag{A-4}$$

where $dv(\omega)$ is called the random amplitude. Let us determine the properties of dv by examining (A-1). First, we require from (A-1a) that

$$\langle dv(\omega)\rangle = 0. \tag{A-5}$$

Next, consider the correlation function (covariance)

$$\langle f(t_1)f^*(t_2)\rangle = \int_{-\infty}^{\infty}\int_{-\infty}^{\infty} \exp(-i\omega_1 t_1 + i\omega_2 t_2)\langle dv(\omega_1)\, dv^*(\omega_2)\rangle. \tag{A-6}$$

According to (A-1b), this must be a function of $t_1 - t_2$ only. Therefore, we require that

$$\langle dv(\omega_1)\, dv^*(\omega_2)\rangle = 0 \qquad \text{if} \quad \omega_1 \neq \omega_2, \tag{A-7a}$$

and that

$$\langle dv(\omega_1)\, dv^*(\omega_2)\rangle = W(\omega_1)\, d\omega_1 \qquad \text{when} \quad \omega_1 = \omega_2. \tag{A-7b}$$

We can compactly represent (A-7a) and (A-7b) by writing

$$\langle dv(\omega_1)\, dv^*(\omega_2)\rangle = W(\omega_1)\, \delta(\omega_1 - \omega_2)\, d\omega_1\, d\omega_2. \tag{A-8}$$

Using (A-8), we have

$$\langle f(t_1)f^*(t_2)\rangle = \int_{-\infty}^{\infty} \exp[-i\omega(t_1 - t_2)]W(\omega)\, d\omega. \tag{A-9}$$

The relationship (A-8) means that if $f(t)$ is expressed in spectral representation (A-4), then the spectra $dv(\omega)$ at different frequencies must be uncorrelated as shown in (A-7a). The transform (A-9) is an ordinary Fourier transform, because the correlation function satisfies Dirichlet's condition. $W(\omega)$ is called the spectral density of the random function $f(t)$ and represents the amount of the "power density" at different frequencies. For example, the variance is a sum of spectral densities over the entire frequency range:

$$\langle |f|^2\rangle = \int_{-\infty}^{\infty} W(\omega)\, d\omega. \tag{A-10}$$

Since (A-9) is an ordinary Fourier transform, we take an inverse transform and obtain

$$W(\omega) = \frac{1}{2\pi}\int_{-\infty}^{\infty} B_f(\tau)e^{i\omega\tau}\, d\omega \tag{A-11}$$

where $\tau = t_1 - t_2$. The transform pair (A-9) and (A-11) is, of course, a statement of the Wiener–Khinchin theorem.

In the measurement of temporal spectra, it is customary to use the definition of W_T:

$$B_f(\tau) = \int_0^{\infty} W_T(f)\cos(2\pi f\tau)\, df, \qquad W_T(f) = 4\int_0^{\infty} B_f(\tau)\cos(2\pi f\tau)\, d\tau. \tag{A-12}$$

W_T differs from W by 4π:

$$W_T(f) = 4\pi W(\omega). \tag{A-13}$$

A-2 STATIONARY REAL RANDOM FUNCTION

If $f(t)$ is a real function, then we require that

$$f(t) = f^*(t). \tag{A-14}$$

Taking the complex conjugate of (A-4), we obtain

$$f^*(t) = \int e^{i\omega t}\, dv^*(\omega). \tag{A-15}$$

By letting $\omega \to -\omega$ and noting (A-14), we get for a real $f(t)$

$$dv(\omega) = dv^*(-\omega). \tag{A-16}$$

A-3 HOMOGENEOUS COMPLEX RANDOM FUNCTION

We can easily extend the one-dimensional spectral representation just discussed to the three-dimensional spectral representation. We write

$$f(\mathbf{r}) = \int e^{i\mathbf{K}\cdot\mathbf{r}}\, dv(\mathbf{K}) \tag{A-17}$$

and the conditions on the random amplitude $dv(\mathbf{K})$ are

$$\langle dv(\mathbf{K})\rangle = 0, \qquad \langle dv(\mathbf{K}_1)\, dv^*(\mathbf{K}_2)\rangle = d\mathbf{K}_1\, d\mathbf{K}_2 \Phi(\mathbf{K}_1)\, \delta(\mathbf{K}_1 - \mathbf{K}_2) \tag{A-18}$$

where

$$\begin{aligned} \mathbf{K} &= K_x \hat{\mathbf{x}} + K_y \hat{\mathbf{y}} + K_z \hat{\mathbf{z}}, \\ \delta(\mathbf{K}_1 - \mathbf{K}_2) &= \delta(K_{x1} - K_{x2})\, \delta(K_{y1} - K_{y2})\, \delta(K_{z1} - K_{z2}) \\ d\mathbf{K}_1 &= dK_{x1}\, dK_{y1}\, dK_{z1}, \qquad d\mathbf{K}_2 = dK_{x2}\, dK_{y2}\, dK_{z2}. \end{aligned}$$

We then have the Wiener–Khinchin theorem

$$\begin{aligned} B_f(\mathbf{r}) &= \langle f(\mathbf{r}_1) f^*(\mathbf{r}_2)\rangle = \int \Phi(\mathbf{K}) e^{i\mathbf{K}\cdot\mathbf{r}}\, d\mathbf{K} \\ \Phi(\mathbf{K}) &= \frac{1}{(2\pi)^3} \int B_f(\mathbf{r}) e^{-i\mathbf{K}\cdot\mathbf{r}}\, d\mathbf{r}, \end{aligned} \tag{A-19}$$

where $\mathbf{r} = \mathbf{r}_1 - \mathbf{r}_2$ and $d\mathbf{r} = dx\, dy\, dz$.

We can also express $f(\mathbf{r})$ in a two-dimensional spectral representation:

$$f(\mathbf{r}) = f(x, \boldsymbol{\rho}) = \int e^{i\boldsymbol{\kappa}\cdot\boldsymbol{\rho}}\, dv(x, \boldsymbol{\kappa}), \tag{A-20}$$

where $\mathbf{r} = x\hat{\mathbf{x}} + \boldsymbol{\rho} = x\hat{\mathbf{x}} + y\hat{\mathbf{y}} + z\hat{\mathbf{z}}$. Then $dv(x, \boldsymbol{\kappa})$ should satisfy

$$\langle dv(x, \boldsymbol{\kappa})\rangle = 0$$

$$\langle dv(x_1, \boldsymbol{\kappa}_1)\, dv^*(x_2, \boldsymbol{\kappa}_2)\rangle = F(x, \boldsymbol{\kappa}_1)\, \delta(\boldsymbol{\kappa}_1 - \boldsymbol{\kappa}_2)\, d\boldsymbol{\kappa}_1\, d\boldsymbol{\kappa}_2, \tag{A-21}$$

where $x = x_1 - x_2$. We then have

$$\begin{aligned} B_f(x, \boldsymbol{\rho}) &= \int F(x, \boldsymbol{\kappa}) e^{i\boldsymbol{\kappa}\cdot\boldsymbol{\rho}}\, d\boldsymbol{\kappa} \\ F(x, \kappa) &= \frac{1}{(2\pi)^2} \int B_f(x, \boldsymbol{\rho}) e^{-i\boldsymbol{\kappa}\cdot\boldsymbol{\rho}}\, d\boldsymbol{\rho}. \end{aligned} \tag{A-22}$$

We can establish the relationship between F and Φ by noting that

$$dv(x, \boldsymbol{\kappa}) = \int \exp(iK_x x)\, dv(\mathbf{K})\, dK_x. \tag{A-23}$$

We then have

$$\begin{aligned} F(x, \boldsymbol{\kappa}) &= \int_{-\infty}^{\infty} \exp(iK_x x)\Phi(\mathbf{K})\, dK_x, \\ \Phi(\mathbf{K}) &= \frac{1}{2\pi} \int_{-\infty}^{\infty} \exp(-iK_x x) F(x, \boldsymbol{\kappa})\, dx, \end{aligned} \tag{A-24}$$

where $\mathbf{K} = K_x\hat{\mathbf{x}} + K_y\hat{\mathbf{y}} + K_z\hat{\mathbf{z}} = K_x\hat{\mathbf{x}} + \boldsymbol{\kappa}$.

A-4 HOMOGENEOUS AND ISOTROPIC RANDOM FUNCTION

If $f(\mathbf{r})$ is isotropic in the two-dimensional yz plane, (A-22) becomes

$$\begin{aligned} B_f(x, \rho) &= 2\pi \int_0^{\infty} J_0(\kappa\rho) F(x, \kappa)\kappa\, d\kappa \\ F(x, \kappa) &= \frac{1}{2\pi} \int_0^{\infty} J_0(\kappa\rho) B_f(x, \rho)\rho\, d\rho, \end{aligned} \tag{A-25}$$

where we have expressed (A-22) in the cylindrical coordinate system and performed the integration with respect to the angle.

If $f(\mathbf{r})$ is isotropic in three dimensions, then (A-24) becomes

$$\Phi(K) = \frac{1}{2\pi} \int_{-\infty}^{\infty} \exp(-iK_x x) F(|x|, \kappa)\, dx. \tag{A-26}$$

Noting that this should hold for any K_x, we let $K_x = 0$ and obtain the following useful formula applicable to homogeneous and isotropic random functions:

$$\Phi(\kappa) = \frac{1}{\pi} \int_0^\infty F(|x|, \kappa)\, dx. \tag{A-27}$$

For a homogeneous and isotropic random function in three dimensions, we have the following relationships: Since Φ is a function of K only, we can write

$$\Phi(K) = \frac{1}{(2\pi)^3} \int_0^\infty r^2\, dr \int_0^\pi \sin\theta\, d\theta \int_0^{2\pi} d\phi B_f(r) \exp(iKr\cos\theta),$$

where use is made of $K_1 x + K_2 y + K_3 z = \mathbf{K} \cdot \mathbf{r}$, and $\mathbf{r}$ is expressed in the spherical coordinate system with $\mathbf{K}$ as the z axis. Then we get the relationship between $\Phi(K)$ and $B_f(r)$:

$$\Phi(K) = \frac{1}{2\pi^2 K} \int_0^\infty B_f(r) r \sin(Kr)\, dr. \tag{A-28}$$

Conversely,

$$B_f(r) = \frac{4\pi}{r} \int_0^\infty \Phi(K) K \sin(Kr)\, dK. \tag{A-29}$$

We also note that the one-dimensional spectrum along x is given by

$$V(K_1) = \frac{1}{2\pi} \int_{-\infty}^\infty B_f(x) \exp(iK_1 x)\, dx, \tag{A-30}$$

which we write as

$$V(K_1) = \frac{1}{\pi} \int_0^\infty B_f(x) \cos K_1 x\, dx,$$

and thus, noting that

$$\frac{dV(K)}{dK} = -\frac{1}{\pi} \int_0^\infty B_f(x) K \sin Kx\, dx,$$

we can express the three-dimensional spectrum $\Phi(K)$ in terms of the one-dimensional spectrum $V(K)$:

$$\Phi(K) = -\frac{1}{2\pi K} \frac{dV(K)}{dK}. \tag{A-31}$$

For example, if

$$B_f(r) = \langle |f|^2 \rangle \exp(-|r|/a), \tag{A-32}$$

then

$$V(K) = \frac{\langle |f|^2 \rangle a}{\pi(1 + K^2 a^2)},$$

and thus

$$\Phi(K) = -\frac{1}{2\pi K}\frac{d}{dK}V(K) = \frac{\langle |f|^2 \rangle a^3}{\pi^2(1 + K^2 a^2)^2}.$$

If

$$B_f(r) = \langle |f|^2 \rangle \exp(-r^2/a^2), \tag{A-33}$$

then

$$V(K) = \frac{\langle |f|^2 \rangle a}{2\sqrt{\pi}} \exp\left(-\frac{K^2 a^2}{4}\right), \qquad \Phi(K) = \frac{\langle |f|^2 \rangle a^3}{8\pi\sqrt{\pi}} \exp\left(-\frac{K^2 a^2}{4}\right).$$

A-5 HOMOGENEOUS AND REAL RANDOM FUNCTION

As an extension of (A-16), we have

$$dv(\mathbf{K}) = dv^*(-\mathbf{K}) \tag{A-34}$$

for a homogeneous random function.

A-6 STATIONARY AND HOMOGENEOUS RANDOM FUNCTION

Let us consider a random function $f(\mathbf{r}, t)$ which is stationary in time and homogeneous in space. We then have

$$f(\mathbf{r}, t) = \int \exp(i\mathbf{K} \cdot \mathbf{r} - i\omega t)\, dv(\mathbf{K}, \omega), \tag{A-35}$$

and dv satisfies the conditions

$$\begin{aligned} &\langle dv(\mathbf{K}, \omega) \rangle = 0 \\ &\langle dv(\mathbf{K}_1, \omega_1)\, dv^*(\mathbf{K}_2, \omega_2) \rangle = U(\mathbf{K}_1, \omega_1)\, \delta(\omega_1 - \omega_2) \\ &\qquad \times \delta(\mathbf{K}_1 - \mathbf{K}_2)\, d\omega_1\, d\omega_2\, d\mathbf{K}_1\, d\mathbf{K}_2. \end{aligned} \tag{A-36}$$

We then have

$$\begin{aligned} B(\mathbf{r}, \tau) &= \int d\mathbf{K}\, d\omega \exp(i\mathbf{K} \cdot \mathbf{r} - i\omega\tau) U(\mathbf{K}, \omega) \\ &= \int d\mathbf{K}\, e^{i\mathbf{K} \cdot \mathbf{r}} \Phi(\mathbf{K}, \tau) = \int d\omega\, e^{-i\omega\tau} W(\mathbf{r}, \omega), \end{aligned} \tag{A-37}$$

where $U(\mathbf{K}, \omega)$ is the four-dimensional spectral density, $\Phi(\mathbf{K}, \tau)$ the time-varying spatial spectral density, and $W(\mathbf{r}, \omega)$ the spatially varying temporal spectral densiiy:

$$\Phi(\mathbf{K}, \tau) = \int d\omega e^{-i\omega\tau} U(\mathbf{K}, \omega), \qquad W(\mathbf{r}, \omega) = \int d\mathbf{K} e^{i\mathbf{K} \cdot \mathbf{r}} U(\mathbf{K}, \omega). \tag{A-38}$$

A-7 "FROZEN-IN" RANDOM FUNCTION

If the time variation of a random function $f(\mathbf{r}, t)$ is caused only by a simple translation of the spatial variation with a velocity $\mathbf{V}$, this is called the frozen-in condition. Mathematically, we have

$$f(\mathbf{r}, t) = f(\mathbf{r} - \mathbf{V}t, 0). \tag{A-39}$$

Then the correlation function becomes

$$\begin{aligned} B(\mathbf{r}, \tau) &= \langle f(\mathbf{r}_1 - \mathbf{V}t_1, 0) f(\mathbf{r}_2 - \mathbf{V}t_2, 0)\rangle \\ &= B(\mathbf{r} - \mathbf{V}\tau) = \int d\mathbf{K} \exp[i\mathbf{K} \cdot (\mathbf{r} - \mathbf{V}\tau)]\Phi(\mathbf{K}). \end{aligned} \tag{A-40}$$

Comparing with (A-37), we get

$$U(\mathbf{K}, \omega) = \delta(\omega - \mathbf{K} \cdot \mathbf{V})\Phi(\mathbf{K}), \qquad \Phi(\mathbf{K}, \tau) = \Phi(\mathbf{K}) e^{-i\mathbf{K} \cdot \mathbf{V}\tau}. \tag{A-41}$$

If the velocity $\mathbf{V}$ consists of the constant velocity $\mathbf{U}$ and the fluctuating velocity $\mathbf{V}_f$,

$$\mathbf{V} = \mathbf{U} + \mathbf{V}_f, \tag{A-42}$$

then (A-40) must be averaged over the fluctuation $\mathbf{V}_f$. We obtain

$$\begin{aligned} B(\mathbf{r}, \tau) &= \int d\mathbf{K} \exp[i\mathbf{K} \cdot (\mathbf{r} - \mathbf{U}\tau)]\Phi(\mathbf{K})\langle \exp(-i\mathbf{K} \cdot \mathbf{V}_f \tau)\rangle \\ &= \int d\mathbf{K} \exp[i\mathbf{K} \cdot (\mathbf{r} - \mathbf{U}\tau)]\Phi(\mathbf{K})\chi(-\mathbf{K}\tau), \end{aligned} \tag{A-43}$$

where $\chi(-\mathbf{K}\tau)$ is the characteristic function of the velocity fluctuation $\mathbf{V}_f$. In this case, we have

$$\Phi(\mathbf{K}, \tau) = \Phi(\mathbf{K}) e^{-i\mathbf{K} \cdot \mathbf{U}\tau} \chi(-\mathbf{K}\tau). \tag{A-44}$$

It is also possible to relate $W(\mathbf{r}, \omega)$ at $\mathbf{r} = 0$ to $\Phi(K)$ when f is homogeneous and isotropic. We have

$$W(0, \omega) = \int d\mathbf{K}\, \delta(\omega - \mathbf{K} \cdot \mathbf{V})\Phi(K) = \frac{2\pi}{v} \int_{|\omega/v|}^{\infty} \Phi(K) K\, dK. \tag{A-45}$$

This is obtained by writing $d\mathbf{K} = K^2\, dK \sin\theta\, d\theta\, d\phi$, choosing $\mathbf{V} = v\hat{\mathbf{z}}$, and integrating with respect to θ and ϕ. From (A-45), we also get

$$\Phi(K) = -\frac{v^2}{2\pi K} \left[\frac{d}{d\omega} W(0, \omega)\right]_{\omega = vK}. \tag{A-46}$$

APPENDIX B □ STRUCTURE FUNCTIONS

Many random processes encountered in practice can be approximated reasonably well by stationary or homogeneous random functions. However, this approximation is valid only within a limited time or spatial distance. At greater time intervals or spatial distances, the average quantities can no longer be constant and the process is not strictly stationary nor homogeneous. An example of this is wind velocity in a turbulent atmosphere. The mean wind velocity can be considered constant only within a limited time.

Kolmogorov in his study of turbulence proposed an important extension of stationary or homogeneous random functions. It was noted that velocity fields in turbulence are not strictly homogeneous since the average field cannot be constant over the different portions of the turbulent medium. However, if we consider the velocities at two different points and take the difference, this difference is almost homogeneous over a wide range of spatial distance. If we denote a nonhomogeneous random function of this type by $f(\mathbf{r})$, then the function $[f(\mathbf{r}+\mathbf{r}_d)-f(\mathbf{r})]$ is homogeneous even though $f(\mathbf{r})$ itself is not homogeneous.

Mathematically, such a random function is called the random process with stationary increments for a function of time and the locally homogeneous random function for a function of position. The description of such random processes can be made most conveniently in terms of the "structure function" discussed in this appendix (see Yaglom, 1962, p. 86; Panchev, 1971, p. 67).

B-1 STRUCTURE FUNCTION AND RANDOM PROCESS WITH STATIONARY INCREMENTS

Consider a random function of time $f(t)$, which may be wind velocity, humidity, temperature, pressure, or the refractive index in the turbulent atmosphere. $f(t)$ is not strictly stationary, but the difference $f(t+\tau)-f(t)$ is considered stationary. Mathematically, we express this situation by the following two basic characteristics:

(a) The ensemble average of $f(t + \tau) - f(t)$ is independent of t and depends only on τ:

$$\langle (f(t + \tau) - f(t)) \rangle = \text{a function of } \tau \text{ only.} \tag{B-1a}$$

(b) The ensemble average of the square of the magnitude of $f(t + \tau) - f(t)$ is a function of τ only:

$$D_f(\tau) = \langle | f(t + \tau) - f(t) |^2 \rangle. \tag{B-1b}$$

The quantity defined in (B-1b) is called the structure function. The average quantity in (B-1a) is in fact proportional to τ:

$$\langle (f(t + \tau) - f(t)) \rangle = c_1 \tau.\dagger \tag{B-2}$$

A random process that satisfies the two conditions (B-1a) and (B-1b) is called the process with stationary increments.

As an example, consider the one-dimensional Brownian motion of a free particle. The distance $x(t)$ traveled by a particular particle is not a stationary process, but if we take the difference $x(t + \tau) - x(t)$, this becomes stationary.

From the definition (B-1b) it is obvious that the structure function has less information about the process than the correlation function. In fact, if we know the correlation function

$$B_f(t_1, t_2) = \langle f(t_1) f^*(t_2) \rangle, \tag{B-3}$$

we can always obtain the structure function:

$$D_f(\tau) = B_f(t + \tau, t + \tau) + B_f(t, t) - B_f(t + \tau, t) - B_f(t, t + \tau). \tag{B-4}$$

However, we cannot in general construct the correlation function B_f from the structure function.

The stationary process may be considered a special case of the process with stationary increments. In this case, we have a one-to-one correspondence between the structure function and the correlation function:

$$D_f(\tau) = 2B_f(0) - B_f(\tau) - B_f^*(\tau), \qquad D_f(\infty) = 2B_f(0)$$
$$\text{Re } B_f(\tau) = \tfrac{1}{2}[D_f(\infty) - D_f(\tau)]. \tag{B-5}$$

† Taking the derivative of (B-1a), we get $(d/dt)\langle f(t + \tau)\rangle - (d/dt)\langle f(t)\rangle = 0$, and thus $(d/dt)\langle f(t)\rangle$ is constant. Therefore, $\langle f(t)\rangle = c_1 t + c_2$ where c_1 and c_2 are constant. From this we get $\langle f(t + \tau) - f(t)\rangle = c_1 \tau$.

B-2 SPECTRAL REPRESENTATION OF THE STRUCTURE FUNCTION

An indefinite integral of a stationary process $g(t)$

$$\int^{t} g(t)\,dt \tag{B-6}$$

constitutes a process with stationary increments, and conversely the derivative of a process $f(t)$ with stationary increments is a stationary process.

If we express $g(t)$ in spectral representation (Yaglom, 1962, p. 88)

$$g(t) = \int_{-\infty}^{\infty} e^{i\lambda t}\,dZ(\lambda), \tag{B-7}$$

where $\langle dZ(\lambda)\,dZ^*(\lambda')\rangle = W(\lambda)\,\delta(\lambda - \lambda')\,d\lambda\,d\lambda'$ and take an indefinite integral, we obtain the spectral representation of the random process with the stationary increments. By taking the limit of integration from $t = 0$ to t, we get

$$f(t) - f(0) = \int_{-\infty}^{\infty} \frac{e^{i\lambda t} - 1}{i\lambda}\,dZ(\lambda). \tag{B-8}$$

We rewrite this in a more conventional form:

$$f(t) = f(0) + \int_{-\infty}^{\infty} (1 - e^{i\lambda})\,d\phi(\lambda) \tag{B-9}$$

where

$$d\phi(\lambda) = -\frac{dZ(\lambda)}{i\lambda}, \qquad \langle d\phi(\lambda)\,d\phi^*(\lambda')\rangle = \Phi(\lambda)\,\delta(\lambda - \lambda')\,d\lambda\,d\lambda', \tag{B-10}$$

and $\phi(\lambda) = W(\lambda)/\lambda^2$ is the "spectral density" of the random process $f(t)$.

From (B-9) we obtain the spectral representation of the structure function:

$$D_f(\tau) = \langle |f(t+\tau) - f(t)|^2\rangle = 2\int_{-\infty}^{\infty} (1 - \cos\lambda\tau)\Phi(\lambda)\,d\lambda. \tag{B-11}$$

To show this, we write

$$f(t+\tau) - f(t) = \int_{-\infty}^{\infty} (e^{i\lambda t} - e^{i\lambda(t+\tau)})\,d\phi(\lambda)$$

$$[f(t+\tau) - f(t)]^* = \int_{-\infty}^{\infty} (e^{-i\lambda' t} - e^{-i\lambda'(t+\tau)})\,d\phi^*(\lambda')$$

and make use of (B-10).

We can obtain $\Phi(\lambda)$ in (B-11) in terms of $D_f(\tau)$. Taking the derivative of (B-11) with respect to τ and using the Fourier sine transform,† we obtain

$$\lambda\Phi(\lambda) = \frac{1}{8\pi}\int_0^\infty \sin \lambda t \frac{d}{d\tau} D_f(\tau)\, d\tau. \tag{B-12}$$

If the process is stationary, the correlation function $B_f(\tau)$ is given by

$$B_f(\tau) = \int_{-\infty}^{\infty} e^{i\lambda\tau}\Phi(\lambda)\, d\lambda. \tag{B-13}$$

Substituting this into (B-5) we also obtain (B-11), and therefore (B-11) is applicable to the stationary process as well as the process with the stationary increments.

In general, however, there is an essential difference between the correlation function and the structure function. The spectral representation (B-13) of the correlation function exists only for the stationary process, whereas the spectral representation (B-11) of the structure function can exist both for the stationary process and the process with stationary increments. To clarify this point, consider the behavior of the spectrum $\Phi(\lambda)$ near $\lambda = 0$. As $\lambda \to 0$, the convergence of the integral (B-13) requires that

$$\phi(\lambda) \sim \lambda^n, \qquad n > -1 \quad \text{as} \quad \lambda \to 0, \tag{B-14}$$

whereas the convergence of the integral (B-11) requires that

$$\Phi(\lambda) \sim \lambda^n, \qquad n > -3 \quad \text{as} \quad \lambda \to 0. \tag{B-15}$$

Therefore, we conclude that if the spectrum satisfies (B-14), then the process is stationary and both the correlation function and the structure function exist. But if (B-15) holds and (B-14) does not, then the process is with the stationary increments and the structure function exists but the correlation function does not exist. This consideration is important in the theory of wave propagation in turbulent media.

B-3 LOCALLY HOMOGENEOUS AND ISOTROPIC RANDOM FUNCTION

The three-dimensional equivalent of the process with stationary increments is called the locally homogeneous random function. We consider a random function $f(\mathbf{r})$. Note that $f(\mathbf{r})$ is not strictly homogeneous, but $f(\mathbf{r}_1 + \mathbf{r}) - f(\mathbf{r}_1)$ is homogeneous.

The structure function $D_f(\mathbf{r})$ is given by

$$D_f(\mathbf{r}) = \langle | f(\mathbf{r}_1 + \mathbf{r}) - f(\mathbf{r}_1)|^2 \rangle. \tag{B-16}$$

† Fourier sine transform pair: $F(\lambda) = \int_0^\infty f(\tau) \sin \lambda\tau\, d\tau$, $f(\tau) = (1/2\pi) \int_0^\infty F(\lambda) \sin \lambda\tau\, d\lambda$ (see Sneddon, 1951, p. 18).

If the process is stationary, D_f can be obtained from the correlation function $B_f(\mathbf{r})$ and vice versa:

$$D_f(\mathbf{r}) = 2B_f(0) - B_f(\mathbf{r}) - B_f^*(\mathbf{r}), \qquad \mathrm{Re}\, B_f(\mathbf{r}) = \tfrac{1}{2}[D_f(\mathbf{r}) - D_f(\infty)]. \tag{B-17}$$

If the process is locally homogeneous and isotropic then the structure function depends only on the magnitude of $\mathbf{r}$, not its direction:

$$D_f(|\mathbf{r}|) = D_f(r) = \langle |f(\mathbf{r}_1 + \mathbf{r}) - f(\mathbf{r}_1)|^2 \rangle. \tag{B-18}$$

The spectral representation of the locally homogeneous random function $f(\mathbf{r})$ is obtained as a three-dimensional extension of the process with stationary increments. We write the spectral representation of the random function $f(\mathbf{r})$ and the structure function $D_f(\mathbf{r})$:

$$f(\mathbf{r}) = f(0) + \iiint_{-\infty}^{\infty} (1 - e^{i\mathbf{K}\cdot\mathbf{r}})\, d\phi(\mathbf{K})$$

$$D_f(\mathbf{r}) = 2 \iiint_{-\infty}^{\infty} (1 - \cos \mathbf{K}\cdot\mathbf{r})\Phi(\mathbf{K})\, d\mathbf{K} \tag{B-19}$$

$$\langle d\phi(\mathbf{K})\, d\phi^*(K') \rangle = \Phi(\mathbf{K})\, \delta(\mathbf{K} - \mathbf{K}')\, d\mathbf{K}\, d\mathbf{K}'.$$

For an isotropic and locally homogeneous process, the structure function becomes a function of $|\mathbf{r}|$ only and the spectral density Φ becomes a function of $|\mathbf{K}|$ only. In this case, we obtain

$$D_f(r) = 8\pi \int_0^{\infty} \left(1 - \frac{\sin Kr}{Kr}\right) \Phi(K) K^2\, dK. \tag{B-20}$$

The three-dimensional spectrum $\Phi(K)$ and the one-dimensional spectrum $V(K)$ are related to each other through

$$\Phi(K) = -\frac{1}{2\pi K} \frac{\partial V(K)}{\partial K}. \tag{B-21}$$

This is obtained by noting that the structure function along one of the coordinate axes, say the x axis, is given by

$$D_f(x) = 4 \int_0^{\infty} (1 - \cos Kx) V(K)\, dK \tag{B-22}$$

and integrating (B-22) by parts.† We also note here that because of Eq. (B-12), $V(K)$ can be obtained from $D_f(x)$:

$$KV(K) = \frac{1}{8\pi} \int_0^\infty \sin Kx \frac{d}{dx} D_f(x)\, dx. \tag{B-23}$$

We can also write $f(\mathbf{r})$ in a two-dimensional spectral representation in the following manner (Tatarski, 1961, p. 22):

$$f(x, \boldsymbol{\rho}) = f(x, 0) + \iint_{-\infty}^{\infty} [1 - \exp(i\boldsymbol{\kappa} \cdot \boldsymbol{\rho})]\, d\phi(x, \boldsymbol{\kappa}) \tag{B-24}$$

where $\kappa = K_2 \hat{\mathbf{y}} + K_3 \hat{\mathbf{z}}$ and $\boldsymbol{\rho} = y\hat{\mathbf{y}} + z\hat{\mathbf{z}}$, and

$$\langle d\phi(x, \boldsymbol{\kappa})\, d\phi^*(x', \boldsymbol{\kappa}') \rangle = F(|x - x'|, \boldsymbol{\kappa})\, \delta(\boldsymbol{\kappa} - \boldsymbol{\kappa}')\, d\boldsymbol{\kappa}\, d\boldsymbol{\kappa}'. \tag{B-25}$$

The structure function is then given by

$$D_f(x, \boldsymbol{\rho}) = D_f(x, 0) + \iint_{-\infty}^{\infty} [1 - \cos(\boldsymbol{\kappa} \cdot \boldsymbol{\rho})] F(|x|, \boldsymbol{\kappa})\, d\boldsymbol{\kappa}.$$

The two-dimensional spectrum F is related to the three-dimensional spectrum Φ through

$$F(|x|, \boldsymbol{\kappa}) = \int_{-\infty}^{\infty} \cos(K_1 x) \Phi(K_1, K_2, K_3)\, dK_1$$

$$\Phi(K_1, K_2, K_3) = \frac{1}{2\pi} \int_{-\infty}^{\infty} \cos(K_1 x) F(|x|, \boldsymbol{\kappa})\, dx. \tag{B-26}$$

If $f(\mathbf{r})$ is locally homogeneous and isotropic in the yz plane, then we have

$$D_f(x, \rho) = D_f(x, 0) + 4\pi \int_0^\infty [1 - J_0(\kappa\rho)] F(|x|, \kappa) \kappa\, d\kappa. \tag{B-27}$$

We note that if $f(\mathbf{r})$ is homogeneous and isotropic, the spectrum must behave as

$$\Phi(K) \to K^n, \qquad n > -3 \quad \text{as} \quad K \to 0. \tag{B-28}$$

If $f(\mathbf{r})$ is locally homogeneous and isotropic, the spectrum must satisfy

$$\Phi(K) \to K^n, \qquad n > -5 \quad \text{as} \quad K \to 0. \tag{B-29}$$

†$D_f(x) = D_f(r) = 4[K - (\sin Kx)/x]V(K)\,|_0^\infty - \int_0^\infty 4[K - (\sin Kx)/x](\partial V/\partial K)\, dK$. This should be compared with (B-21).

B-4 KOLMOGOROV SPECTRUM

As an example, consider the Kolmogorov spectrum. The structure function $D_n(r)$ of the refractive index fluctuation is given by

$$D_n(r) = \langle |n_1(\mathbf{r} + \mathbf{r}_1) - n_1(\mathbf{r}_1)|^2 \rangle$$

$$= C_n{}^2 r^{2/3} \qquad \text{for} \quad L_0 \gg r \gg l_0 \tag{B-30a}$$

$$= C_n{}^2 l_0^{2/3} (r/l_0)^2 \qquad \text{for} \quad r \ll l_0 \tag{B-30b}$$

where $n = \langle n \rangle (1 + n_1)$ and L_0 and l_0 are the outer and inner scales of turbulence.

If $D_n(r) = C_n{}^2 r^p$, then we have

$$V_n(K) = \frac{1}{2\pi K} \int_0^\infty \sin Kx \frac{d}{dx} D_n(x)\, dx = \left[\frac{\Gamma(p+1)}{2\pi} \sin \frac{\pi p}{2}\right] C_n{}^2 K^{-(p+1)}. \tag{B-31}$$

The three-dimensional spectrum $\Phi(K)$ can be obtained from this by using (B-21)

$$\Phi_n(K) = \frac{1}{2\pi K} \frac{\partial}{\partial K} V_n(K) = \left[\frac{\Gamma(p+2)}{4\pi^2} \sin \frac{\pi p}{2}\right] C_n{}^2 K^{-(p+3)} \tag{B-32}$$

which for $p = \frac{2}{3}$ becomes

$$\Phi_n(K) = 0.033 C_n{}^2 K^{-11/3}. \tag{B-33}$$

Equation (B-33) represents the spectrum $\Phi_n(K)$ in the range $2\pi/L_0 \ll K \ll 2\pi/l_0$. In the range $K \gg 2\pi/l_0$, we expect that $\Phi_n(K)$ has small magnitude and that its structure function should reduce to (B-30b) for $r \ll l_0$. One way to realize this condition was suggested by Tatarski. He assumed a form:

$$\Phi_n(K) = 0.033 C_n{}^2 K^{-11/3} \exp(-K^2/K_m{}^2) \tag{B-34}$$

for $K > 2\pi/L_0$ and adjusted the value of K_m such that the structure function becomes (B-30b) for $r \ll l_0$.

To determine K_m, we substitute (B-34) into (B-20) and evaluate the following integral for $r \ll l_0$.

$$D_f(r) = 8\pi \int_0^\infty \left(1 - \frac{\sin Kr}{Kr}\right) \Phi_n(K) K^2\, dK.$$

For $r \ll l_0$, we can use the approximation

$$1 - \frac{\sin Kr}{Kr} \simeq \frac{K^2 r^2}{6}$$

and obtain

$$D_f(r) = C_n{}^2\{(5/27\pi)[\Gamma(\tfrac{2}{3})]^2 \sin(\pi/3)K_m^{4/3}\}r^2. \tag{B-35}$$

We choose K_m so that (B-35) becomes equal to (B-30b). Thus we get

$$K_m l_0 = \{(5/27\pi)[\Gamma(\tfrac{2}{3})]^2 \sin(\pi/3)\}^{-3/4} = 5.91.$$

Therefore, we conclude that the spectrum $\Phi_n(K)$ to correspond to the structure function (B-30a) and (B-30b) is

$$\Phi_n(K) = 0.033 C_n{}^2 K^{-11/3} \exp(-K^2/K_m{}^2) \qquad \text{for} \quad K > 2\pi/L_0 \tag{B-36}$$

and $K_m = 5.91/l_0$.

APPENDIX C □ TURBULENCE AND REFRACTIVE INDEX FLUCTUATIONS

C-1 LAMINAR FLOW AND TURBULENCE

There are two distinct states of motion for a viscous fluid: laminar and turbulent. For example, if a fluid passes through a pipe of diameter l with an average velocity of v, we observe by means of coloring dye that if the velocity is low, the streamline is smooth and distinct. This is a laminar flow. As the velocity is increased, a point may be reached where the streamline is no longer smooth and the fluid undergoes irregular and random motion. This is called the turbulence (Sommerfeld, 1950; Lumley and Panofsky, 1964; Friedlander and Topper, 1961; Yaglom and Tatarski, 1967; Landau and Lifshitz, 1959; Saxton, 1969; Panchev, 1971; Humphreys, 1964; Vinnichenko *et al.*, 1973; Tennekes and Lumley, 1972; Olsen *et al.*, 1971; Abramovich, 1963; Batchelor, 1953, 1971; Beran, 1968; Monin and Yaglom, 1971).

Reynolds made the first systematic investigation of turbulence in 1883. Using the similarity principle, he deduced that if a nondimensional number

$$\mathrm{Re} = vl/\nu \tag{C-1}$$

is smaller than a certain critical number Re_{cr}, then the flow is laminar, and if Re exceeds Re_{cr}, the motion becomes turbulent. v and l are the characteristic velocity and dimension of the flow such as the average velocity and the diameter of a pipe, and ν is the kinematic viscosity. This number Re is called the Reynolds number, and Re_{cr}, the critical Reynolds number.

We can get some insight into the meaning of the Reynolds number by considering the following: If a turbulence is created within a pipe of dimension l and velocities of eddies in the pipe are on the order of v, then the characteristic time associated with the eddy is on the order of $\tau = l/v$. The kinetic energy of the turbulence per unit mass of the fluid per unit time is, therefore, on the order of

$$v^2/\tau = v^3/l. \tag{C-2}$$

On the other hand, the energy of the turbulence is dissipated due to viscosity. Obviously, if the turbulence is to be sustained, the kinetic energy must be much greater than the energy dissipation. The energy dissipation per unit mass per unit time† is on the order of

$$\nu\, v^2/l^2. \tag{C-3}$$

The ratio of the kinetic energy and the energy dissipation is recognized as the Reynolds number.

The critical Reynolds number Re_{cr} is not a universal constant. Its value depends not only on the geometry of the structure but on how the turbulence is introduced.‡

C-2 DEVELOPED TURBULENCE

When the Reynolds number is close to the critical Reynolds number, the characteristics of the turbulence depend on the initial conditions. However, when Re is much greater than Recr, the influences of the initial conditions disappear, the fluid motion is almost completely random and irregular, and its characteristics can be described only in statistical terms.

In 1941, Kolmogorov advanced an important theory of local structure of turbulence with a very large Reynolds number, and subsequently, this became the basis for all the contemporary theories of turbulence.

According to Kolmogorov, the energy is introduced into a turbulence

† For a laminar flow, the shear stress (force per unit area) due to the rate of shear strain (velocity gradient $\partial V_x/\partial y$) is given by $\mu\, \partial V_x/\partial y$, where μ is the coefficient of viscosity. In turbulence, the shear stress is dimensionally given by $\mu\, v/l$. The energy dissipation per unit mass per unit time is therefore $(\mu v/l)(1/\rho)(v/l) = \nu\, v^2/l^2$, where ρ is the density of the fluid and $\nu = \mu/\rho$ (Somerfeld, 1950, p. 113).

Typical values of μ, ρ, and ν at 20°C are:

	Air	Water
μ (kg/m-sec)	1.81×10^{-5}	0.001
ρ (kg/m^3)	1.21	998.0
ν (m^2/sec)	1.5×10^{-5}	10^{-6}

‡ If a small velocity fluctuation $\mathbf{v}_1$ is introduced into a system, it is possible in principle to solve the Navier–Stokes equation: $\partial \mathbf{V}/\partial t + (\mathbf{V} \cdot \text{grad})\mathbf{V} = -\nabla P/\rho + \nu\nabla^2\mathbf{V}$, together with the continuity equation $\nabla \cdot \mathbf{V} = 0$ and the initial and boundary conditions. In general, $\mathbf{v}_1$ has the time dependence $\exp(-i\omega t)$ and if $Re > Re_{cr}$, the imaginary part of ω, ω_i, is positive and the fluctuation increases with time. If $Re < Re_{cr}$, ω_i is negative and the fluctuation decays in time.

Some typical critical Reynolds numbers are (Landau and Lifshitz, 1959)
flow across a cylinder, 34;
flow between moving plates (or large rotating cylinders), 1500;
flow between two stationary plates (or pipe), 7700.

as a result of the variation of the average velocity. For example, near the ground, the average horizontal wind velocity varies as a function of height and this variation creates turbulence of a size approximately equal to the height z (see Section C-10). The size at which the energy enters into the turbulence is called the outer scale of turbulence and is designated by L_0. The eddies of sizes greater than L_0 are generally anisotropic. For example, in the case of atmospheric turbulence near the ground, the eddies of sizes greater than $L_0 \simeq z$ may be spread considerably in the horizontal plane. The eddies of size smaller than the outer scale L_0 are, however, in general isotropic.

Let us consider the kinetic energy of the eddies of size L_0. Let V_0 be the velocity associated with the eddies of this size. As given in (C-1), the kinetic energy per unit mass per unit time is on the order of $V_0{}^3/L_0$. The energy dissipation per unit mass per unit time is on the order of $\nu V_0{}^2/L_0{}^2$ and since the Reynolds number Re is very large, the kinetic energy is very much greater than the dissipation. Since the dissipation is negligible, almost all the kinetic energy may be transferred to eddies of smaller size. Let $V_1, V_2, \ldots, V_n$ be the velocities of the eddies of sizes $L_1, L_2, \ldots, L_n$, where $L_0 > L_1 > L_2 > \cdots > L_n$. Then the kinetic energies per unit mass per unit time for eddies of all sizes must be approximately the same.

$$V_0{}^3/L_0 \simeq V_1{}^3/L_1 \simeq V_2{}^3/L_2 \simeq V_3{}^3/L_3 \simeq \cdots \simeq V_n{}^3/L_n. \qquad \text{(C-4a)}$$

However, as the size becomes smaller, the dissipation $\nu V_n{}^2/L_n{}^2$ increases† until the smallest size l_0 of the eddies is reached where the kinetic energy is on the same order as the energy dissipation ε:

$$V_0{}^3/L_0 \simeq V_1{}^3/L_1 \simeq \cdots \simeq V_l{}^3/l_0 \simeq \nu V_l{}^2/l_0{}^2 \simeq \varepsilon. \qquad \text{(C-4b)}$$

At this size l_0, all the energy is dissipated into heat and practically no energy is left for eddies of size smaller than l_0. This size l_0 is called the inner scale of turbulence. The energy dissipation per unit mass per unit time ε is an important quantity in turbulence theory and is called the energy dissipation rate.

From (C-4b), we obtain an important property of the velocity V of the eddy of size L. If the eddy size is between the outer scale L_0 and the inner scale l_0, then the velocity fluctuation V depends only on the size L and the energy dissipation rate ε, as is evident from (C-4b):

$$V \sim (\varepsilon L)^{1/3}. \qquad \text{(C-5)}$$

It is now possible to deduce from this the form of the structure function

† Note that $V_n \propto L_n^{1/3}$ and therefore $\nu V_n{}^2/L_n{}^2 \sim \nu L_n^{-1/3}$.

for the velocity fluctuation. It must be isotropic and should have a dimension of $(\varepsilon L)^{2/3}$:

$$D_v(r) = C(\varepsilon r)^{2/3} \qquad \text{for} \quad l_0 \ll r \ll L_0 \tag{C-6a}$$

where C is a dimensionless constant.†

Equation (C-6a) was first obtained by Kolmogorov and Obukhov and is known as the "two-thirds law."

For $r \ll l_0$, from (C-4b), we deduce

$$D_v = C'(\varepsilon/v)r^2 \tag{C-6b}$$

where C' is known to be $\frac{1}{15}$ for $D_r(r)$ (Tatarski, 1961, p. 32).

The energy dissipation rate ε is already shown in (C-4b) to be on the order of vV_l^2/l_0^2. Also, since this is the energy which enters into the turbulence at large size, we expect that ε should be related to the gradient of the average wind velocity $\langle V \rangle$. We can therefore write‡

$$\varepsilon \cong K_m(\text{grad}\langle V \rangle)^2. \tag{C-7}$$

The constant K_m is called the coefficient of eddy viscosity. It is important to realize that the dissipation rate ε is not only related to the smallest size l_0 through (C-4b), but it is related to the behavior of the average quantity through (C-7).

Since our main objectives are the development of the fluctuation characteristics of the refractive index, we will not pursue the detailed account of the fluctuations of the velocity field (Lumley and Panofsky, 1964).

C-3 SCALAR QUANTITIES CONSERVED IN A TURBULENCE AND NEUTRAL, STABLE, AND UNSTABLE ATMOSPHERE

The index of refraction n of the earth's atmosphere in the troposphere (height < 17 km) is given by§

$$n - 1 = \frac{77.6}{T}(P + 4810e/T) \times 10^{-6} \tag{C-8a}$$

† Since the velocity $\mathbf{V}(\mathbf{r})$ is a vector field, we need to consider the structure function between different components. However, it can be shown that all the components can be deduced from the structure function for the component of $\mathbf{v}$ along $\mathbf{r}$: $D_r(r) = \langle [v_r(\mathbf{r}_1 + \mathbf{r}) - v_r(\mathbf{r}_1)]^2 \rangle$. For this case, C in (C-6a) is 1.2 (Tatarski, 1961, pp. 30, 192).

‡ More generally, ε depends also on the buoyancy, as discussed in Section C-6 [see (C-33)].

§ Considering the effects of dry air (pressure P_d), water vapor (partial pressure e), and CO_2 (pressure P_c), the index of refraction n is given by $(n-1)10^6 = k_1(P_d/T) + K_2(e/T) + K_3(e/T^2) + K_4(P_c/T)$ (Bean and Dutton, 1968, p. 4). Under an appropriate assumption about CO_2 content, and using the total pressure $P = P_d + e$, (C-8) results.

where T is the absolute temperature in degrees Kelvin, P the pressure in millibars, and e the water vapor pressure in millibars.†

Equation (C-8) is considered valid for frequencies ranging from 1 MHz to at least 30 GHz and perhaps up to 72 GHz. For optical propagation, the humidity effect is negligible and the index of refraction is approximately given by

$$n - 1 = (77.6/T)P \times 10^{-6}. \tag{C-8b}$$

In turbulent atmospheres, P, T, and e in (C-8) undergo irregular variations, and in general these quantities do not necessarily follow the motions of the turbulence. However, some quantities, which we will indicate shortly, preserve themselves in a volume element in space as it moves about in turbulence. Since the characteristics of such quantities are conserved in turbulent motion, they are called the *conservative additives*. Furthermore, as is usually the case, if they do not exchange energy with turbulence, they are called *passive*. We, therefore, speak of *conservative passive additives*.

Which quantities are conservative passive additives? In general, there are two quantities which may be regarded "conservative": potential temperature and specific humidity. Let us examine these two quantities for the earth's atmosphere.

Let us assume that no evaporation or condensation of water vapor is taking place in the atmosphere. Under such a condition, we can reasonably expect that as a mass of air moves about, no heat is added to or withdrawn from the air within an appreciable time (minutes). Thus, we can regard the thermodynamic process of turbulence as *adiabatic*.

Consider a mass of air at T and p at a height z. If we bring this air to the earth's surface ($P = P_0$ and $z = 0$) adiabatically, then the temperature T becomes θ. This temperature θ is called the potential temperature (Humphrey, 1964, p. 34). The potential temperature can be obtained in the following manner:

Since the pressure P in the atmosphere is caused by the gravitational force, it satisfies a hydrostatic equation:

$$dP = -\rho g \, dz \tag{C-9}$$

where ρ is the air density and g the gravitational acceleration. The pressure P and the density ρ are related by the equation of state. For the gas in 1 mole, it is given by

$$P = \rho R T \tag{C-10}$$

where R is the universal gas constant [8.314 J/mole K]. For an adiabatic

† 1 millibar = 10^{-3} dyne/cm^2 = 1 mbar; 1 standard atmosphere pressure = 1013.2 mbar.

process, we have $p\rho^{-\gamma} = \text{const}$ where $\gamma = C_p/C_v$ is the ratio of the specific heat at constant pressure to constant volume. We therefore get

$$\frac{dT}{T} = \left(\frac{\gamma - 1}{\gamma}\right)\frac{dP}{P}. \tag{C-11}$$

Using the hydrostatic equation, we obtain

$$\frac{dT}{dz} = -\left(\frac{\gamma - 1}{\gamma}\right)\frac{g}{R} = -\frac{g}{C_p} = -\alpha_a \tag{C-12}$$

where we used the relationship $C_p - C_v = R$.

For the standard atmosphere,

$$C_p = 0.240\ (\text{cal}/^\circ\text{C}) = 0.240 \times 10^3 \times 4.184\ (\text{J/kg} - \text{K})$$

$$C_v = 0.171\ (\text{cal}/^\circ\text{C}) \qquad \text{and} \qquad g = 9.81\ (\text{J/kg} - \text{m})$$

and therefore $\alpha_a = 0.98$ K/100 m. Integrating this from z to 0, we get the potential temperature θ:

$$\theta - T = -\alpha_a z\Big|_z^0 = \alpha_a z, \qquad \theta = T + \alpha_a z. \tag{C-13}$$

α_a is called the adiabatic rate of the decrease of the temperature.

It is clear that as long as the potential temperature of the atmosphere is the same at all heights (the temperature T decreases with height at the rate of α_a), then the air, even if displaced, has the same temperature and pressure as the surrounding air; therefore, it will neither rise nor fall, and the atmosphere is said to be "neutral." On the other hand, if the potential temperature increases with height (T decreases at a rate less than α_a), then the air, if displaced up or down, becomes cooler or warmer than the surrounding air and thus it returns to its original position, and the atmosphere is "stable." If the potential temperature decreases with height, the air will not return to its original position and thus the atmosphere is "unstable."

From the foregoing, we note that it is the variation of the potential temperature, not the temperature itself, which governs the motion of the air in the turbulence. Only when we consider a small range of height can we neglect the effect of the adiabatic rate α_a and treat the temperature as a "conservative additive."

The water vapor pressure e is not conservative, but the specific humidity q, which is the mass of water vapor per unit mass of air, should approximately conserve itself in the turbulent motion of the air. The specific humidity q is given by

$$q = 0.622e/P \tag{C-14}$$

where 0.622 is the ratio of the molecular weight of water vapor to that of air.

C-4 FLUCTUATIONS OF THE INDEX OF REFRACTION

From the foregoing discussions on conservative passive additives, it is clear that the variations of these additives, the potential temperature θ and the specific humidity q, are directly related to the motion of the air. We therefore express the index of refraction n in terms of θ and q:

$$n - 1 = \frac{77.6P}{\theta - \alpha_a z}\left(1 - \frac{7733q}{\theta - \alpha_a z}\right) \times 10^{-6}. \tag{C-15}$$

Assuming the adiabatic process (C-11), the fluctuations of the index of refraction are given by

$$\delta n = \frac{\partial n}{\partial q}\partial\theta + \frac{\partial n}{\partial q}\delta q$$

$$\frac{\partial n}{\partial \theta} = \frac{77.6P}{(\gamma - 1)T^2}\left(1 + \frac{15466q}{T}\right) \times 10^{-6}, \qquad \frac{\partial n}{\partial q} = \frac{77.6 \times 7733}{T^2} \times 10^{-6}. \tag{C-16}$$

Equation (C-16) shows that the behaviors of these additives θ and q in turbulence are directly related to the fluctuation characteristics of the index of refraction in turbulence.

C-5 STRUCTURE FUNCTIONS OF A CONSERVATIVE SCALAR AND THE INDEX OF REFRACTION FLUCTUATION

Let us consider the fluctuation characteristics of the potential temperature θ in a developed turbulence. Let us write θ as a sum of the average $\langle\theta\rangle$ and the fluctuation θ'.

$$\theta = \langle\theta\rangle + \theta'. \tag{C-17}$$

The behavior of the fluctuation θ' can be deduced from the behaviors of the velocity field in Section C-3. Let us consider the eddies of the size L_0, the outer scale of the turbulence. These eddies carry with them the concentration of the additive θ. Thus, in analogy with the velocity field, the amount of inhomogeneity θ' associated with the size L_0 per unit time should be on the order of

$$\theta'^2/\tau_0 = V_0\theta'^2/L_0. \tag{C-18}$$

The amount of inhomogeneity disappearing per unit time is on the order of

$$D\,\theta'^2/L_0{}^2 \tag{C-19}$$

where D is called the coefficient of molecular diffusion. D plays a role similar to that of the kinematic viscosity ν for the velocity field, and the value of D is very close to that of ν (Tatarski, 1961, p. 45).† For the eddy size between L_0 and l_0, the dissipation (C-19) is negligible compared with the amount of inhomogeneity θ' in (C-18) and the inhomogeneity given in (C-18) is transferred from the larger size to the smaller size. Finally, at the inner scale l_0, the inhomogeneity disappears due to the molecular diffusion.

Thus, we write in analogy with (C-4b)

$$\frac{V_0\theta_0'^2}{L_0} \simeq \frac{V_1\theta_1'^2}{L_1} \simeq \frac{V_2\theta_2'^2}{L_2} \simeq \cdots \simeq \frac{V_l\theta_l'^2}{l_0} \simeq D\frac{\theta_l'^2}{l_0{}^2} = N_\theta \tag{C-20}$$

where N_θ is the rate of dissipation of the fluctuations of the scalar θ'. From (C-20), we note that between L_0 and l_0, we have

$$V\theta'^2/L \simeq N_\theta. \tag{C-21}$$

But, since in turbulence, V is given by [see (C-5)]

$$V \simeq (\varepsilon L)^{1/3}, \tag{C-22}$$

we get

$$\theta'^2 \sim N_\theta L^{2/3}/\varepsilon^{1/3}. \tag{C-23}$$

It is now possible to write the structure function for the fluctuation of the scalar θ. From (C-23), we deduce

$$D_\theta(r) = C_\theta{}^2 r^{2/3}, \qquad C_\theta{}^2 = bN_\theta/\varepsilon^{1/3} \qquad \text{for} \quad l_0 \ll r \ll L_0, \tag{C-24}$$

and b is considered to be a universal constant. The exact magnitude of b is not well known; it has been estimated to be in the range 1.5–3.5, and 2.8 is suggested by Monin and Yaglom (1971). C_θ is called the structure constant of the scalar θ.

For $r \ll l_0$, we can write

$$D_\theta(r) = C_\theta{}^2 l_0^{2/3}(r/l_0)^2. \tag{C-25}$$

Equations (C-24) and (C-25) are the fundamental equations for the structure function of the fluctuation of the scalar θ.

Since the fluctuation of the index of refraction is directly proportional to the fluctuation of the conservative additive, we can also write the structure function for the index of refraction:

$$D_n(r) = \begin{cases} C_n{}^2 r^{2/3} & \text{for} \quad l_0 \ll r \ll L_0 \\ C_n{}^2 l_0^{2/3}(r/l_0)^2 & \text{for} \quad r \ll l_0 \end{cases} \tag{C-26}$$

† For temperature, $D \simeq 1.9 \times 10^{-5}$ m^2/sec; for atmospheric water vapor, $D = 2.0 \times 10^{-5}$ m^2/sec; and for atmospheric CO_2, $D = 1.4 \times 10^{-5}$ m^2/sec. Compare these with the air kinematic viscosity $\nu = 1.5 \times 10^{-5}$ m^2/sec.

and $C_n^2 = bN_n/\varepsilon^{1/3}$. N_n is the rate of dissipation of the fluctuation of the index of refraction, and C_n is the structure constant of the index of refraction fluctuation.

The structure constants C_θ and C_n represent the intensity of the turbulence and their determinations are of critical importance. To do this, we need to know more about the two important quantities ε and N, which are discussed in the following sections.

C-6 THE ENERGY DISSIPATION RATE ε AND THE ENERGY BUDGET OF ATMOSPHERIC TURBULENCE

In order to investigate the method to calculate ε, it is necessary first to study the general properties of the atmospheric turbulence. These properties are most conveniently expressed in a form of the energy budget equation. Let E be the mean kinetic energy of turbulence per unit mass. The change of E per unit time is caused by (a) M, the rate of production of turbulent energy by wind shear, (b) B, the rate of production of energy by buoyancy, and (c) ε, the energy dissipation.

Thus, we write†

$$dE/dt = M + B - \varepsilon. \tag{C-27}$$

The energy production M is caused by the rate of the change of wind velocity with elevation and is given by

$$M = K_{\mathrm{m}}[(\partial V_x{}^2/\partial z) + (\partial V_y/\partial z)^2] \tag{C-28}$$

where V_x and V_y are two orthogonal horizontal components of the wind velocity, and K_{m} is the coefficient of "eddy viscosity." The buoyancy term B depends on the vertical temperature distribution. If the potential temperature θ is constant (neutral atmosphere), B is negative and it tends to diminish the turbulence. If θ decreases with height (unstable atmosphere), B is positive and it tends to increase the turbulent energy. B is approximately given by

$$B = -K_{\mathrm{h}} \frac{g}{\theta} \frac{\partial \theta}{\partial z} \tag{C-29}$$

where g is the gravitational acceleration and K_{h} is the "eddy coefficient of heat conduction." K_{m} and K_{h} are of the same order and are often assumed to be equal.

The ratio of the shear energy production M to the buoyancy B is called

† To this may be added a transport term T, which describes the energy exported and dissipated elsewhere. But this is normally negligible (Panofsky, 1969a).

the flux Richardson number R_f. It is approximately equal to the gradient form of the Richardson number R_i:

$$R_f = -\frac{B}{M}, \qquad R_i = \frac{(g/\theta)\,\partial\theta/\partial z}{(\partial V_x/\partial z)^2 + (\partial V_y/\partial z)^2}. \tag{C-30}$$

We can now rewrite (C-27) as

$$dE/dt = M(1 - R_f) - \varepsilon. \tag{C-31}$$

If R_f is negative, then both the wind shear and buoyancy are feeding energy to the turbulence. If R_f is large and positive, energy is withdrawn by buoyancy so much that turbulence does not develop. We can, therefore, speak of the "critical Richardson number" R_{icr}, such that the turbulence will sustain only if $R_i < R_{icr}$. The critical Richardson number is considered to be between 0.15 and 0.5. We can therefore state the region in which turbulence occurs in terms of the following two conditions:

$$R_i < R_{icr}, \qquad R_e > R_{ecr}. \tag{C-32}$$

For the turbulence to maintain itself, the energy dissipation rate ε must be balanced by the sum of the wind shear term M and the buoyancy term B. Thus, we obtain

$$\varepsilon = M + B = M(1 - R_f). \tag{C-33}$$

In free atmosphere (see Section C-9), the values of the dissipation rate (in m^2/sec^3) for turbulence of different strengths are approximately (Vinnichenko and Dutton, 1969):

$\varepsilon > 100 \times 10^{-3}$	severe turbulence
$100 \times 10^{-3} > \varepsilon > 12 \times 10^{-3}$	moderate turbulence
$12 \times 10^{-3} > \varepsilon > 3 \times 10^{-3}$	light turbulence
$\varepsilon < 3 \times 10^{-3}$	almost no turbulence.

C-7 THE RATE OF DISSIPATION OF THE FLUCTUATION N

Just as the dissipation rate ε is balanced by the turbulence production rate $M + B$ to sustain turbulence, the rate of dissipation N should be approximately equal to the rate of production of the fluctuation of the scalar. The fluctuation is produced by the gradient of the mean value of the scalar. Therefore, we write

$$N_\theta = K(\mathrm{grad}\langle\theta\rangle)^2 \simeq D\,\theta_l'^2/l_0{}^2 \tag{C-34}$$

where K is the coefficient of "turbulent diffusion." K and K_m have the same dimension (L^2/T) and are numerically of the same order.

The values of N_θ for the temperature fluctuations in free atmosphere at altitudes of approximately 18 km range generally from 10^{-6} to 10^{-4} $(°C)^2$/sec for light turbulence, from 10^{-5} to 10^{-3} $(°C)^2$/sec for moderate turbulence, and from 10^{-4} to 10^{-3} $(°C)^2$/sec for severe turbulence (Vinnichenko and Dutton, 1969).

C-8 CALCULATION OF THE STRUCTURE CONSTANT

As was shown in (C-24), the structure constant $C_\theta{}^2$ for a scalar quantity θ such as the refractive index or temperature can be obtained if the energy dissipation rate ε and the dissipation rate N_θ of the fluctuation of the scalar θ are known [see (C-24)]:

$$C_\theta{}^2 = bN_\theta/\varepsilon^{1/3}, \tag{C-35}$$

where N_θ and ε are given in Sections C-7 and C-6, respectively.

Alternatively we can calculate $C_\theta{}^2$ in terms of the outer scale of turbulence L_0. To do this, consider the structure function $D_\theta(r)$, which is the mean square of the difference of fluctuations at two points. Between two points, there should be some difference of the average values of the scalar $\langle\theta\rangle$. In the inertial subrange, the mean square of this difference $[r\ \mathrm{grad}\langle\theta\rangle]^2$ is much smaller than $D_\theta(r)$, the mean square of the difference of fluctuation. However, at the distance of the outer scale L_0, these two quantities should become comparable:

$$D_\theta(L_0) \sim [L_0\ \mathrm{grad}\langle\theta\rangle]^2. \tag{C-36}$$

Therefore, we write

$$b(N_\theta/\varepsilon^{1/3})L_0^{2/3} = b[L_0\ \mathrm{grad}\langle\theta\rangle]^2 \tag{C-37}$$

where a constant b is placed on the right-hand side to simplify the definition of L_0. Noting that $N_\theta = K(\mathrm{grad}\langle\theta\rangle)^2$ from (C-34), we obtain

$$L_0 = [K/\varepsilon^{1/3}]^{3/4}. \tag{C-38}$$

Since $D_\theta(r) = C_\theta{}^2 r^{2/3}$ and $D_\theta(L_0) = b[L_0\ \mathrm{grad}\langle\theta\rangle]^2$, we have

$$C_\theta{}^2 = bL_0^{4/3}(\mathrm{grad}\langle\theta\rangle)^2. \tag{C-39}$$

Equations (C-35) and (C-39) are two alternative ways of calculating the structure constant $C_\theta{}^2$. We will make use of (C-39) to calculate $C_n{}^2$ in Section C-10.

C-9 BOUNDARY LAYER, FREE ATMOSPHERE, LARGE- AND SMALL-SCALE TURBULENCE

In this section, we digress somewhat from the description of the refractive index fluctuation in turbulence and give a brief classification of different turbulences.

In the earth's atmosphere, we can classify the turbulence into two general cases. The turbulence appearing near the ground is strongly dependent on the height from the surface and the surface condition, and is called the turbulence in the "boundary layer." Turbulence also appears in the atmosphere at high altitude and we speak of turbulence in "free atmosphere."

Turbulence in the atmosphere may also be classified into small-scale and large-scale turbulences. The large-scale turbulence has eddies of the order of hundreds and thousands of kilometers and is responsible for global weather conditions. The small-scale turbulence eddies are of the order of millimeters to hundreds of meters and are responsible for fluctuations of the index of refraction, which we are considering in this appendix.

C-10 THE STRUCTURE CONSTANT FOR THE INDEX OF REFRACTION IN THE BOUNDARY LAYER†

As indicated in Section C-6, the turbulence is caused by the wind shear and the buoyancy. For the turbulence in the boundary layer, it is therefore necessary to investigate the change of the average horizontal wind velocity with height. Owing to the viscosity, the wind velocity on ground is zero, and for the lowest few hundred meters, the velocity is known to have the logarithmic profile with height:‡

$$V(z) = (V_*/0.4)\ln(z/z_0) \tag{C-40}$$

where V_* is called the friction velocity and is constant, and z_0 depends on the surface roughness. Let us calculate M, the rate of energy production given by (C-28):

$$M = K_m(\partial V/\partial z)^2. \tag{C-41}$$

Using (C-40), we get

$$\partial V/\partial z = V_*/0.4z \tag{C-42}$$

From experimental data, however, a small correction needs to be introduced. We thus write

$$\partial V/\partial z = (V_*/0.4z)\phi \tag{C-43}$$

† For free atmosphere, see Vinnichenko and Dutton (1969) and Hufnagal and Stanley (1964), as well as Panofsky (1968, 1969a,b).

‡ Except for small z, where the velocity is almost linear with z (Landau and Lifshitz, 1959).

where ϕ is unity for neutral atmosphere. Some empirical expressions for ϕ are given in Eqs. (C-47). The eddy viscosity coefficient K_m is known to be equal to

$$K_m = 0.4V_* z/\phi. \qquad \text{(C-44)}$$

Using (C-43) and (C-44), we obtain

$$M = (V_*{}^3/0.4z)\phi. \qquad \text{(C-45)}$$

Now, if we take into account the buoyancy B, we can write

$$\varepsilon = M + B = (V_*{}^3/0.4z)\phi_\varepsilon \qquad \text{(C-46)}$$

where ϕ_ε contains ϕ as well as the effect of B.

In terms of the Richardson number R_i, the following expressions for ϕ and ϕ_ε seem to fit existing data well (Panofsky, 1968, 1969a):

For unstable air,

$$\phi_\varepsilon = (1 - 18R_i)^{-1/4} - R_i, \qquad \phi = (1 - 18R_i)^{-1/4}. \qquad \text{(C-47a)}$$

For stable air,

$$\phi_\varepsilon = \frac{1 - 0.7R_i}{1 - 7R_i}, \qquad \phi = \frac{R_i}{1 - 7R_i}. \qquad \text{(C-47b)}$$

The turbulent diffusion coefficient K is close to K_m and is given approximately by

$$K = K_m/\phi = 0.4V_* z/\phi^2. \qquad \text{(C-48)}$$

Substituting (C-46) and (C-48) into (C-38), we get the outer scale of turbulence:

$$L_0 = 0.4z\phi^{-3/2}\phi_\varepsilon^{-1/4}. \qquad \text{(C-49)}$$

Using (C-39), we obtain

$$C_\theta{}^2 = b(0.4z)^{4/3}\phi^{-2}\phi_\varepsilon^{-1/3}(\partial\langle\theta\rangle/\partial z). \qquad \text{(C-50)}$$

We note here that the outer scale of turbulence L_0 is approximately equal to $0.4z$, which confirms the earlier conjecture that the outer scale is on the order of height near the ground.

If we neglect the effect of buoyancy, we can set ϕ and ϕ_ε to be unity. In general, ϕ and ϕ_ε are not much different from unity.

The structure constant C_n of the index of refraction can be calculated using (C-50):

$$C_n{}^2 = b(0.4z)^{4/3}\phi^{-2}\phi_\varepsilon^{-1/3}(\partial\langle n\rangle/\partial z)^2. \qquad \text{(C-51)}$$

The gradient of the mean index of refraction is given by (C-16):

$$\frac{\partial\langle n\rangle}{\partial z} = \frac{\partial n}{\partial \theta}\frac{\partial \theta}{\partial z} + \frac{\partial n}{\partial q}\frac{\partial q}{\partial z}. \tag{C-52}$$

Let us assume that the humidity variation is negligible and therefore

$$\frac{\partial\langle n\rangle}{\partial z} \approx \frac{\partial n}{\partial \theta}\frac{\partial \theta}{\partial z}. \tag{C-53}$$

$\partial n/\partial\theta$ is already given in (C-16). To calculate $\partial\theta/\partial z$, we note (Tatarski, 1961, p. 190) that the potential temperature θ is also logarithmic with height:

$$\theta(z) = \text{const} + \theta_* \ln(z/z_0), \qquad \theta_* = \text{const}. \tag{C-54}$$

Therefore,

$$\frac{\partial\theta}{\partial z} = \frac{\theta_*}{z} \tag{C-55}$$

To determine θ_*, we may measure temperatures T_1 and T_2 at two different heights z_1 and z_2. Then we get

$$\theta(z_1) = T_1 + \alpha_a z_1 \qquad \text{and} \qquad \theta(z_2) = T_2 + \alpha_a z_2. \tag{C-56}$$

Substituting this into (C-54) and taking the difference, we obtain

$$\frac{\partial\theta}{\partial z} = \frac{1}{z}\left[\frac{T_1 - T_2 + \alpha_a(z_1 - z_2)}{\ln(z_1/z_2)}\right]. \tag{C-57}$$

Substituting this into (C-53), we get $\partial\langle n\rangle/\partial z$. Using (C-53), we get C_n^2 from (C-51).

C-11 THE STRUCTURE CONSTANT C_n FOR FREE ATMOSPHERE

The structure constant C_n has been determined only for a few selected heights. A model profile for C_n for optical propagation has been proposed by Hufnagel (Hufnagel and Stanley, 1964). The values of C_n for microwaves appear to be considerably higher than that of optical waves (see Fig. C-1). Some experimental data (Ochs and Lawrence, 1972; Consortini and Ronchi, 1972) are available on temperature and C_n^2 profiles over land and ocean to 3 km above the surface.

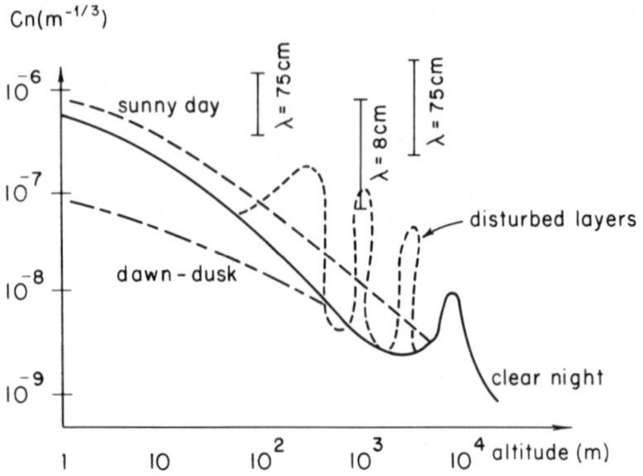

FIG. C-1 Structure constant C_n of the refractive index fluctuation in the atmosphere as a function of altitude.

C-12 RELATION BETWEEN THE STRUCTURE CONSTANT C_n AND THE VARIANCE OF THE INDEX OF REFRACTION FLUCTUATION

Since turbulence is, in general, locally homogeneous, the structure function $D_n(r)$ exists, but the correlation function $B_n(r)$ does not exist. This is equivalent to saying that the spectrum $\Phi_n(K)$ of the index of refraction can be described in the inertial and dissipation ranges $K \gg 1/L_0$, but it is anisotropic and not known for the input range $K < 1/L_0$.

However, if the spectrum is modified so that it is isotropic and finite in the input range, then the turbulence can be considered homogeneous and the correlation functions exist.

One such modification has already been suggested. We know that if

$$\begin{aligned} D_n(r) &= 2[B_n(0) - B_n(r)] \\ &= 2\langle n_1{}^2\rangle\left[1 - \frac{1}{2^{\nu-1}\Gamma(\nu)}\left(\frac{r}{L_0}\right)^{\nu} K_\nu\left(\frac{r}{L_0}\right)\right], \end{aligned} \tag{C-58}$$

then

$$\Phi_n(K) = \frac{\Gamma(\nu + \frac{3}{2})}{\pi\sqrt{\pi}\,\Gamma(\nu)} \frac{\langle n_1{}^2\rangle L_0{}^3}{(1 + K^2 L_0{}^2)^{\nu+3/2}}. \tag{C-59}$$

To make (C-59) correspond to the Kolmogorov spectrum of $K^{-11/3}$ for

$K \gg 1/L_0$, we use $\nu = \frac{1}{3}$. Furthermore, the structure function $D_n(r)$ for small r is reduced to

$$D_n(r) = \frac{2\Gamma(1-\nu)}{\Gamma(1+\nu)} \frac{\langle n_1{}^2 \rangle}{2^{2\nu}} \left(\frac{r}{L_0}\right)^{2/3}, \dagger \tag{C-60}$$

which must be equal to $C_n{}^2 r^{2/3}$. Therefore, we obtain the relationship between the structure constant C_n and the variance $\langle n_1{}^2 \rangle$ in terms of the outer scale of turbulence:

$$C_n{}^2 = \frac{2\Gamma(\frac{2}{3}) L_0^{-2/3}}{\Gamma(\frac{4}{3}) 2^{2/3}} \langle n_1{}^2 \rangle = 1.91 \langle n_1{}^2 \rangle L_0^{-2/3}. \tag{C-61}$$

The variance of the index of refraction fluctuation $\langle n_1{}^2 \rangle$ is on the order of 10^{-12}.

† Note that

$$K_\nu(z) \sim \frac{(\pi/2)}{\sin \nu\pi} \left[\left(\frac{z}{2}\right)^{-\nu} \frac{1}{\Gamma(1-\nu)} - \left(\frac{z}{2}\right)^{\nu} \frac{1}{\Gamma(\nu+1)} \right] \quad \text{for} \quad |Z| \ll 1.$$

APPENDIX D □ SOME USEFUL MATHEMATICAL FORMULAS†

D-1 KUMMER FUNCTION ${}_1F_1(a, c; z)$

$$\int_0^\infty \exp(-a^2t^2)t^{\mu-1}J_\nu(bt)\,dt$$

$$= \frac{\Gamma((\nu+\mu)/2)(b/2a)^\nu}{2a^\mu\Gamma(\nu+1)}\,{}_1F_1\left(\frac{\nu+\mu}{2}, \nu+1; -\frac{b^2}{4a^2}\right),$$

$$\operatorname{Re}(\mu+\nu) > 0, \quad \operatorname{Re} a^2 > 0. \tag{D-1}$$

As $|z| \to 0$,

$${}_1F_1(a, b; z) \to 1 + (a/b)z + \cdots. \tag{D-2}$$

As $|z| \to \infty$,

$${}_1F_1(a, b; z) \to \frac{\Gamma(b)}{\Gamma(a)}e^z z^{a-b}, \qquad \operatorname{Re} z > 0$$

$$\to \frac{\Gamma(b)}{\Gamma(b-a)}(-z)^{-a}, \qquad \operatorname{Re} z < 0. \tag{D-3}$$

$${}_1F_1(a, b; z) = e^z\,{}_1F_1(b-a, b; -z). \tag{D-4}$$

D-2 CONFLUENT HYPERGEOMETRIC FUNCTION $\psi(a, b; z)$

$$\int_0^\infty e^{-zt}t^{a-1}(1+t)^{b-a-1}\,dt = \Gamma(a)\psi(a, b, z), \qquad \operatorname{Re} a > 0, \quad \operatorname{Re} z > 0. \tag{D-5}$$

† Magnus and Oberhettinger (1954), Abramowitz and Stegun (1964).

As $|z| \to 0$

$$\psi(a, b; z) \to \frac{\Gamma(b-1)}{\Gamma(a)} z^{1-b}, \qquad \text{Re } b \geq 1, \quad b \neq 1$$

$$\to \frac{(-1)}{\Gamma(a)}\left[\ln z + \frac{\Gamma'(a)}{\Gamma(a)}\right], \qquad b = 1$$

$$\to \frac{\Gamma(1-b)}{\Gamma(1+a-b)}, \qquad \text{Re } b < 1. \tag{D-6}$$

As $|z| \to \infty$

$$\psi(a, b; z) \to z^{-a}. \tag{D-7}$$

D-3 OTHER INTEGRALS

$$\int_0^\infty \frac{t^{\nu+1} J_\nu(at)\, dt}{(t^2 + z^2)^{\mu+1}} = \frac{a^\mu z^{\nu-\mu}}{2^\mu \Gamma(\mu+1)} K_{\nu-\mu}(az),$$

$$a > 0, \quad \text{Re } z > 0, \quad -1 < \text{Re } \nu < \text{Re}(2\mu + \tfrac{3}{2}).\dagger \tag{D-8}$$

$$\int_0^\infty J_\nu(at) \exp(-p^2 t^2) t^{\nu+1}\, dt = \frac{a^\nu}{(2p^2)^{\nu+1}} \exp\left(-\frac{a^2}{4p^2}\right). \tag{D-9}$$

$$\int_0^\infty t^{-(p+1)} \sin^2 t\, dt = \frac{2^{p-2}\pi}{\sin(\pi p/2)\Gamma(p+1)}, \qquad 0 < p < 2. \tag{D-10}$$

$$\int_{-\infty}^\infty \exp(-at^2 + bt)\, dt = \left(\frac{\pi}{a}\right)^{1/2} \exp\left(\frac{b^2}{4a}\right), \qquad \text{Re } a > 0. \tag{D-11}$$

† Gradshteyn and Ryzhik (1965, p. 686).

REFERENCES

Ables, J. G., M. M. Komesaroff, and P. A. Hamilton (1970). Pulse broadening of PSR 0833-45 by interstellar scattering, *Astrophys. Lett.* **6**, 147–150.

Abramovich, G. N. (1963). "The Theory of Turbulent Jet." MIT Press, Cambridge, Massachussetts.

Abramowitz, M., and I. A. Stegun, eds. (1964). "Handbook of Mathematical Functions." U.S. Govt. Printing Office, Washington, D.C.

Ackley, M. H. (1971). Phase variations in atmospheric optical propagation, *J. Opt. Soc. Am.* **61**, 1279–1284.

Ament, W. S. (1953). Toward a theory of reflection by a rough surface, *Proc. IRE* **41**, 142–146.

Anderson, R. C., and E. V. Browell (1972). First and second order backscattering from clouds illuminated by finite beams, *Appl. Opt.* **11**, 1345–1351.

Backus, G., and F. Gilbert (1970). Uniqueness in the inversion of inaccurate gross earth data, *Phil. Trans. Roy. Soc. London Ser. A* **266**, 123–192.

Bahar, E. (1971). Radio-wave method for geophysical prospecting, *J. Geophys. Res.* **70**, 1921–1928.

Baker, D. W. (1970). Pulsed ultrasonic doppler blood-flow sensing, *IEEE Trans. Sonics and Ultrasonics* **SU-17**, 170–184.

Banakh, V. A., G. M. Krekov, V. L. Mironov, S. S. Khmelevtsov, and R. Sh. Tsvik (1974). Focused-laser beam scintillations in the turbulent atmosphere, *J. Opt. Soc. Am.* **64**, 516–518.

Barabanenkov, Y. N. (1968). Application of the smooth-perturbation method to the solution of general equations of multiple wave-scattering theory, *Sov. Phys. JETP* **27**, 6, 954–959.

Barabanenkov, Y. N. (1969). On the spectral theory of radiation transport equations, *Sov. Phys. JETP* **29**, 4, 679–684.

Barabanenkov, Y., and V. Finkel'berg (1968a). Radiation transport equation for correlated scatterers, *Sov. Phys. JETP* **26**, 587–591.

Barabanenkov, Y., and V. Finkel'berg (1968b). Optical theorem in the theory of multiple wave scattering, *Radiophys. Quantum Electron.* **11**, 407–410.

Barabanenkov, Y. N., Y. A. Kravtsov, S. M. Rytov, and V. I. Tatarski (1971). Status of the theory of propagation of waves in a randomly inhomogeneous medium, *Sov. Phys. Usp.* **13**, 551–580.

Barrick, D. E. (1972). First order theory and analysis of MF/HF/VHF scatter from the sea, *IEEE Trans. Antennas Propagat.* **AP-20**, 2–10.

Barrick, D. E., and W. H. Peake (1968). A review of scattering from surfaces with different roughness scales, *Radio Sci.* **3**, 865–868.

Barrick, D. E. (1971). Theory of HF and VHF propagation across the rough sea—Parts I and II, *Radio Sci.* **6**, 517–533.

Basharinov, A. E., A. S. Gurvich, L. T. Tuchkov, and K. S. Shifrin (1970). The terrestrial thermal radio-emission field, *Izv. Atmos. Ocean Phys.* **6**, 366–380.

Batchelor, G. K. (1953). "Theory of Homogeneous Turbulence." Cambridge Univ. Press, London and New York.

Batchelor, G. K. (1971). "Scientific Papers of G. I. Taylor," Vol. 4. Cambridge Univ. Press, London and New York.

Bates, R. H. T. (1969). Rayleigh Hypothesis, the Extended-Boundary-Condition and Point-Matching, *Electron. Lett.* **5**, 654–655.

Bean, B. R., and E. J. Dutton (1968). "Radio Meteorology." Dover, New York.

Beard, C. I. (1961). Coherent and incoherent scattering of microwaves from the ocean, *IRE Trans. Antennas Propagat.* **AP-9**, 5, 740–783.

Beard, C. I. (1962). Statistics of phase quadrature components of microwave field transmitted through a random medium, *IRE Trans. Antennas Propagat.* **AP-10**, 721–731.

Beard, C. I. (1967). Behavior of non-Rayleigh statistics of microwave forward scatter from a random water surface, *IEEE Trans. Antennas Propagat.* **AP-15**, 5, 649–657.

Beard, C. I., T. H. Kays, and V. Twersky (1965). Scattered intensities for random distributions—microwave data and optical applications, *Appl. Opt.* **4**, 1299.

Beard, C. I., T. H. Kays, and V. Twersky (1967). Scattering by random distribution of spheres vs. concentration, *IEEE Trans. Antennas Propagat.* **AP-15**, 1, 99–118.

Beard, C. I., W. T. Kreiss, and W. G. Tank (1969). A multi wavelength line-of-sight experiment for remote atmospheric sensing, *Proc. IEEE* **57**, 4, 446–458.

Beckmann, P., and A. Spizzichino (1963). "The Scattering of Electromagnetic Waves from Rough Surfaces." Pergamon, Oxford.

Bendat, J. S., and A. G. Piersol (1966). "Measurement and Analysis of Random Data." Wiley, New York.

Bendat, J. S., and A. G. Piersol (1971) "Random Data", Interscience, New York.

Beran, M. J. (1968). "Statistical Continuum Theories." Wiley (Interscience), New York.

Beran, M. J. (1970). Propagation of a finite beam in a random medium, *J. Opt. Soc. Am.* **60**, 4, 518–520.

Beran, M. J. (1975). Coherence equations governing propagation through random media, *Radio Sci.* **10**, 1, 15–22.

Beran, M. J., and T. L. Ho (1969). "Propagation of the fourth-order coherence function in a random medium (a non-perturbative formulation), *J. Opt. Soc. Am.* **59**, 9, 1134–1138.

Beran, M. J., and G. B. Parrent, Jr. (1964). "Theory of Partial Coherence." Prentice-Hall, Englewood Cliffs, New Jersey.

Beran, M. J., and A. M. Whitman (1971). Asymptotic theory for beam propagation in a random medium, *J. Opt. Soc. Am.* **61**, 8, 1044–1050.

Booker, H. G. (1956a). Turbulence in the ionosphere with applications to meteor-trails, radio-star scintillation, auroral radar echoes and other phenomena, *J. Geophys. Res.* **61**, 4, 673–704.

Booker, H. G. (1956b). A theory of scattering by nonisotropic irregularities with application to radar reflections from the aurora, *J. Atmos. Terrestrial Phys.* **8**, 204–221.

Booker, H. G. (1958). The use of radio stars to study irregular refraction of radio waves in the ionosphere, *Proc. IRE* **46**, 294–313.

Booker, H. G., and W. E. Gordon (1950a). Outline of a theory of radio scatterings in the troposphere, *J, Geophys. Res.* **55**, 3, 241–246.

Booker, H. G., and W. E. Gordon (1950b). A theory of radio scattering in the troposphere, *Proc. IRE* **38**, 401–412.

Borisov, B. D., V. M. Sazanovich, and S. S. Khmelevtosov (1969). Investigation of the fluctuations in the angles of arrival of laser radiation in the surface layer of the atmosphere, *Iz. V. U. k. Zav. Fiz.* **12**, 1, 103–106.

Born, M., and E. Wolf (1964). "Principles of Optics." MacMillan, New York.

Bouricius, G. M. B., and S. F. Clifford (1970). Experimental study of atmospherically induced phase fluctuations in an optical signal, *J. Opt. Soc. Am.* **60**, 1484–1489.

Bradley, L. C., and J. Herrmann (1974). Phase compensation for thermal blooming, *Appl. Opt.* **13**, 331–334.

✓Bremmer, H. (1964). Random volume scattering, *J. Res. Nat. Bur. Std.* **68D**, 967–981.

Bremmer, H. (1973). General remarks concerning theories dealing with scattering and diffraction in random media, *Radio Sci.* **8**, 511–534.

Bridges, W. B., P. T. Brunner, S. P. Lazzara, T. A. Nussmeier, T. R. O'Meara, J. A. Sanguinet, and W. P. Brown, Jr. (1974). Coherent optical adaptive techniques, *Appl. Opt.* **13**, 291–300.

Brinkworth, B. J. (1973). Pulsed-lidar reflectance of clouds, *Appl. Opt.* **12**, 2, 427–428.

Brookner, E. (1970). Atmosphere propagation and communication channel model for laser wavelengths, *IEEE Trans.* **COM-18**.

Brown, H. E., and S. F. Clifford (1973). Spectral broadening of an acoustic pulse propagating through turbulence, *JASA* **54**, 36–39.

Brown, W. P. Jr. (1966). Validity of the Rytov approximation in optical propagation calculation, *J. Opt. Soc. Am.* **56**, 8, 1045–1052.

Brown, W. P. Jr. (1967). Validity of the Rytov Approximation, *J. Opt. Soc. Am.* **57**, 12, 1539–1543.

Brown, W. P. (1971). Second moment of a wave propagating in a random medium, *J. Opt. Soc. Am.* **61**, 1051–1059.

Brown, W. P. (1972a). Moment equations for waves propagated in random media, *J. Opt. Soc. Am.* **62**, 1, 45–54.

Brown, W. P. (1972b). Fourth moment of a wave propagating in a random medium, *J. Opt. Soc. Am.* **62**, 966–971.

Bucher, E. A., and R. M. Lerner (1973). Experiments on light pulse communication and propagation through atmospheric clouds, *Appl. Opt.* **12**, 2401–2414.

Buck, A. L. (1967). Effects of the atmosphere on laser beam propagation, *Appl. Opt.* **6**, 703–708.

Budden, K. G., and B. J. Uscinski (1970). The scintillation of extended radio sources when the receiver has a finite bandwidth, *Proc. Roy. Soc. London* **A316**, 315–339; **A321**, 15–40 (1971); **A330**, 65–67 (1972).

Bugnolo, D. (1960). Transport equation for the spectral density of a multiple scattered electromagnetic field, *J. Appl. Phys.* **31**, 1176–1182.

Bugnolo, D. (1961). Radio star scintillation and multiple scattering in the ionosphere, *IRE Trans.* **AP-9**, 89–96.

Bugnolo, D. (1972). Critique of electromagnetic turbulent-plasma scattering theories, *Phys. Rev.* **A6**, 477–484.

Burrows, M. L. (1969a). Equivalence of the Rayleigh solution and the extended-boundary-condition solution for scattering problems, *Electron. Lett.* **5**, 277–278.

Burrows, M. L. (1969b). Example of the generalized-function validity of the Rayleigh hypothesis, *Electron. Lett.* **5**, 694–695.

Buser, R. G. (1971). Interferometric determination of the distance dependence of the phase structure function for near-group horizontal propagation at 6328Å, *J. Opt. Soc. Am.* **61**, 488–491.

Carlson, F. P. (1969). Application of optical scintillation measurements to turbulence diagnostics, *J. Opt. Soc. Am.* **59**, 1343–1347.

Carlson, F. P., and A. Ishimaru (1969). The propagation of spherical waves in locally homogeneous random media, *J. Opt. Soc. Am.* **59**, 319–327.

Chernov, L. A. (1960). "Wave Propagation in a Random Medium." McGraw-Hill, New York.

Chow, L. C., and C. L. Tien (1976). Inversion techniques for determining the droplet size distribution in clouds: numerical examination, *Appl. Opt.* **15**, 378–383.

Clifford, S. F. (1971). Temporal frequency spectral for a spherical wave propagating through atmospheric turbulence, *J. Opt. Soc. Am.* **61**, 1282–1292.

Clifford, S. F., and J. W. Strohbehn (1970). The theory of microwave line-of-sight propagation through a turbulent atmosphere, *IEEE Trans. Antennas Propagat.* **AP-18**, 264–274.

Clifford, S. F., G. R. Ochs, and R. S. Lawrence (1974), Saturation of optical scintillation by strong turbulence, *J. Opt. Soc. Am.* **64**, 148–154.

Clifford, S., G. Ochs, and T.-I. Wang (1975). Optical wind sensing by observing the scintillations of a random scene, *Appl. Opt.* **14**, 2844–2850.

Cohen, A. (1975). Cloud-base water content measurement using single wavelength laser-radar data, *Appl. Opt.* **14**, 2873–2877.

Collett, E., R. Alferness, and T. Forbes (1973). Log-intensity correlations of a laser beam in a turbulent medium, *Appl. Opt.* **12**, 1067–1070.

Collin, R. E. (1971). A comparison of selective summation techniques for the coherent Green's function in random media, *Radio Sci.* **6**, 991–995.

Consortini, A., L. Ronchi, and E. Moroder (1973). Role of the outer scale of turbulence in **11**, 1205–1211.

Consortini, A., L. Ronchi, and E. Moroder (1973). Role of the outer scale of turbulence in atmospheric degradation of optical images, *J. Opt. Soc. Am.* **63**, 1246–1248.

Coulman, C. E. (1966). Dependence of image quality on horizontal range in a turbulent atmosphere, *J. Opt. Soc. Am.* **56**, 9, 1232–1238.

Crane, R. K. (1971). Propagation phenomena affecting satellite communication systems operating in the centimeter and millimeter wavelength bands, *Proc. IEEE* **59**, 173–188.

Crane, R. K. (1977). Ionospheric scintillation, *Proc. IEEE* **65**, 180–199.

Crombie, D. D. (1955). Doppler spectrum of sea echo at 13.56 Me/s, *Nature* (*London*) **175**, 4459, 681–682.

Crosignani, B., P. DiPorto, and M. Bertolotti (1975). "Statistical Properties of Scattered Light". Academic Press, New York.

Dabberdt, W. F. (1973). Slant-path scintillation in the planetary boundary layer, *Appl. Opt.* **12**, 1536–1543.

Dabberdt, W. F., and W. B. Johnson (1973). Analysis of multiwavelength observations of optical scintillation, *Appl. Opt.* **12**, 1544–1551.

Davenport, W. B. Jr., and W. L. Root (1958). "Random Signals and Noise." McGraw-Hill, New York.

Davis, J. I. (1966). Consideration of atmospheric turbulence in laser systems design, *Appl. Opt.* **5**, 1, 139–147.

Deitz, P. H., and N. J. Wright (1969). Saturation of scintillation magnitude in near-earth optical propagation, *J. Opt. Soc. Am.* **59**, 527–535.

Derr, V. E. (1972). "Remote Sensing of the Troposphere". U.S. Govt. Printing Office, Washington, D.C.

Derr, V. E., and C. G. Little (1970). A comparison of remote sensing of the clear atmosphere by optical, radio and acoustic radar techniques, *Appl. Opt.* **9**, 1976–1992.

DeSanto, J. A. (1971). Scattering from a periodic corrugated structure, *J. Math. Phys.* **12**, 1913–1923.

DeSanto, J. A. (1973). Scattering from a random rough surface: Diagram methods for elastic media, *J. Math. Phys.* **14**, 1566–1573.

Deschamps, G. A., and H. S. Cabayan (1972). Antenna synthesis and solution of inverse problems by regularization methods, *IEEE PGAP* **AP-20**, 3, 268–274.

de Wolf, D. (1968). Saturation of irradiance fluctuations due to turbulent atmosphere, *J. Opt. Soc. Am.* **58**, 461–466.

de Wolf, D. (1969). Are strong irradiance fluctuations log normal or Rayleigh distributed, *J. Opt. Soc. Am.* **59**, 1455–1460.

de Wolf, D. (1971). Electromagnetic reflection from an extended turbulent medium: Cumulative forward-scatter single-backscatter approximation, *IEEE Trans. Antennas Propagat.* **AP-19**, 254–292.

de Wolf, D. (1972). Discussion of radiative transfer methods applied to electromagnetic reflection from turbulent plasma, *IEEE Trans. Antennas Propagat.* **AP-20**, 805–807.
de Wolf, D. A. (1973). Strong irradiance fluctuations in turbulent air: Plane wave, *J. Opt. Soc. Am.* **63**, 171–179.
de Wolf, D. A. (1974). Strong irradiance fluctuation in turbulent air, III, diffraction cutoff, *J. Opt. Soc. Am.* **64**, 360–365.
Dickey, F. M., C. King, J. C. Holtzman, and R. K. Moore (1974). Moisture dependency of radar backscatter from irrigated and nonirrigated fields at 400 MHz and 13.3 GHz. *IEEE Trans. Geosci. Electron.* **GE-12**, 19–22.
Dolin, L. (1966). Propagation of a narrow light beam in a medium with strongly anisotropic scattering, *Radiophys. Quantum Electron.* **9**, 40–47.
Dolin, L. (1968). Equations for the correlation functions of a wave beam in a randomly inhomogeneous medium, *Izv. Vuz. Radiofiz.* **11**, 6, 840–849.
Dowling, J. A., and P. M. Livingston (1973). Behavior of focused beams in atmospheric turbulence, *J. Opt. Soc. Am.* **63**, 846–858.
Du Castel, F. (1966). "Tropospheric Radiowave Propagation Beyond the Horizon". Pergamon, Oxford.
Dunphy, J. R., and J. P. Kerr (1974). Atmospheric beam-wander cancellation by a fast-tracking transmitter, *J. Opt. Soc. Am.* **64**, 7, 1015–1016.
Duntley, S. Q. (1974). Underwater visibility and photography, *In* "Optical Aspects of Oceanography" (N. G. Jerlov and E. Steemann Nielsen, eds.), Chapter 7. Academic Press, New York.
Dyson, F. J. (1975). Photon noise and atmospheric noise in active optical systems, *J. Opt. Soc. Am.* **65**, 551–558.
Edenhofer, P., J. N. Franklin, and C. H. Papas (1973). A new inversion method in electromagnetic wave propagation, *IEEE Trans. Antennas Propagat.* **AP-21**, 260–263.
El-Khamy, S. E., and R. E. McIntosh (1973). Optimum transionospheric pulse transmission, *IEEE Trans. Antennas Propagat.* **AP-21**, 269–273.
England, A. W. (1975). Thermal microwave emission from a scattering layer, *J. Geophys. Res.* **80**, 4484–4496.
Erickson, K. R., F. J. Fry, and J. P. Jones (1974). Ultrasound in medicine—A review, *IEEE Trans. Sonics and Ultrasonics* **SU-21**, 144–170.
Erukhimov, L. M. (1972). The effect of interstellar inhomogeneities on the shape of radio pulses from pulsars, *Radiofizika* **15**, 821–825.
Fante, R. L. (1973). Propagation of electromagnetic waves through turbulent plasma using transport theory, *IEEE Trans. Antennas Propagat.* **AP-21**, 750–755.
Fante, R. L. (1974). Mutual coherence function and frequency spectrum of a laser beam propagating through atmospheric turbulence, *J. Opt. Soc. Am.* **64**, 5, 592–598.
Fante, R. L. (1975). Electromagnetic beam propagation in turbulent media, *Proc. IEEE* **63**, 12, 1669–1692.
Fante, R. L. (1976). Comparison of theories for intensity fluctuations in strong turbulence, *Radio Sci.* **11**, 215–220.
Fante, R. L., and J. L. Poirier (1973). Mutual coherence function of a finite optical beam in a turbulent medium, *Appl. Opt.* **12**, 2247–2248.
Feinstein, D. L., F. E. Butler, K. R. Piech, and A. Leonard (1972). Radiative transport analysis of electromagnetic propagation in isotropic plasma turbulence, *Phys. Fluids* **15**, 1641–1651.
Feizulin, Z., and Y. Kravtsov (1967). Broadening of a laser beam in a turbulent medium, *Radiophys. Quantum Electron.* **10**, 33–35.
Feller, W. (1957). "An Introduction to Probability Theory and Its Applications", Vols. 1 and 2. Wiley, New York.

Feyzulin, Z. I. (1970). Amplitude and phase fluctuations of a confined wave beam propagating in a randomly inhomogeneous medium, *Radio Eng. Electron. Phys.* **15**, 7, 1189–1195.

Fisher, M. J., and F. R. Krause (1967). The cross-beam correlation technique, *J. Fluid Mech.* **28**, 705–717.

Fitsmaurice, M. W., and J. L. Bufton (1969). Measurement of log-amplitude variance, *J. Opt. Soc. Am.* **59**, 462–563.

Fjeldbo, G., and V. R. Eshleman (1969). Atmosphere of Venus as studied with Mariner 5 dual radio-frequency occultation experiment, *Radio Sci.* **4**, 879–897.

Foldy, L. O. (1945). The multiple scattering of waves, *Phys. Rev.* **67**, 3, 4, 107–119.

Fortuin, L. (1970). Survey of literature on reflection and scattering of sound waves at the sea surface, *JASA* **50**, 1209–1228.

Franklin, J. N. (1968). "Matrix Theory". Prentice Hall, Englewood Cliffs, New Jersey.

Franklin, J. N. (1970). Well-posed stochastic extensions of ill-posed linear problems, *J. Math. Anal. Appl.* **31**, 682–716.

Fried, D. L. (1964). Limiting resolution looking down through the atmosphere, *J. Opt. Soc. Am.* **56**, 10, 1380–1384.

Fried, D. L. (1966). Optical resolution through a randomly inhomogeneous medium for very long and very short exposures, *J. Opt. Soc. Am.* **56**, 10, 1372–1379.

Fried, D. L. (1967a). Optical heterodyne detection of an atmospherically distorted signal wave front, *Proc. IEEE* **55**, 57–66.

Fried, D. L. (1967b). Aperture averaging of scintillation, *J. Opt. Soc. Am.* **57**, 169–175.

Fried, D. L. (1967c). Scintillation of a ground-to-space laser illuminator, *J. Opt. Soc. Am.* **57**, 8, 980–983.

Fried, D. L. (1967d). Test of the Rytov approximation, *J. Opt. Soc. Am.* **57**, 2, 268–269.

Fried, D. L. (1968). Diffusion analysis for the propagation of mutual coherence, *J. Opt. Soc. Am.* **58**, 961–969.

Fried, D. L. (1977). Topical issue on adaptive optics, *J. Opt. Soc. Am.* **67**, 421–422.

Fried, D. L., and J. D. Cloud (1966). Propagation of an infinite plane wave in a randomly inhomogeneous medium, *J. Opt. Soc. Am.* **56**, 12, 1667–1676.

Fried, D. L., and J. B. Seidman (1967). Laser-beam scintillation in the atmosphere, *J. Opt. Soc. Am.* **57**, 181–185.

Fried, D. L., G. E. Mevers, and M. P. Keister, Jr. (1967). Measurements of laser-beam scintillation in the atmosphere, *J. Opt. Soc. Am.* **57**, 787–797.

Friedlander, S. K. and L. Topper (1961). "Turbulence." Wiley (Interscience), New York.

Frisch, V. (1968). Wave propagation in random media, *in* "Probabilistic Methods in Applied Mathematics", (A. T. Barucha-Reid, ed.), Vol. 1, pp. 76–198. Academic Press, New York.

Fung, A. K. (1968). Mechanisms of polarized and depolarized scattering from a rough dielectric surface, *J. Franklin Inst.* **2**, 125–133.

Fung, A. K. (1970). Scattering and depolarization of electromagnetic waves from a slightly rough dielectric layer, *Am. J. Phys.* **48**, 127–136.

Fung, A. K., and H. L. Chan (1969). Backscattering of waves by composite rough surfaces, *IEEE Trans. Antenna Propagat.* **AP-17**, 590–597.

Furuhama, Y. (1976). Probability distribution of irradiance fluctuations of beam waves in a weakly turbulent atmosphere, *Radio Sci.* **11**, 763–774.

Furuhama, Y., and M. Fukushima (1973). Measurements of the log-irradiance distribution of laser wave propagated through the turbulent atmosphere, *Boundary-Layer Meterol.* **4**, 433–447.

Furutsu, K. (1972). Statistical theory of wave propagation in a random medium and the irradiance distribution function, *J. Opt. Soc. Am.* **62**, 240–254.

Furutsu, K. (1975). Multiple scattering of waves in a medium of randomly distributed particles and derivation of the transport equation, *Radio Sci.* **10**, 29–44.

Gallager, R. G. (1964). Characterization and measurement of time- and frequency-spread channels, MIT Lincoln Lab. TR 352, Reference Documentation Center, AD443715.

Gardner, C. S., and M. A. Plonus (1975). The effects of atmospheric turbulence on the propagation of pulsed laser beams, *Radio Sci.* **10**, 129–137.

Gebhardt, F. G. (1976). High power laser propagation, *Appl. Opt.* **15**, 1479–1493.

Gebhardt, F. G., and S. A. Collins, Jr. (1969). Log-amplitude mean for laser-beam propagation in the atmosphere, *J. Opt. Soc. Am.* **59**, 1139–1148.

Gelfand, I. M., and S. V. Fomin (1963). "Calculus of Variations." Prentice-Hall, Englewood Cliffs, New Jersey.

Gnedenko, B. V. (1962). "Theory of Probability." Chelsea, New York.

Gnedin, U. N., and A. Z. Dolginov (1964). Theory of multiple scattering, *Sov. Phys. JETP* **18**, 3, 784–791.

Golitsyn, G. S. (1970). A similarity approach to the general circulation of planetary atmospheres, *ICARUS* **13**, 1–24.

Goodman, J. W. (1968) "Fourier Optics." McGraw-Hill, New York.

Goodman, J. W. (1976), Some fundamental properties of speckle, *J. Opt. Soc. Am.* **66**, 1145–1149.

Gracheva, M. E., and A. S. Gurvich (1965). Strong fluctuations in the intensity of light propagated through the atmosphere close to the earth, *Izv. Vuz. Radiofiz.* **8**, 717–724.

Gradshteyn, I. S., and I. M. Ryzhik (1965), "Tables of Integrals, Series, and Products," Academic Press, New York.

Granatstein, V. L., M. Rhinewine, A. M. Levine, D. L. Feinstein, M. J. Mazurowski, and K. R. Piech (1972). Multiple scattering of light from a turbid medium, *Appl. Opt.* **11**, 1217–1225.

Greenwood, D. P., and D. L. Fried (1976). Power spectra requirements for wave-front-compensative systems, *J. Opt. Soc. Am.* **66**, 193–206.

Guinard, N. W., and J. C. Daley (1970). An experimental study of a sea clutter model, *Proc. IEEE* **58**, 543–550.

Gurvich, A. (1969a). An estimate of the parameters of small-scale turbulence in the atmosphere of Venus obtained from fluctuations or radio signal from Venus 4 and Mariner 5, *Izv. Acad. Sci. USSR Atmos. Oceanic Phys.* **5**, 11, 1172–1178.

Gurvich, A. S. (1969b). Light intensity fluctuations in divergent beams, *Izv. Vuz Radiofiz.* **12**, 1, 147–149.

Gurvich, A. S., and V. I. Tatarski (1975). Coherence and intensity fluctuations of light in the turbulent atmosphere, *Radio Sci.* **10**, 1, 3–14.

Gurvich, A. S., M. A. Kallistrotova, and N. S. Time (1968). Fluctuations in the parameters of a light wave from a laser during propagation in the atmosphere, *Radiofizika* **11**, 9, 1360–1370.

Gurvich, A. S., V. I. Kalinin, and D. T. Matveyev (1973). Influence of the internal structure of glaciers on their thermal radio emission, *Atmos. Oceanic Phys.* **9**, 712–717.

Heidbreder, G. R., and R. L. Mitchell (1969). Effect of a turbulent medium on the power pattern of a wavefront-tracking circular apeture, *J. Opt. Soc. Am.* **56**, 12, 1677–1684.

Heneghan, J. M., and A. Ishimaru (1974). Remote determination of the profiles of the atmospheric structure constant and wind velocity along a line-of-sight by a statistical inversion procedure, *IEEE Trans. Antennas Propagat.* **AP-12**, 457–463.

Ho, T. L. (1969). Log-amplitude fluctuations of laser beam in a turbulent atmosphere, *J. Opt. Soc. Am.* **59**, 385–390.

Ho, T. L., and M. J. Beran (1968). Propagation of the Fourth-Order coherence function in a random medium, *J. Opt. Soc. Am.* **58**, 10, 1335–1341.

Hoffman, W. C., ed. (1960). "Statistical Methods in Radio Wave Propagation." Pergamon, Oxford.

Homstad, G. E., J. W. Strohbehn, R. H. Beyer, and J. M. Heneghan (1974). Aperture-averaging effects for weak scintillations, *J. Opt. Soc. Am.* **64**, 2, 162–165.

Hong, S. T., and A. Ishimaru (1976). Two-frequency mutual coherence function, coherence bandwidth, and coherence time of millimeter and optical waves in rain, fog, and turbulence, *Radio Sci.* **11**, 551–559.

Hufnagel, R., and N. Stanley (1964). Modulation transfer function associated with image transmission through turbulent media, *J. Opt. Soc. Am.* **54**, 52–61.

Humphreys, W. J. (1964). "Physics of the Air," Dover, New York.

Ishimaru, A. (1969a). Fluctuations of a beam wave propagating through a locally homogeneous medium, *Radio Sci.* **4**, 4, 295–305.

Ishimaru, A. (1969b). Fluctuations of a focused beam wave for atmospheric turbulence probing, *Proc. IEEE* **57**, 4, 407–419.

Ishimaru, A. (1972). Temporal frequency spectra of multifrequency waves in turbulent atmosphere, *IEEE Trans. Antennas Propagat.* **AP-20**, 10–19.

Ishimaru, A. (1973). A new approach to the problem of wave fluctuations in localized smoothly varying turbulence, *IEEE Trans. Antennas Propagat.* **AP-21**, 47–53.

Ishimaru, A. (1975). Correlation functions of a wave in a random distribution of stationary and moving scatterers, *Radio Sci.* **10**, 1, 45–52.

Ishimaru, A. (1977a). Theory and application of wave propagation and scattering in random media, *Proc. IEEE* **65**, 1030–1061.

Ishimaru, A. (1977b). The beam wave case and remote sensing, *In* "Laser Beam Propagation Through the Atmosphere" (J. W. Strohbehn, ed.), Chapter 3, Topics in Applied Physics. Springer-Verlag, Berlin and New York.

Ishimaru, A., and S. T. Hong (1975). Multiple scattering effects on coherent bandwidth and pulse distortion of a wave propagating in a random distribution of particles, *Radio Sci.* **10**, 6, 637–644.

Izyumov, A. O. (1968). Amplitude and phase fluctuations of a plane monochromatic submillimeter wave in a near-ground layer of moisture-containing turbulent air, *Radio Eng. Electron. Phys.* **13**, 7, 1009–1013.

Jahnke, E., F. Emde, and F. Losch (1960). "Table of Higher Functions." McGraw-Hill, New York.

Janes, H. B., and M. C. Thompson, Jr. (1973). Comparison of observed and predicted phase-front distortion in line-of-sight microwave signals, *IEEE Trans. Antennas Propagat.* **AP-21**, 263–266.

Janes, H. B., M. C. Thompson, Jr., D. Smith, and A. W. Kirkpatric (1970). Comparison of simultaneous line-of-sight signals at 9.6 and 34.52 GHz, *IEEE Trans. Antennas Propagat.* **AP-18**, 447–451.

Kalaschnikov, N. P. (1966). On the theory of the multiple scattering in an inhomogeneous medium, *Il Nuovo Cimento* **XLV**, A, 2, 6458–6460.

Kalaschnikov, N. P., and M. I. Ryazanov (1966). Multiple scattering of electromagnetic waves in an inhomogeneous medium, *Soc. Phys. JETP* **23**, 306–313.

Kaydanovskiy, N. L., and N. A. Smirnova (1968). Resolution limits of radio telescopes and radio interferometers imposed by propagation of waves in the space and in the atmosphere of the earth, *Radio Eng. Electron. Phys.* **13**, 9, 1355–1362.

Keller, J. B. (1964). Stochastic equations and wave propagation in random media, *Proc. Symp. Appl. Math.* **16**, 145–170.

Keller, J. B. (1969). Accuracy and validity of the Born and Rytov approximation, *J. Opt. Soc. Am.* **59**, 1003–1004.

Keller, J. B., and F. C. Karal, Jr. (1966). Effective dielectric constant, permeability, and conductivity of a random media and the velocity and attenuation coefficient of coherent waves, *J. Math. Phys.* **7**, 4, 661–670.

Kennedy, R. S., and E. V. Hoverstein (1968). On the atmosphere as an optical communications channel, *IEEE Trans.* **IT-14**, 716–725.

Kennedy, R. S., and S. Karp (1969). Optical space communication, *Proc. MIT-NASA Workshop* NASA SP-217. U.S. Govt. Printing Office, Washington, D.C.

Kerr, J. R. (1972a). Comments on "Irradiance fluctuations in optical transmission through the atmosphere," *J. Opt. Soc. Am.* **62**, 916.

Kerr, J. R. (1972b). Experiments on turbulence characteristics and multiwavelength scintillation phenomena, *J. Opt. Soc. Am.* **62**, 1040–1049.

Kerr, J. R., and R. Eiss (1972). Transmitter-size and focus effects on scintillations, *J. Opt. Soc. Am.* **63**, 1–8.

Kerr, J. R., R. P. J. Titterton, A. R. Kraemer, and C. R. Cooke (1970). Atmospheric optical communications systems, *Proc. IEEE* **58**, 1691–1709.

Khmelevtsov, S. S. (1973). Propagation of laser radiation in a turbulent atmosphere. *Appl. Opt.* **12**, 2421–2433.

Khmelevtsov, S. S., and R. Sh. Tsvyk (1970). Intensity fluctuations of a laser beam propagating in a turbulent atmosphere. *Radiofizika* **13**, 146.

Kinoshita, Y., T. Asakura, and M. Suzuki (1968). Fluctuation distribution of Gaussian beam propagating through a random medium, *J. Opt. Soc. Am.* **58**, 798–807.

Kyatskin, V. I. (1970). Applicability of the approximation of a Markov random process in problems relating to the propagation of light in a medium with random inhomogeneities, *Sov. Phys. JETP* **30**, 3, 520–523.

Klyatskin, V., and V. Tatarski (1969). Strong fluctuations of a plane light wave moving in a medium with weak random inhomogeneities, *Sov. Phys. JETP* **28**, 346–353.

Klyatskin, V. and V. Tatarski (1970). Parabolic equation approximation for propagation of waves in a medium with random inhomogeneities, *Sov. Phys. JETP* **31**, 335–339.

Knollman, G. C. (1965). Pulse propagation in anisotropic, randomly inhomogeneous media, *J. Appl. Phys.* **36**, 12, 3704–3715.

Komisarov, W. M. (1967). Fluctuations in a bounded light beam propagating in a randomly inhomogeneous medium, *Sov. Phys. JETP* **25**, 3, 467–469.

Kon, A. I. (1970). Focusing of light in a turbulent medium, *Radiofizika* **13**, 61, 43–50.

Kon, A. I., and V. I. Feizulin (1970). Fluctuations in the parameters of spherical waves propagating in a turbulent atmosphere, *Izv. VUZ Radiofiz* **13**, 71–74.

Kon, A. I., and V. I. Tatarskii (1965). Parameter fluctuations of a space-limited light beam in a turbulent atmosphere, *Izv. VUZ Radiofiz.* **8**, 870–875.

Korff, D., G. Dryden, and R. P. Leavitt (1975). Isoplanicity: the translation invariance of the atmospheric Green's function, *J. Opt. Soc. Am.* **65**, 1321–1330.

Krishen, K. (1970). Scattering of electromagnetic waves from a layer with rough front and plane back (small perturbation method by Rice), *IEEE Trans. Antennas Propagat.* **AP-19**, 573–576.

Krischen, K. (1971). Correlation of radar backscattering cross sections with ocean wave height and wind velocity, *J. Geophys. Res.* **76**, 6528–6539.

Krishen, K. (1973). Detection of oil spills using a 13.3 GHz radar scatterometer, *J. Geophys. Res.* **78**, 1952–1963.

Kuriksha, A. A. (1968). Optimal discrimination of extended radiation sources in the presence of turbulent scatter, *Radio Eng. Elect. Phys.* **13**, 5, 671–678.

Laing, M. B. (1971). The upwind/downwind dependence of the Doppler spectra of radar sea echo, *IEEE Trans. Antennas Propagat.* **AP-19**, 5, 712–714.

Landau, L. D., and E. M. Lifshitz (1959). "Fluid Mechanics." Pergamon, Oxford.

Lang, K. R. (1971a). Interstellar scintillation of pulsar radiation, *Astrophys. J.* **164**, 249–264.

Lang, K. R. (1971). Pulse broadening due to angular scattering in the interstellar medium, *Astrophys. Lett.* **7**, 175–178.

Lawrence, R. S. (1972a). Irradiance fluctuations in optical transmission through the atmosphere, *J. Opt. Soc. Am.* **62**, 701.

Lawrence, R. S. (1972b). Remote sensing by optical line-of-sight propagation, *In* "Remote Sensing of the Troposphere" by V. E. Derr, U.S. Govt. Printing Office, Washington, D.C.

Lawrence, R. S., and J. W. Strohbehn (1970). A survey of clear-air propagation effects relevant to optical communications. *Proc. IEEE* **58**, 10, 1523–1545.

Lawrence, R. S., G. R. Ochs, and S. F. Clifford (1972). Use of scintillations to measure average wind across a light beam, *Appl. Opt.* **11**, 2, 239–243.

Lax, M. (1951). Multiple scattering of waves, *Rev. Mod. Phys.* **23**, 4, 287–310.

Leader, J. C. (1971). The relationship between the Kirchhohh approach and small perturbation analysis in rough surface scattering theory, *IEEE Trans. Antennas Propagat.* **AP-19**, 786–788.

Lee, L. C. (1974). Wave propagation in a random medium: A complete set of the moment equations with different wave numbers, *J. Math. Phys.* **15**, 9, 1431–1435.

Lee, L. C., and J. R. Jokipii (1975a). Strong scintillations in astrophysics II, A theory of temporal broadening of pulses, *Astrophys. J.* **201**, 532–543.

Lee, L. C., and J. R. Jokipii (1975b). Strong scintillations in astrophysics I, The Markov approximation, its validity and application to angular broadening, *Astrophys. J.* **196**, 695–707.

Lee, M. H., J. F. Holmes, and J. R. Kerr (1976). Statistics of sparkle propagation through the turbulent atmosphere, *J. Opt. Soc. Am.* **66**, 1164–1172.

Lee, R. W. (1974). Remote probing using spatially filtered apertures, *J. Opt. Soc. Am.* **64**, 10, 1295–1303.

Lee, R. W., and J. C. Harp (1969). Weak scattering in random media, with applications to remote probing, *Proc. IEEE* **57**, 375–406.

Lee, R. W., and T. Waterman, Jr. (1966). A large antenna array for millimeter wave propagation studies, *Proc. IEEE* **54**, 454–458.

Levi, L. (1968). "Applied Optics." Wiley, New York.

Liu, C. H., A. W. Wernik, and K. C. Yeh (1974). Propagation of pulse trains through a random medium, *IEEE Trans. Antennas Propagat.* **AP-20**, 624–627.

Livingston, P. M., P. H. Deitz, and E. C. Alcaraz (1970). Light propagation through a turbulent atmosphere: Measurements of the optical-filter function, *J. Opt. Soc. Am.* **60**, 7, 925–935.

Lotova, N. A. (1975). Current ideas concerning the spectrum of the irregularities in the planetary plasma, *Sov. Phys. Usp.* **18**, 4, 292–300.

Lumley, J. L., and J. A. Panofsky (1964). "The Structure of Atmospheric Turbulence." Wiley (Interscience), New York.

Lutomirski, R. F., and R. G. Buser (1973). Mutual coherence function of a finite optical beam and application to coherent detection, *Appl. Opt.* **12**, 2153–2160.

Lutomirski, R. F., and H. T. Yura (1969). Aperture-averaging factor of a fluctuating light signal, *J. Opt. Soc. Am.* **59**, 1247–1248.

Lutomirski, R., and H. Yura (1971a). Wave structure function and mutual coherence function of an optical wave in a turbulent atmosphere, *J. Opt. Soc. Am.* **61**, 482–487.

Lutomirski, R., and H. Yura (1971b). Propagation of a finite optical beam in an inhomogeneous medium, *Appl. Opt.* **10**, 1652–1658.

Magnus, W., and F. Oberhettinger (1954). "Formulas and Theorems for Functions of Mathematical Physics," Chelsea, New York.

Mandics, P. A., and R. W. Lee (1969). On a limitation of multifrequency atmospheric probing, *Proc. IEEE* **57**, 4, 685–686.

Mandics, P. A., R. W. Lee, and A. T. Waterman, Jr. (1973). Spectra of short-term fluctations of line-of-sight signals: Electromagnetic and acoustic, *Radio Sci.* **8**, 185–201.

Mandics, P. A., J. C. Harp, R. W. Lee, and A. T. Waterman, Jr. (1974). Multifrequency coherences of short-term fluctuations of line-of-sight signals—Electromagnetic and acoustic, *Radio Sci.* **9**, 723–732.

Mano, K. (1969). Mutual power spectrum for propagation through random media: Generalization of the Beran result, *J. Opt. Soc. Am.* **59**, 318–384.
Marcuvitz, N. (1974). On the theory of plasma turbulence, *J. Math. Phys.* **15**, 869–879.
Mercier, R. P. (1962). Diffraction by a screen causing large random phase fluctuations, *Proc. Cambridge Phil. Soc.* **58**, 382–400.
Mertens, L. E. (1970). "In-Water Photography." Wiley, New York.
Middleton, D. (1960). "Introduction to Statistical Communication Theory." McGraw-Hill, New York.
Millar, R. F. (1969). On the Rayleigh assumption in scattering by a periodic surface, *Proc. Cambridge Phil. Soc.* **65**, 773–791.
Millar, R. F. (1971). On the Rayleigh assumption in scattering by a periodic surface—II, *Proc. Cambridge Phil. Soc.* **69**, 217–225.
Millar, R. F., and M. L. Burrows (1969). Rayleigh hypothesis in scattering problems, *Electron. Lett.* **5**, 416–417.
Milton, J. E., R. C. Anderson, and E. V. Browell (1972). Lidar reflectance of fair-weather cumulus clouds at 0.903μ, *Appl. Opt.* **11**, 3, 697–698.
Mintzer, D. (1953). Wave propagation in a randomly inhomogeneous medium, *J. Acoust. Soc. Am.* **25**, 5, 922–927.
Mitchell, R. L. (1968). Permanence of the log-normal distribution, *J. Opt. Soc. Am.* **58**, 9, 1267–1272.
Mittra, R. (1973). "Computer Techniques for Electromagnetics." Pergamon, Oxford.
Molyneux, J. E. (1971). Propagation of the N-th order coherence function in a random medium: The governing equations, *J. Opt. Soc. Am.* **60**, 2, 248–256.
Monin, A. S., and A. M. Yaglom (1971). "Statistical Fluid Mechanics." MIT Press, Cambridge, Massachusetts.
Muchmore, R. B., and A. D. Wheelon (1955). Line-of-sight propagation phenomena I. ray treatment, *Proc. IRE* **43**, 1437–1449.
Muchmore, R. B., and A. D. Wheelon (1963). Frequency correlation of line-of-sight signal scintillations, *IEEE Trans. Antennas Propagat.* **AP-11**, 46–51.
Muller, R. A., and A. Buffington (1974). Real-time correction of atmospherically degraded telescope images through image sharpening, *J. Opt. Soc. Am.* **64**, 1200–1210.
Murata, K. (1966). Instruments for the measuring of optical transfer functions, *Progr. Opt.*, **5**, 201–248.
Norton, K. A., P. L. Rice, H. B. Janes, and A. P. Barsis (1955). The rate of fading in propagation through a turbulent atmosphere, *Proc. IRE* **43**, 1341–1353.
Novikov, E. A. (1965). Functionals and the random-force method in turbulence theory, *Sov. Phys, JETP* **20**, 5, 1290–1294.
Ochs, G. R., and R. S. Lawrence (1969). Saturation of laser-beam scintillation under conditions of strong atmospheric turbulence, *J. Opt. Soc. Am.* **59**, 226–227.
Ochs, G. R., and R. S. Lawrence (1972). Temperature and C_n^2 profiles measured over land and ocean to 3 km above the surface, NOAA Technical Rep. ERL 251-WPL22. U.S. Govt. Printing Office, Washington, D.C.
Ochs, G. R., R. R. Bergman, and J. R. Snyder (1969). Laser-beam scintillation over horizontal paths from 5.5 to 145 kilometers, *J. Opt. Soc. Am.* **59**, 231–234.
Ochs, G. R., T-I. Wang, R. S. Lawrence, and S. F. Clifford (1976). Refractive turbulence profiles measured by one-dimensional spatial filtering of scintillations, *Appl. Opt.* **15**, 2504–2510.
Olsen, T. H., A. Goldburg, and M. Rogers (1971). "Aircraft Wake Turbulence and its Detection." Plenum Press, New York.
O'Neill, E. L. (1963). "Introduction to Statistical Optics." Addison-Wesley, Reading, Massachusetts.

Panchev, S. (1971). "Random Functions and Turbulence." Pergamon, Oxford.

Panofsky, H. A. (1968). The structure constant for the index of refraction in relation to the gradient index of refraction in the surface layer, *J. Geophys. Res.* **73**, 18, 6047–6049.

Panofsky, H. A. (1969a). Spectra of atmospheric variables in the boundary layer (review paper), *Radio Sci.* **4**, 12, 1101–1110.

Panofsky, H. A. (1969b). Spectrum of temperature, *Radio Sci.* **4**, 12, 1143–1146.

Pao, Y.-H., and A. Goldburg (1969). "Clear Air Turbulence and its Detection." Plenum Press, New York.

Peake, W. H., D. L. Barrick, A. K. Fung, and H. L. Chan (1970). Comments on "Backscattering of waves by composite rough surfaces," *IEEE Trans. Antennas Propagat.* **AP-18**, 716–728.

Pearson, J. E. (1976). Atmospheric turbulence compensation using coherent optical adaptive techniques, *Appl. Opt.* **15**, 622–631.

Pearson, J. E., W. B. Bridges, S. Hansen, T. A. Nussmeier, and M. E. Pedinoff (1976a). Coherent optical adaptive techniques: design and performance of an 18-element visible multidither COAT system, *Appl. Oct.* **15**, 611–621.

Pearson, J. E., S. A. Kokorowski, and M. E. Pedinoff 1976b). Effects of speckle in adaptive optical systems, *J. Opt. Soc. Am.* **66**, 1261–1267.

Pekeris, C. L. (1947). Note on the scattering of radiation in an inhomogeneous medium, *Phys. Rev.* **71**, 4, 268–269.

Phillips, D. L. (1961). A technique for the numerical solution of certain integral equations of the first kind, *JACM* **9**, 84–97.

Phillips, O. M. (1969). "The Dynamics of the Upper Ocean." Cambridge Univ. Press, London and New York.

Plonus, M. A., H. H. Su, and C. S. Gardner (1972). Correlation and structure functions for pulse propagation in a turbulent atmosphere, *IEEE Trans. Antennas Propagat.* **AP-20**, 801–805.

Pokasov, V. V., and S. S. Khmelevtsov (1968). Effect of atmospheric turbulence upon the spatial coherence of laser radiation, *Izv. Vuz. Fiz.* **11**, 5, 82–86.

Porcello, L. J. (1970). Turbulence-induced phase errors in synthetic-aperture radars, *IEEE Trans. Aerospace Electron. Syst.* **AES-6**, 5, 636–644.

Post, M. J. (1976). Limitations of cloud droplet size distribution by Backus-Gilbert inversion of optical scattering data, *J. Opt. Soc. Am.* **66**, 483–486.

Primmerman, C. A., and D. G. Fouche (1976). Thermal-blooming compensation: experimental observations using a deformable mirror system, *Appl. Opt.* **15**, 990–995.

Prokhorov, A. M., F. V. Bunkin, K. S. Gochelashvily, and V. I. Shishov (1975). Laser irradiance propagation in turbulent media, *Proc. IEEE* **63**, 790-811.

Rice, S. O. (1951). Reflection of electromagnetic waves from slightly rough surfaces, *Comm. Pure Appl. Math.* **4**, 351–378.

Rino, C. L. (1976). Ionspheric scintillation theory—A mini review, *IEEE Trans. Antennas Propagat.* **AP-24**, 912–915.

Roddier, C., and F. Roddier (1973). Correlation measurements on the complex amplitude of stellar plane waves perturbed by atmospheric turbulence, *J. Opt. Soc. Am.*, **63**, 6, 661–663.

Ruck, G. T., D. E. Barrick, W. D. Stuart, and C. K. Krichbaum (1970). "Radar Cross Section Handbook." Vols. I and II. Plenum Press, New York.

Rumsey, V. H. (1975). Scintillations due to a concentrated layer with a power-law turbulence spectrum, *Radio Sci.* **10**, 107–114.

Ryde, J. W. (1931). The scattering of light by turbid media—Part I, *Proc. Roy. Soc. London Ser. A* **131**, 451–464.

Ryde, J. W., and B. S. Cooper (1931). The scattering of light by turbid media—Part II, *Proc. Roy. Soc. London Ser. A* **131**, 464–475.

Ryzhov, Y. A., V. V. Tamoikin, and V. I. Tatarski (1965). Spatial dispersion of inhomogeneous media, *Sov. Phys. JETP* **21**, 433–438.

Salpeter, E. E. (1967). Interplanetary scintillations, *Astrophys. J.* **147**, 433–448.

Sancer, M. I., and A. D. Varvatsis (1970). Saturation calculation for light propagation in the turbulent atmosphere, *J. Opt. Soc. Am.* **60**, 654–659.

Saxton, J. A., ed. (1969). *Proc. Colloq. Spectra Meteorol. Variables, Stockholm, June, Radio Sci.* **4**, 12.

Schmeltzer, R. A. (1967). Means, variances, and covariances for laser beam propagation through a random medium, *Quart. Appl. Math.* **24**, 339–354.

S'edin, V. Y., S. S. Khmelevstov, and M. F. Nebol'sin (1970). Intensity fluctuations in a pulsed laser beam propagating up to 9.8 km in the atmosphere, *Izv. VUZ Radiofiz.* **13**, 1, 44–49.

Sergeyenko, T. N. (1970). Synthesis of a phase-manipulated signal from its autocorrelation function, *Radio Eng. Electron Phys.* **15**, 3, 399–404.

Shapiro, J. H. (1976a). Propagation medium limiations on phase-compensated atmospheric function, *Radio Eng. Electron. Phys.* **15**, 3, 399–405.

Shapiro, J. H. (1976b). Diffraction-limited atmospheric imaging of extended objects, *J. Opt. Soc. Am.* **66**, 469–477.

Shapiro, J. H. (1977). Imaging and optical communication through atmospheric turbulence, *In* "Laser Beam Propagation through the Atmosphere" (J. W. Strohbehn, ed.), Topics in Applied Physics. Springer-Verlag, Berlin and New York.

Shen, L. (1970). Remote probing of atmosphere and wind velocity by millimeter waves, *IEEE Trans. Antennas Propagat.* **AP-18**, 493–497.

Shifrin, Y. S. (1971). "Statistical Antenna Theory." Golem Press, Boulder, Colorado.

Shirokova, T. A. (1963). The variation in the form of a pulse caused by the effect of random inhomogeneities in a medium, *Sov. Phys.* **9**, 1, 78–81.

Shishov, V. I. (1968). Theory of wave propagation in random media, *Izv. Vuz. Radiofiz.* **11**, 6, 866–875.

Shishov, V. I. (1974). Effect of refraction on scintillation characteristics and average pulse shape of pulsars, *Sov. Astron.* **17**, 598–602.

Silverman, R. A. (1957a). Fading of radio waves scattered by dielectric turbulence, *J. Appl. Phys.* **28**, 4, 506–511.

Silverman, R. (1957b). Locally stationary random processes, *IRE Trans.* **IT-3(3)**, 182–187.

Silverman, R. A. (1958). Scattering of plane waves by locally homogeneous dielectric noise, *Proc. Cambridge Phil. Soc.* **54**, 530–537.

Sneddon, I. N. (1951). "Fourier Transforms." McGraw-Hill, New York.

Snyder, H. S., and W. T. Scott (1949). Multiple scattering of fast charged particles, *Rev. Phys.* **76**, 220–225.

Sommerfeld, A. (1950). "Mechanics of Deformable Bodies." Academic Press, New York.

Sreenivasiah, I., A. Ishimaru, and S. T. Hong (1976). Two-frequency mutual coherence function and pulse propagation in a random medium: An analytic solution to the plane wave case, *Radio Sci.* **11**, 10, 775–778.

Staras, H. (1952). Scattering of electromagnetic energy in a randomly inhomogeneous atmosphere, *J. Appl. Phys.* **23**, 10, 1152–1156.

Staras, H. (1955). Forward scattering of radio waves by anisotropic turbulence, *Proc. IRE* **43**, 1374–1380.

Stogryn, A. (1970). The brightness temperature of a vertically structured medium, *Radio Sci.* **5**, 1397–1406.

Stott, P. (1968). A transport theory for the multiple scattering of electromagnetic wave by a turbulent plasma, *J. Phys. A.* **1**, 675–689.

Strand, O. N., and E. R. Westwater (1968). Statistical estimation of the numerical solution of a Fredholm integration equation of the first kind, *JACM* **15**, 100–114.

Stratonvich, R. L. (1963). "Topics in the Theory of Random Noise," Vols. 1 and 2. Gordon and Breach, New York.

Strohbehn, J. W. (1968). Line-of-sight wave propagation through the turbulent atmosphere, *Proc. IEEE* **56**, 1301–1318.

Strohbehn, J. W. (1971). Optical propagation through the turbulent atmosphere, *Progr. Opt.* **9**, 75–122.

Strohbehn, J. W., ed. (1977). Laser beam propagation in the atmosphere, "Topics in Applied Physics." Springer-Verlag, Berlin and New York.

Strohbehn, J. W., T. Wang, and J. P. Speck (1975). On the probability distribution of line-of-sight fluctuations of optical signals, *Radio Sci.* **10**, 59–70.

Strohbehn, J. W., and S. F. Clifford (1967). Polarization and angle-of-arrival fluctuations for a plane wave propagated through a turbulent medium, *IEEE Trans. Antennas Propagat.* **AP-15**, 416–421.

Swift, C. T., ed. (1977). "Special Issue on Radio Oceanography" *IEEE Trans. Antenna Propagat.* **AP-25**, 1–3.

Szymanski, H. A. (1967). "Raman Spectroscopy," Plenum Press, New York.

Tai, C-T. (1972). Kirchhoff theory: Scalar, vector or dyadic, *IEEE Trans. Antenna Propagat.* **AP-20**, 114–115.

Tatarski, V. I. (1961). "Wave Propagation in a Turbulent Medium." McGraw-Hill, New York.

Tatarski, V. I. (1971). The Effects of the Turbulent Atmosphere on Wave Propagation, U.S. Dept. of Commerce, TT-68-50464, Springfield, Virginia.

Tatarski, V. I., and M. Gertsenshtein (1963). Propagation of waves in a medium with strong fluctuations of the refractive index, Sov. Phys. *JETP* **17**, 458–463.

Tavis, M. T., and H. T. Yura (1976). Short-term average invariance profile of an optical beam in a turbulent medium, *Appl. Opt.* **15**, 2922–2931.

Taylor, L. S. (1972). Scintillation of randomized electromagnetic fields, *J. Math. Phys.* **13**, 590–595.

Tennekes, H., and J. O. Lumley (1972). "A First Course in Turbulence." MIT Press, Cambridge, Massachusetts.

Tihonov, A. N. (1963). Regularization of incorrectly posed problems, *Dokl. Akad. Nauk SSSR* **153**, 49; *Soviet Math. Dokl.* **4**, 1624–1627.

Titterton, P. J. (1973). Scintillation and transmitter-aperture averaging over vertical paths, *J. Opt. Soc. Am.* **63**, 439–444.

Torrieri, D. J., and L. S. Taylor (1972). Irradiance fluctuations in optical transmission through the atmosphere, *J. Opt. Soc. Am.* **62**, 145–147.

Tsang, L., and J. A. Kong (1976). Thermal microwave emission from half-space random media, *Radio Sci.* **11**, 599–609.

Turchin, V. F., V. P. Kozlov, and M. S. Malkevich (1971). The use of mathematical-statistics methods in the solution of incorrectly posed problems, *Sov. Phys. Usp.* **13**, 681–840.

Twersky, V. (1957). On multiple scattering and reflection of waves by rough surfaces, *IRE Trans. Antennas Propagat.* **5**, 81.

Twersky, V. (1964). On propagation in random media of discrete scatterers, *Proc. Am. Math. Soc. Symp. Stochas. Proc. Math. Phys. Eng.* **16**, 84–116.

Twersky, V. (1967). Theory of microwave measurements on higher statistical moments of randomly scattered fields, "Electromagnetic Scattering" (R. L. Rowell and R. S. Stein, eds.), pp. 579–695. Gordon and Breach, New York.

Twersky, V. (1970a). Interference effects in multiple scattering by large, low-refracting, absorbing particles, *J. Opt. Soc. Am.* **60**, 908–914.

Twersky, V. (1970b). Absorption and multiple scattering by biological suspensions, *J. Opt. Soc. Am.* **60**, 1084–1093.

Twersky, V. (1973). Multiple scattering of sound by a periodic line of obstacles, *J. Acoust. Soc. Am.* **53**, 1, 96–112.

Twersky, V. (1975). Transparency of pair-correlated, random distributions of small scatterers, with applications to the cornea, *J. Opt. Soc. Am.* **65**, 524–530.

Ulaby, F. T., and P. P. Batilivala (1976). Diurnal variations of radar backscatter from a vegetation canopy, *IEEE Trans. Antennas Propagat.* **AP-24**, 11–17.

Uscinski, B. J. (1974). The propagation and broadening of pulses in weakly irregular media, *Proc. Roy. Soc. London*, **A336**, 379–392.

Valenzuela, G. R. (1967). Depolarization of EM waves by slightly rough surfaces, *IEEE Trans. Antennas Propagat.* **AP-15**, 4, 552–557.

Valenzuela, G. R. (1968a). Scattering of electromagnetic waves from a slightly rough surface moving with uniform velocity, *Radio Sci.* **3**, 12.

Valenzuela, G. R. (1968b). Scattering of electromagnetic waves from a tilted slightly rough surface, *Radio Sci.* **3**, 11, 1057–1066.

Valenzuela, G. R. (1970). The effective reflection coefficients in forward scatter from a dielectric slightly rough surface, *Proc. IEEE* **58**, 8, 1279.

Valenzuela, G. R., and M. B. Laing (1970). Study of doppler spectra of radar sea echo, *J. Geophys. Res.* **75**, 3, 551–563.

Valenzuela, G. R., and M. B. Laing (1972). Nonlinear energy transfer in gravity-capillary wave spectra, with applications, *J. Fluid Mech.* **54**, 507.

Villars, F., and V. F. Weisskopf (1955). On the scattering of radio waves by turbulent fluctuations of the atmosphere, *Proc. IRE* **43**, 1232–1239.

Vinnichenko, N. K., and J. A. Dutton (1969). Empirical studies of atmospheric structure and spectra in the free atmosphere (review paper), *Radio Sci.* **4**, 12, 1115–1126.

Vinnichenko, N. K., N. Z. Pinus, S. M. Shmeter, and G. N. Shur (1973). "Turbulence in the Free Atmosphere." Consultant Bureau, New York.

Wait, J. R. (1971a). "Electromagnetic Probing in Geophysics." Golem Press, Boulder, Colorado.

Wait, J. R. (1971b). Perturbation analysis for reflection from two-dimensional periodic sea waves, *Radio Sci.* **6**, 387–391.

Wait, J. R. (1972). Remote sensing of observables in geophysics, *In* "Future Directions of Electromagnetics of Continuous Media" (Proc. NSF Workshop), (D. J. Angelakos and W. K. Kahn, eds.). U.S. Govt. Printing Office, Washington, D.C.

Wang, T., and J. W. Strohbehn (1974a). Log-normal paradox in atmospheric scintillations, *J. Opt. Soc. Am.* **64**, 5, 583–591.

Wang, T., and J. W. Strohbehn (1974b). Perturbed log-normal distribution of irradiance fluctuations, *J. Opt. Soc. Am.* **64**, 7, 994–999.

Wang, T., S. F. Clifford, and G. R. Ochs (1974). Wind and refractive-turbulence sensing using crossed laser beams, *Appl. Opt.* **13**, 11, 2602–2608.

Watson, K. (1969). Multiple scattering of electromagnetic waves in an underdense plasma, *J. Math. Phys.* **10**, 688–702.

Watson, K. (1970). Electromagnetic wave scattering within a plasma in the transport approximation, *Phys. Fluids* **13**, 2514–2523.

Wells, P. N. T. (1969). "Physical Principles of Ultrasonic Diagnosis." Academic Press, New York.

Wenzel, A. R., and J. B. Keller (1971). Propagation of acoustic waves in a turbulent medium, *J. Acoust. Soc. Am.* **50**, 911–920.

Westwater, E. R., and A. Cohen (1973). Application of Backus-Gilbert inversion technique to determination of aerosol size distributions from optical scattering measurements, *Appl. Opt.* **12**, 1340–1348.

Wheelon, A. D. (1959). Radio wave scattering by tropospheric irregularities, *J. Res. Nat. Bur. Std.* (*Radio Propagat.*) **63D**, 205–233.

Wheelon, A. D., and R. B. Muchmore (1955). Line-of-sight propagation phenomena II. Scattered components, *Proc. IRE* **43**, 1450–1458.

Whitman, A. M., and M. J. Beran (1970). Beam spread of laser light propagating in a random medium, *J. Opt. Soc. Am.* **60**, 12, 1595–1602.

Williamson, I. P. (1972). Pulse broadening due to multiple scattering in the interstellar medium, *Mon. Notices Roy. Astron. Soc.* **157**, 55–71.

Williamson, I. P. (1973). *Mon. Notices Roy. Astron. Soc.* **163**, 345–356.

Wolf, E. (1976). New theory of radiative energy transfer in free electromagnetic fields, *Phys. Rev. D* **13**, 869–886.

Woo, R. (1975a). Multifrequency technique for studying interplanetary scintillations, *Astrophys. J.* **201**, 1, Pt. 1, 238–248.

Woo, R. (1975b). Observation of turbulence in the atmosphere of Venus using Mariner 10 Radio Occultation measurement, *J. Atmos. Sci.* **32**, 1084–1090.

Woo, R., and A. Ishimaru (1973). Remote sensing of the turbulence characteristics of a planetary atmosphere by radio occultation of a space probe, *Radio Sci.* **8**, 2, 103–108.

Woo, R., and A. Ishimaru (1974). Effects of turbulence in a planetary atmosphere on radio occultation, *IEEE Trans. Antennas Propagat.* **AP-22**, 566–573.

Woo, R., and F. C. Yang (1976). Measurements of electron density irregularity in the ionosphere of Jupiter by Pioneer 10, *J. Geophys. Res.* **81**, 3417–3422.

Woo, R., A. Ishimaru, and W. B. Kendall (1974). Observations of small-scale turbulence in the atmosphere of Venus by Mariner 5, *J. Atmos. Sci.* **31**, 6, 1698–1706.

Woo, R., F-C Yang, K. W. Yip, and W. B. Kendall (1976a). Measurements of large-scale density fluctuations in the solar wind using dual-frequency phase scintillations, *Atrophys. J.* **210**, 2, Pt. 1, 568–574.

Woo, R., F-C Yang, and A. Ishimaru (1976b). Structure of density fluctuations near the sun deduced from Pioneer 6 spectral broadening measurements, *Atrophys. J.* **210**, 2, Pt. 1, 593–602.

Wright, J. W. (1968). A new model for sea clutter, *IEEE Trans. Antennas Propagat.* **AP-16**, 2, 217–223.

Wyant, J. C. (1975). Use of an AC heterodyne lateral shear interferometer with real time wavefront correction systems, *Appl. Opt.* **14**, 2622–2626.

Yaglom, A. M. (1962). "Stationary Random Functions." Prentice-Hall, Englewood Cliffs, New Jersey.

Yaglom, A. M., and V. I. Tatarski (1967). "Atmospheric Turbulence and Radio Wave Propagation." "NAUKA" Publishing House, Moscow.

Yates, H. W. (1970). A general discussion of remote sensing of the atmosphere, *Appl. Opt.* **9**, 1971–1975.

Young, A. T. (1974). Seeing: Its cause and cure, *Astrophys. J.* **189**, 587–604.

Yura, H. T. (1969). Optical propagation through a turbulent medium, *J. Opt. Soc. Am.* **59**, 111–113.

Yura, H. T. (1972). Mutual coherence function of a finite cross section optical beam propagating in a turbulent medium, *Appl. Opt.* **11**, 6, 1399–1406.

Yura, H. T. (1973a). Imagining in clear ocean water, *Appl. Opt.* **12**, 1061–1066.

Yura, H. T. (1973b). Propagation of finite cross-section laser beams in sea water, *Appl. Opt.* **12**, 1, 108–115.

Yura, H. T. (1973c). Optical beam spread in a turbulent medium: Effect of the outer scale of turbulence, *J. Opt. Soc. Am.* **63**, 107–109.

Yura, H. T. (1974a). Physical model for strong optical-amplitude fluctuations in a turbulent medium, *J. Opt. Soc. Am.* **64**, 59–67.

Yura, H. T. (1974b). Temporal-frequency spectrum of an optical wave propagating under saturation conditions, *J. Opt. Soc. Am.* **64**, 357–359.

Zavody, A. M. (1974). Effect of scattering by rain on radiometric measurements at millimeter wavelengths, *Proc. IEE* **121**, 257–263.

Zipfel, G. G., and J. A. DeSanto (1972). Scattering of a scalar wave from a random rough surface: A diagrammatic approach, *J. Math. Phys.* **13**, 1903–1911.

INDEX

T

U

V

W

A
B
C 8
D 9
E 0
F 1
G 2
H 3
I 4
J 5